Study Guide for

Gillespie, Baird, Humphreys, and Robinson

CHEMISTRY

Second Edition

Prepared by

Joseph D. Laposa
McMaster University

Edward A. Robinson
University of Toronto

N. Colin Baird
University of Western Ontario

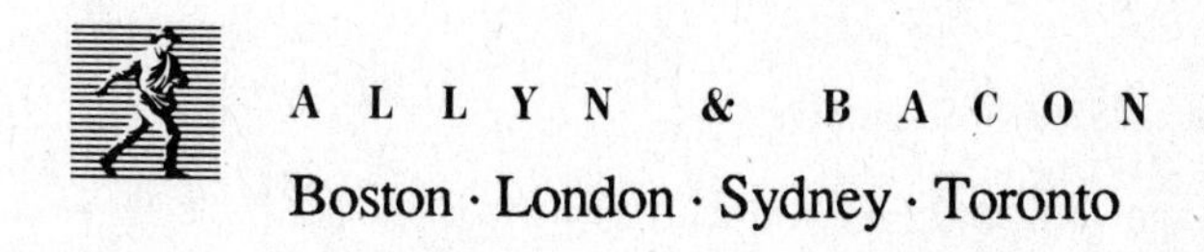

Boston · London · Sydney · Toronto

A Division of Simon & Schuster, Inc.
160 Gould Street
Needham Heights, Massachusetts 02194

ISBN 0-205-11798-8

Printed in the United States of America

10 9 8 7 6 5 4 3 2 1 95 94 93 92 91 90 89

CONTENTS

PREFACE

What the Study Guide Contains

This **Study Guide to accompany Chemistry** is intended to be used along with the second edition of **Chemistry** by Gillespie, Humphreys, Baird and Robinson. The Study Guide is not meant to replace the textbook, but to supplement it. Each chapter corresponds to a chapter in the text, and contains a **summary review** of the principal topics in the chapter, a set of simple **review questions**, a list of **learning objectives** for the chapter, **problem solving strategies** for the more difficult topics, and a **self test** with answers.

How to Use the Study Guide

There is no one way that works best for all students attempting to master chemistry. In addition to attending the formal lectures and performing the laboratory experiments, the following approach, using the Study Guide, may be useful.

1. Read the chapter in the text.
2. Read the summary review in the Study Guide, and refer to the list of important terms at the end of each chapter of the text, as necessary.
3. Try all the review questions in the Study Guide for that chapter.
4. Examine all the learning objectives in the Study Guide.
5. Reread all the worked examples in the text, along with the problem solving strategies in the Study Guide. Note that the section numbering in the Study Guide is the same as that in the text.
6. Study all the worked examples in the Study Guide for the chapter.
7. Take the self test in the Study Guide and evaluate your test performance. Use the text and the Study Guide to review again any areas that still are giving you trouble.
8. Work problems at the end of each chapter in the text. Selected odd-numbered ones have answers at the back of the text.

General Review and Study Hints

Each of you has to work out for yourself the appropriate amount of time spent outside formal classes so that each subject taken, including chemistry, receives adequate attention on a regular basis. This is a personal matter, and should be done for the semester, term, or academic year so that the amount of work needed to be done before quizzes, tests, or final examinations does not suddenly become impossible.

Study habits are important. Not only is a steady application of effort on a regular basis far more effective than occasional blitzes prompted by panic, but there is a practical limit to what can be achieved in any one study session. Concentrated study at the eleventh hour usually induces greater panic and confusion rather than an increased understanding. Insight into and comprehension of any subject, especially one as complex as chemistry, is a cumulative process. The solution of numerical problems requires repeated practice. As well, success in learning both general classes of chemical reactions and specific reactions only comes with repetition. Steady effort as part of a regular schedule is needed.

Equally important as time for study is time for relaxation and leisure. You will soon discover your own natural period for study, usually three or four hours for most students, beyond which little of lasting value is achieved. It is unwise to push yourself into a state of mental exhaustion. Nevertheless, increased confidence comes as a result of steady practice, as you will soon discover.

Time must also be used economically. Thus, for example, in problem solving, be prepared to give up and seek advice if after twenty minutes or so you have made no progress -- almost all of the problems in the text are straightforward **once you grasp the principles** involved. Similarly, when you encounter difficulties in your private study, make a note of the problem and seek help as soon as possible from your professor, tutor or fellow students.

Templates for making an Octahedron and a Tetrahedron, see text page 219

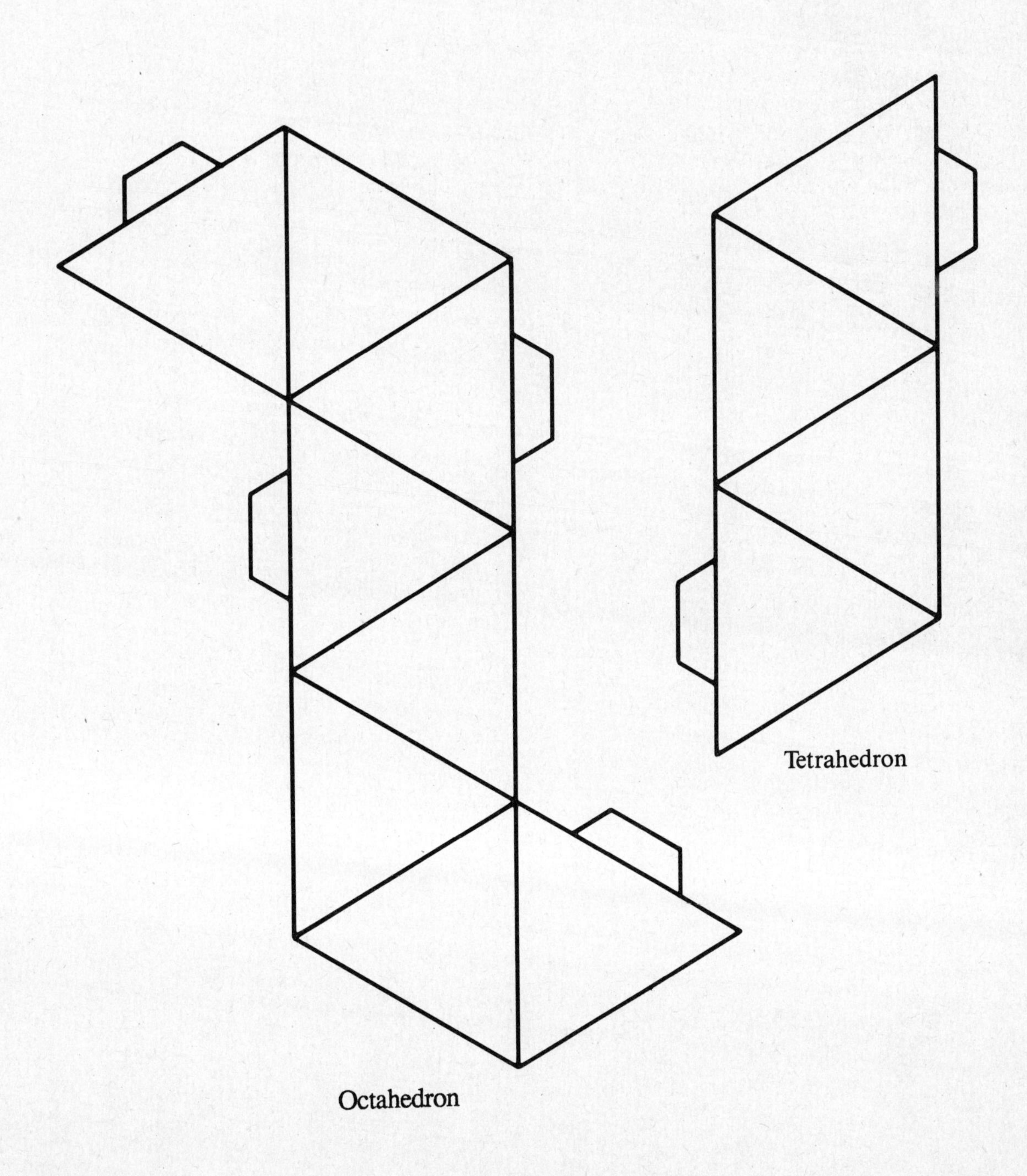

CHAPTER 1

STRUCTURE OF MATTER

SUMMARY REVIEW

Matter has mass and occupies space, while light and heat are forms of **energy.** The many varieties of matter can occur as solids, liquids and gases (vapors) and are called **substances. Chemistry** is concerned with the composition, structure and properties of substances, with **reactions,** which are processes in which substances are changed into other substances, and with the energy changes that accompany reactions. An **element** cannot be broken down into other substances; a substance composed of two or more elements, combined in fixed proportions, is a **compound.** Each of the more than 100 known elements is composed of **atoms.** Each kind of atom is the smallest distinguishable part of an element and is indivisible by chemical means. Each element is given a unique symbol, e.g. hydrogen, H, sulfur, S, potassium, K. The composition of a compound is expressed by a **formula** which gives the relative numbers of atoms of each kind it contains. The **simplest** formula that gives the correct ratio of atoms is the **empirical formula.** A **molecular formula** gives the actual number of atoms that are strongly bonded together in a molecule -- for example, water consists of H_2O molecules, ethane of C_2H_6 molecules, white phosphorus of P_4 molecules and nitrogen of N_2 molecules.

Many molecules consist of relatively small numbers of atoms; others contain large numbers and are called macromolecules or polymers and contain repeat patterns of smaller groups of atoms. Substances in which molecules cannot be discerned, such as diamond, C, graphite, C, magnesium, Mg, and sodium chloride, NaCl, have an infinite network of atoms forming a single giant molecule.

In chemical reactions, atoms in **reactant** substances are rearranged to give new **product** substances.

Atoms in molecules have specific relative locations in space, or **geometries.** Experimental determination of structure gives the **geometric arrangement** of the atoms, the exact **interatomic distances,** and the **angles** formed by groups of three atoms. For example water consists of H_2O molecules with two O-H distances **(bond lengths)** of 97 pm and an HOH angle **(bond angle)** of 104.5°

Substances are either "pure", or they are **mixtures** of two or more substances. Each pure substance has characteristic **physical properties** and **chemical properties** that together distinguish it from all others. Mixtures are either **homogeneous** (with uniform properties throughout) or **heterogeneous** (with different properties in various parts of a sample).

The **SI System** (International System of Units) is based on the metric system. Its base units are enlarged or reduced by means of prefixes - some of the more important of which are: kilo, **k**, 10^3; deci, **d**, 10^{-1}; centi, **c**, 10^{-2}; milli, **m**, 10^{-3}; micro, **μ**, 10^{-6}, nano, **n**, 10^{-9}; pico, **p**, 10^{-12}. The derived unit of **volume** is the cubic decimeter (dm^3) which is commonly called the **liter, L.** A common pratical unit is the **mL.**

$$1\ L = 10^{-3} m^3 = 1\ dm^3; \qquad 1\ mL = 10^{-3} L = 1\ cm^3$$

Density is mass per unit volume, with the usual units **g cm^{-3}** or **g mL^{-1}** for liquids and solids, and g L^{-1} for gases.

$$\text{density} = \frac{\text{mass}}{\text{volume}}$$

The three states of matter are **solid, liquid** and **gas.** When a gas is cooled it **condenses** to a liquid and then **freezes** to a solid. A **gas** is a fluid with no definite volume or shape, which expands to fill any container of any volume or shape. A **liquid** is also fluid but has a fixed volume. A **solid** has both a fixed volume and shape. Such **macroscopic properties** are explained at the **microscopic** (atomic and molecular) level in terms of **intermolecular forces.**

In a solid the molecules are packed together in a regular array and the intermolecular forces are relatively strong. An increase in temperature provides the molecules with increased energy, so that they can vibrate and oscillate with increasing amplitude. At the melting point the molecules break free from their fixed positions and the solid melts. In the resulting liquid, molecules move relative to each other but are still held together by relatively strong intermolecular forces. As the temperature is further increased, the molecular motions become sufficiently violent that some molecules leave the surface as vapor (gas). With continued heating, all of the liquid becomes gas. In the gas, the molecules move sufficiently fast and are so far apart that the intermolecular forces become negligible.

A **solution** is a homogeneous mixture of two or more substances (components). The component in largest amount usually is called the **solvent;** the other components are **solutes.** **Aqueous solutions** have water as the solvent. **Concentrations** of solutions are measured in terms of the amount of solute in a given amount of solution (or solvent):

$$(\text{mass }\%)_{\text{solute}} = \frac{\text{grams of solute}}{\text{100 grams of solution}}$$

For very dilute solutions, or trace impurities, concentrations are often expressed as parts per million (ppm):

$$1\ \text{ppm} = 1\ \text{g in } 10^6\ \text{g}$$

A **saturated solution** contains the maximum concentration of solute that can dissolve at a given temperature; this concentration is called its **solubility.** A saturated solution is one that is in dynamic equilibrium with pure undissolved solid solute at a given temperature, or would be if any solid solute were present.

Separation of mixtures into their components is achieved by a variety of physical methods. **Crystallization** involves formation of crystals from a saturated solution as the temperature is lowered, or as some of the solvent evaporates. **Distillation** involves heating a liquid to form vapor, which subsequently can be condensed back to liquid by cooling. In a mixture of liquids, the different liquids have different rates of vaporization, and give a vapor which is enriched in the lowest boiling component. Efficient separation of liquids is achieved by multiple distillations using a fractionating column that achieves this in a single step. **Chromatographic** separation of substances makes use of their different tendencies to be adsorbed from solution onto paper, other solids, or the surface of an inert liquid.

REVIEW QUESTIONS

1. What is the definition of matter?
2. What are the essential phenomena with which the study of chemistry is concerned?
3. What are the three states of matter?
4. What is an atom?
5. What is an element?
6. What is a chemical reaction?
7. Write the symbols for the following elements: hydrogen, oxygen, nitrogen, fluorine, sodium, potassium, iron and lead.
8. What is the chemical formula for methane?
9. What information does the formula for hydrogen peroxide, H_2O_2, provide?
10. Give two examples of substances that are elements, and two that are compounds.
11. What is meant by the term pure substances?
12. How could it be shown that a particular substance was pure?
13. Cite four physical properties that are characteristic of a sample of pure water.
14. What is a mixture?
15. What characteristics differentiate a heterogeneous mixture from a homogeneous mixture?
16. Classify each of the following mixtures as heterogeneous or homogeneous: milk, concrete, a mixture of alcohol and water, air, household bleach, rain water, soil.
17. How can a mixture be distinguished from a pure substance?
18. What is meant by the structure of a molecule?
19. What information is given by the empirical formula of a substance?
20. What are the empirical and molecular formulas of ethane?
21. What are the empirical and molecular formulas of sulfur, phosphorus and hydrogen peroxide?
22. What is the empirical formula of sucrose, $C_{12}H_{22}O_{11}$?
23. Describe the molecular shapes (geometries) of the following molecules: water, ammonia, P_4, methane.
24. Construct a tetrahedron inside a cube.
25. Write the following numbers in scientific notation: 0.0004, 4000, 4 million, 4 billion.
26. What are the base units of length, mass and time in SI?
27. What unit conversion factor is used to convert 1 hour to seconds?
28. What unit conversion factors are necessary to convert a velocity in miles per hour to meters per second?

29. The velocity of light is 3.00×10^8 m s^{-1}. What is this velocity in miles per hour?
30. Express the following in basic SI units and in scientific notation: 2.54 cm, 2.54 mg, 1.54 pm.
31. In what units is the density of a substance normally given?
32. If 100.0 mL of alcohol (ethanol) has a mass of 86.1 g at 16°C, what is its density at this temperature?
33. If the density of benzene is 0.879 g cm^{-3} at 20°C, what is the mass of 1.00 L of benzene at 20°C?
34. What macroscopic properties distinguish ice from liquid water, and liquid water from gaseous water vapor?
35. In terms of its molecular interactions, how does solid bromine differ from liquid bromine, and liquid bromine from gaseous bromine?
36. What is a solution?
37. 5.00 g sodium chloride, NaCl, are dissolved in 100 g of water. Which substance is defined as the solute, and which as the solvent?
38. What is meant by the statement that water and ethylene glycol (anti-freeze) are miscible?
39. What is meant by the statement that water and gasoline are immiscible?
40. A solution contains 5.00 g of sodium chloride dissolved in 100.00 g of water. What is the mass % of sodium chloride?
41. 100 g of water contains 1.5 mg of an impurity. What is the concentration of the impurity in ppm?
42. In what units is the solubility of a substance in a liquid expressed?
43. What is a saturated solution?
44. The solubility of sugar (sucrose) in water at 25°C is 67.9 mass %. What is the concentration of water saturated with sugar at 25°C?
45. How can the potassium nitrate in a solution containing 80 g of potassium nitrate and 10 g of potassium chloride in 100 g of water at 100°C be obtained as the pure solid?
46. How can a solid be further purified by recrystallization?
47. How is distilled water obtained from tap water?
48. How can a mixture of water and glycol be separated?
49. How does fractional distillation differ from ordinary distillation?
50. How are nitrogen and oxygen obtained from air?
51. Why is it possible to separate a mixture of volatile substances by means of gas chromatography?

Answers to Selected Review Questions

25. 4×10^{-4}; 4×10^3; 4×10^6; 4×10^9 29. 6.71×10^8 miles hr^{-1}.
30. 2.54×10^{-2} m; 2.54×10^{-6} kg; 1.54×10^{-12} m. 32. 0.816 g mL^{-1}.
33. 879 g. 40. 4.76% 41. 15 ppm. 44. 32.1%.

OBJECTIVES

Be able to:

1. Define the terms chemical reaction, element, compound, atom, molecule.

2. Write the symbols for all the elements discussed in the textbook (see Table 1.1).
3. Distinguish between physical and chemical properties of a substance, between homogeneous and heterogeneous mixtures.
4. Appreciate that atoms in a molecule have a spatial arrangement or structure.
5. Realize that not all substances are composed of molecules.
6. Distinguish between empirical and molecular formulas of a substance.
7. State the molecular formulas for water, carbon monoxide, carbon dioxide, hydrogen, oxygen, nitrogen, hydrogen peroxide, ammonia, methane, carbon disulfide.
8. Use exponential notation.
9. Use SI units, especially those for length and mass.
10. Convert from one set of units to another (preferably by the unit factor method).
11. Describe the three states of matter, both macroscopically and microscopically.
12. Define the terms solvent, solute, solution and express aqueous solution concentrations as mass percentages and as ppm.
13. Define the terms solubility and saturated solution.
14. Describe the separation techniques crystallization, distillation, and chromatography.

PROBLEM-SOLVING STRATEGIES

In this chapter you are introduced to some basic definitions that are a part of the common vocabulary of chemistry. The principles of chemistry are expressed using terms such as element, compound, atom, and molecule, so be sure you understand what they mean.

1.2 Units of Measurement

Scientific Notation

As you learn chemical concepts, you will also be expected to apply them to specific examples, which may involve calculations with very large and/or very small numbers. Scientific (or exponential) notation is very useful in these cases.

Example 1.1

Express 1,930,000,000 and 0.0040009 in scientific notation.

Solution 1.1

We want to write each of the above numbers as the product of a number between 1 and 10 times 10 raised to the appropriate power. We need not be concerned with zeros that are merely used to locate a decimal point. For the first number, all the zeros after the 3 are only there to tell us that the number is one trillion, nine hundred thirty billion. Thus we write 1.93×10^9 since the decimal point is moved nine places to the left. For the second number, the two zeros immediately to the right of the decimal point are just there to tell us where to put the decimal point; this number is then 4.0009×10^{-3} in exponential notation since the decimal point is moved three places to the right.

Conversion of Units

Chemistry is an experimental science, and chemical calculations involve measurements that must be accompanied by proper units. Many times it is necessary to convert from one set of units to another. A reasonable way to begin is to write the definition of the property that is under consideration. Next, write the conversion factor or factors that change the old units into the desired units. Finally, perform the necessary arithmetic. Let's try an example.

Example 1.2

A sample of an iron-containing ore has a density of 6.95 g cm^{-3}. Express the density in terms of kg m^{-3}.

Solution 1.2

Since the density is defined as the mass of a given volume of substance, we write

$$\text{density} = \frac{\text{mass}}{\text{volume}} = \frac{6.95 \text{ g}}{1.00 \text{ cm}^3}$$

The problem requires the mass to be expressed in kilograms, so we write

$$1.000 \text{ kg} = 1.000 \times 10^3 \text{ g}$$

After dividing both sides by 1.000×10^3 g, we obtain

$$\frac{1.000 \text{ kg}}{1.000 \times 10^3 \text{ g}} = 1$$

This is an example of a **unit conversion factor** that relates kilograms and grams. With this factor we can now express 6.95 grams in terms of kilograms, by multiplying 6.95 grams by the unit conversion factor:

$$6.95 \text{ g} \times \frac{1.000 \text{ kg}}{1.000 \times 10^3 \text{ g}}$$

Note that the unit conversion factor has the **desired mass unit in the numerator** and **the original mass unit in the denominator.** When the unit conversion factor is multiplied by the original mass unit, the original mass unit cancels, and the desired mass unit is produced. If we had chosen the wrong unit conversion factor, the correct unit would not have been obtained. The same procedure must be applied to the conversion of cm^3 to m^3.

$$1.00 \text{ m} = 1.00 \times 10^2 \text{ cm.}$$

By raising both sides to the power 3 (including the units!) we obtain the necessary unit conversion factor.

$$(1.00)^3 \text{ m}^3 = (1.00 \times 10^2)^3 \text{ cm}^3 = 1.00 \times 10^6 \text{ cm}^3,$$

or

$$\frac{1.00 \text{ m}^3}{1.00 \times 10^6 \text{ cm}^3} = 1$$

With this unit conversion factor, the 1.00 cm^3 in the density expression can be converted to m^3:

$$1.00 \text{ cm}^3 \times \frac{1.00 \text{ m}^3}{1.00 \times 10^6 \text{ cm}^3} = 1.00 \times 10^{-6} \text{ m}^3$$

Thus a density of 6.95 g cm^{-3} is equivalent to

$$\frac{6.95 \times 10^{-3} \text{ kg}}{1.00 \times 10^{-6} \text{ m}^3} = 6.95 \times 10^3 \text{ kg m}^{-3}$$

Rather than do the problem in 2 separate steps, it is faster to use **two** conversion factors at the same time. Here the plan of attack is to write the original units on the left side, the desired units on the right side, and by inspection multiply the **left** side units by the appropriate unit conversion factors, so that the desired units are produced.

$$\frac{6.95 \text{ g}}{1 \text{ cm}^3} \times \frac{1.000 \text{ kg}}{1.000 \times 10^3 \text{ g}} \times \frac{1.000 \times 10^6 \text{ cm}^3}{1.00 \text{ m}^3} = 6.95 \times 10^3 \text{ kg m}^{-3}$$

Note here that the last unit conversion factor is **inverted**, compared to the factor we used in the **two** step solution. There we wanted to convert cm^3 to m^3. Here we want to change cm^{-3} to m^{-3}. You may use a unit conversion factor or its reciprocal (since both equal 1), depending on which units you want to cancel.

1.4 Solutions

Much of the chemistry discussed in an introductory-level course involves substances dissolved in water, forming solutions. There are several ways to express the concentrations of such solutions; they all describe how much of the substance is dissolved in a given amount of solution. You must first learn the **definitions** of the different concentration expressions, in order to be able to calculate concentrations of solutions. Here are two examples. Notice the use of unit conversion factors in Example 1.4.

Example 1.3

Household bleach is a solution of sodium hypochlorite, NaOCl, in water. What are the mass percentages of NaOCl and water in a bleach solution made from 2.67 g NaOCl and 60.32 g water?

Solution 1.3

Mass percentage is defined as the number of grams of the given substance divided by the total number of grams of all substances in the solution, multiplied by 100. Thus

$$\text{mass \% NaOCl} = \frac{2.67 \text{ g}}{(2.67 + 60.32) \text{ g}} \times 100\% = 4.24 \text{ mass \%}$$

$$\text{mass \% H}_2\text{O} = \frac{60.32 \text{ g}}{(2.67 + 60.32) \text{ g}} \times 100\% = 95.76 \text{ mass \%}$$

As a check, the sum of all the mass percentages must yield 100%. In this problem,

$$\text{mass \% NaOCl} + \text{mass \% H}_2\text{O} = 4.24 + 95.76 = 100.00 \text{ mass \%}$$

Example 1.4

Many communities add about 0.7 ppm of sodium fluoride, NaF, to their municipal water supplies to help reduce tooth decay in children. If an average person drinks 2.00 L of fluoridated water per day, what is his daily intake of NaF? Assume the density of fluoridated water is 1.00 g cm^{-3}.

Solution 1.4

Let us first calculate how many grams of solution correspond to 2.00 L. This is accomplished using the definition of density and a series of unit conversion factors.

$$1\ L = 10^3\ mL \quad \frac{1\ L}{10^3\ mL} = 1 \qquad 1\ mL = 1.00\ cm^3 \quad \frac{1.00\ mL}{1.00\ cm^3} = 1$$

$$\text{density} = \frac{\text{mass}}{\text{volume}} = \frac{\text{g solution}}{\text{cm}^3\ \text{solution}}$$

$$\text{density} \times \text{cm}^3\ \text{solution} = \text{g solution}$$

$$\frac{1.00\ g}{1\ cm^3} \times 2.00\ L \times \frac{10^3\ mL}{1\ L} \times \frac{1.00\ cm^3}{1\ mL} = 2.00 \times 10^3\ \text{g solution}$$

Next, since ppm is a **concentration** unit, defined as

$$\text{ppm} = \frac{\text{g solute}}{10^6\ \text{g solution}}$$

the present solution has a concentration of

$$\frac{0.7\ \text{g NaF}}{10^6\ \text{g solution}}$$

Finally, with the amount of solution calculated above, we have

$$\text{g NaF} = \frac{0.7\ \text{g NaF}}{10^6\ \text{g solution}} \times 2.00 \times 10^3\ \text{g solution}$$

$$= 1.4 \times 10^{-3}\ \text{g NaF}$$

SELF-TEST

Part I True or False

1. Carbon can exist in the form of graphite or in the form of diamond. Both of these are called the element carbon.
2. A pure substance has a fixed, constant composition and a characteristic set of chemical and physical properties.
3. The melting of solid water to liquid water is an example of a chemical change.
4. HOOH is the empirical formula of hydrogen peroxide.
5. The molecule H_2O, water, is composed of a hydrogen atom and two oxygen atoms.
6. Molecules are much farther apart from each other in the gas state than they are in the liquid state.
7. The solubility of a substance is dependent upon temperature.

8. 1.0×10^{-3} g of an impurity is present in 1.00 kg solution. The concentration of the impurity is 1000 ppm.
9. The density of a pure substance is constant; it does not change even if the substance is transformed from a solid to a liquid or gas.
10. Both methane, CH_4, and ammonia, NH_3, are nonplanar.
11. In exponential notation, 0.0028 is 2.8×10^{-2}.

Part II Multiple Choice

12. Which of the following is the symbol for the element tin?
(a) Si (b) Ti (c) Tn (d) Sn (e) In
13. A single hydrogen atom has a mass of 1.67×10^{-27} kg. Express this mass numerically in nanograms.
(a) 1.67×10^{-9} (b) 1.67×10^{-15} (c) 1.67×10^{-18} (d) 1.67×10^{-36}
(e) 1.67×10^{-39}
14. The P-P distance in the P_4 molecule is 221 ppm. What is the distance in angstroms?
(a) 22.1 (b) 221×10^{-10} (c) 2.21 (d) 0.221 (e) 221×10^{8}
15. The solubility of a substance
(a) does not change with temperature
(b) always increases with increasing temperature
(c) always decreases with increasing temperature
(d) may decrease or increase with increasing temperature, depending on the substance.
16. Regular beer is about 4% ethanol, C_2H_5OH, by mass in water. What mass of alcohol is present in a 340 mL bottle of beer, assuming the density of the solution to be 1.00 g cm^{-3}?
(a) 0.01 g (b) 13.6 g (c) 85 g (d) 340 g (e) 1360 g.
17. A carbon atom has a mass 11.91 times larger than the mass of a hydrogen atom. Numerically, what is the mass % of hydrogen in methane, CH_4?
(a) $\frac{4}{11.91 + 4}$ (b) $\frac{4}{11.91}$ (c) $\frac{400}{11.91}$
(d) $\frac{400}{11.91 + 4}$ (e) cannot answer, since mass of a hydrogen atom not given.
18. The density of lead is 11.34 g cm^{-3}. A 15.00 g sample of lead is placed into a 100 mL graduated cylinder originally filled with 50.00 mL water. What will be the final mL reading of the water in the cylinder? (Lead does not react with water).
(a) $50.00 + 15.00$ (b) $\frac{50.00}{15.00} + 11.34$ (c) $50.00 + 11.34$
(d) $50.00 + \frac{11.34}{15.00}$ (e) $50.00 + \frac{15.00}{11.34}$
19. What volume of beer, in mL (see question 16), contains 10.0 g of alcohol?
(a) 0.4 (b) 2.5 (c) 40 (d) 250 (e) 340.

Answers to Self Test

1. T; 2. T; 3. F; 4. F; 5. F; 6. T; 7. T; 8. F; 9. F; 10. T;
11. F; 12. d; 13. b; 14. c; 15. d; 16. b; 17. d; 18. e; 19. d.

CHAPTER 2

STOICHIOMETRY

SUMMARY REVIEW

The atom of smallest mass is the **hydrogen atom:** it contains one electron and a nucleus which consists of a proton, and which accounts for most of the mass of the atom. The electron and proton are held together by an **electrostatic force;** the electron is negatively-charged and the proton carries an equal positive charge. The proton (and the nucleus of any atom) is very small compared to the overall atomic volume. In general, nuclei are positively charged and are surrounded by sufficient electrons to make the atom neutral overall. Because there is no experimental way of determining the paths by which electrons travel around nuclei, the size of an atom is defined in terms of the radius of a sphere drawn so that there is a 90% probability of finding the electrons within the spherical surface.

Three types of forces are recognized in nature: **gravitational, electrical** and **nuclear,** of which only electrical forces are important in chemistry. The **electrostatic force** F between two charges of magnitude Q_1 and Q_2 is given by **Coulomb's law**

$$F = kQ_1Q_2/r^2$$

where r is their distance apart. The SI unit of force is the **newton, N.**

$$1\ N = 1\ kg\ m^2 s^{-2}$$

In any system, energy is conserved **(Law of conservation of energy).** When a negative charge is moved away from a positive charge, its potential energy increases. When the negative charge is then allowed to move freely, it accelerates towards the positive charge; its potential energy decreases (i.e. becomes more negative) and its kinetic energy increases but the sum of potential energy and kinetic energy remains constant. The total energy of any isolated system remains constant; energy cannot be created or destroyed but only changed from one form to another.

The **work w** done when a force **F** displaces an object by distance **d** is given by

$$w = F\ d$$

The **SI unit of work** is the newton meter, N m, or **joule,** J; it is the work done when 1 newton acts through a distance of 1 meter.

$$1\ J = 1\ N\ m = 1\ kg\ m^2 s^{-2}$$

The capacity to do work is called **energy.** Energy and work are both measured in joules.

The **helium atom** has a nucleus surrounded by 2 electrons but has a mass approximately four times that of an H atom. Two protons, to balance the two electrons, account for 2 mass units. The remaining additional 2 mass units are due to the presence of 2 **neutrons** in the He nucleus. (He nucleus = 2 protons + 2 neutrons.) The **neutron** is a fundamental **neutral** particle with a mass which is slightly greater than that of the proton.

In general, the **nuclear charge** of an atom is given by its **atomic number,** Z, the number of protons it contains. The number of electrons outside of the nucleus is also Z in the **neutral atom.** The **nuclear mass** is given by the mass of Z protons and A neutrons. The sum, Z + A, of these nucleons is called the **mass number** or nucleon number. The constitution of any nucleus is indicated by adding to the symbol of the element, the atomic number, Z, as a **subscript** and the mass number, Z + A, as a **superscript**. For example, for atom X, we write

$$^{(Z+A)}_{Z}X$$

The mass of the Z electrons surrounding the nucleus makes only a very small contribution to the overall mass of an atom; the **atomic mass** is almost the same as the mass of the nucleus.

An **element is defined as a substance all of the atoms of which have the same atomic number Z.** Atoms with the same Z but different numbers of neutrons in their nuclei are known as **isotopes** of a given element. **Atomic mass** is measured in terms of the **atomic mass unit, u,** defined as one-twelfth of the mass of one $^{12}_{6}C$ atom.

The mass of one $^{12}_{6}C$ atom is exactly 12 u.

The actual atomic mass of any atom is not exactly the same as the sum of the masses of its constituent particles (protons, neutrons and electrons). A **mass defect** always occurs as a result of the mass that is converted to energy ($E = mc^2$) when an atom is formed from its constituent particles. The amount of energy involved is very large; in contrast, **energy changes that accompany chemical reactions** are relatively small and the corresponding changes in mass are negligible. In a chemical reaction, mass is conserved and the total number of atoms of all kinds remains unchanged in going from reactants to products; **atoms are conserved in chemical reactions.**

On earth, the natural isotopic composition of any element is approximately constant, as a result of the mixing together of substances during geological time. Compounds contain all of the constituent isotopes of their elements in the same abundances that they occur naturally. This makes it possible to draw up a table of **average atomic masses** for routine use in chemical calculations. Average atomic masses may be calculated using relative isotopic abundances and the masses of individual isotopes. Atomic masses in atomic units, **u,** represent the average masses of single atoms. Atomic masses in grams, g, represent the mass of **one mole** of atoms.

1 mole of entities = Avogadro's number of entities, N

$$N = 6.022 \times 10^{23}$$

$$1\ g = 6.022 \times 10^{23}\ u; \quad 1\ u = 1.66 \times 10^{-24}\ g$$

The **empirical formula mass** of a substance is obtained by adding together the masses of all of the atoms in its empirical formula; the **molecular mass** results from adding together the masses of all of the atoms in its molecular formula. For a single molecule the mass units are atomic units, u; for 1 mole of molecules, the units are grams, g. The mass of **one mole** of atoms, molecules or other entities is called the **molar mass.**

The **empirical formula** of a compound may be determined from its experimental elemental compositions (expressed normally as the % of each element by mass) by dividing the number of grams of each element in a given mass of compound (usually 100 g for convenience) by its atomic mass. This gives the ratio of moles of atoms of each kind, which is the same as the ratio of atoms of each kind. This result, when expressed as the simplest whole number ratio of atoms, gives the empirical formula. The **molecular formula** can only be determined when the molar or molecular mass is also known.

A **chemical equation** is a shorthand description of a reaction, and it gives the formulas of all the reactants and products. In a **balanced equation** the same numbers of atoms of each element appear on both sides of the equation. Symbols indicating the physical states of the reactants and products are also often included -- gas (g), liquid (l), solid (s) and aqueous solution (aq). To balance an equation the formulas of all the reactants and products must be known or deduced. The elements that appear least frequently in the equation are balanced first on both sides by balancing the numbers of atoms. It is usually then a simple matter to balance the numbers of atoms of the remaining elements.

A balanced equation is interpreted in a number of ways: (a) as giving the number of product molecules that result from the complete reaction of given numbers of reactant molecules; (b) as giving the number of moles of product molecules that result from complete reaction of given numbers of moles of reactant molecules (or any product thereof). By multiplying numbers of moles by the respective molar masses, the masses of products that result from complete reaction of certain masses of reactants can be calculated.

The reactants of a reaction are not necessarily present in exactly the amounts required by the balanced equation, in which case one reactant will be completely consumed in the reaction, leaving the other(s) in excess. The former is referred to as the **limiting reagent.** Not every reaction goes to completion. In general, because of incomplete reaction and product loss during purification the yields of reactions are not quantitative. The actual yield is expressed as a percentage of the theoretical yield calculated from the balanced equation.

In expressing the **concentrations of solutions,** the **molarity, M,** of a solute is defined as **moles of solute per liter of solution.** The units of molarity are mol L^{-1}, so that the molarity of a particular solute is obtained by dividing the number of moles by the volume in liters. A solution of known **molarity** is prepared by weighing out a known mass of solute (and hence a known number of moles), dissolving it in a minimum amount of solvent in a **volumetric flask** and then adding more solvent (and mixing) until the graduation mark on the neck of the flask is reached, which indicates a known, predetermined volume.

REVIEW QUESTIONS

1. How is the elemental composition of a substance normally expressed?
2. What is the lightest element in the periodic table?

3. Of what elementary particles is a hydrogen atom composed?
4. What elementary particles constitute the nuclei of atoms?
5. How does the relative size of a nucleus compare to the size of an atom?
6. What determines the approximate size of an atom, and why cannot the size be precisely determined?
7. What is force?
8. What is the SI unit of force?
9. What are the three types of forces that are known in nature?
10. Which of these forces is important in chemistry?
11. State Coulomb's law in words and as a mathematical expression.
12. What is the definition of work? What is the unit of work?
13. What is energy? What unit is used to measure energy?
14. What is kinetic energy? How is it related to mass and velocity?
15. What is potential energy?
16. State the law of conservation of energy.
17. Of what elementary particles, and in what numbers, is a He atom composed?
18. What is the atomic number, Z, of an atom?
19. What is the mass number (nucleon number) of an atom?
20. Why are the chemical properties of an atom related to its atomic number Z?
21. Why are atoms with the same atomic number Z atoms of the same element?
22. What is an isotope?
23. What is the unit of atomic mass?
24. Why are the masses of atoms less than the sum of the masses of their constituent elementary particles?
25. What is meant by the mass equivalent of energy?
26. Why can we say that in a chemical reaction the total mass of the reactants and products is conserved?
27. What are the isotopes of hydrogen? oxygen?
28. What is meant by average atomic mass?
29. Why can the average atomic masses of the atoms of most elements be considered as essentially constant over the earth?
30. How is the average atomic mass of the atoms of an element related to the masses of its isotopes?
31. Why are some atomic masses quoted to more significant figures than others?
32. How many ${}^{12}_{6}C$ atoms are there in exactly 12 g of ${}^{12}_{6}C$?
33. How many molecules are there in 1 mole of hydrogen molecules?
34. What is the value of the atomic mass unit u expressed in terms of Avogadro's constant?
35. What is meant by the empirical formula of a substance?
36. Why is the molecular formula of a substance not necessarily the same as its empirical formula?
37. What is the molar mass of sulfuric acid, H_2SO_4?
38. How are the empirical formulas of organic compounds containing only carbon and hydrogen, or carbon, hydrogen and oxygen, determined experimentally?
39. What are the empirical formula mass, the molecular mass, and the molar mass of C_4H_{10}?

40. Define the term molarity of a solute in solution; what are the units of molar concentration, M?
41. How would you make up a 0.1 M solution of sodium chloride in water, if you were supplied with solid NaCl, distilled water and a 500 mL volumetric flask?
42. How many grams of carbon (graphite) are required to react completely with 2.00 g of oxygen (O_2) to give CO_2, and what is the maximum mass of CO_2 that could be obtained?
43. What is meant by the limiting reagent in a chemical reaction?
44. What information is required before a balanced equation can be written for a particular reaction?
45. Why is the actual yield of a product normally less than the theoretical yield?

Answers to Selected Review Questions

37. 98.08 g mol^{-1} 39. 29.06 u; 58.12 u; 58.12 g mol^{-1}
41. 2.92 g NaCl and sufficient water to make up to 500 mL
42. 0.751 g graphite, 2.75 g CO_2

OBJECTIVES

Be able to:

1. State the basic types of forces present in nature and recognize the one important force in chemistry.
2. Give the number of electrons, protons and neutrons in an atom from its mass number and atomic number.
3. Define the term isotope and realize that molecules containing different isotopes have similar but not identical properties.
4. Define the term unified atomic mass unit.
5. Calculate average atomic masses from isotope abundance data, or abundances from the average mass.
6. Define the term work and give the SI units of force, work and energy.
7. State the laws of conservation of mass and energy.
8. Calculate the molecular mass, the formula mass, and the empirical formula mass of a substance.
9. Relate an atomic or molecular mass (in unified atomic mass units) to a molar mass (in grams).
10. Calculate the mass percentage composition of a compound from its formula.
11. Determine the empirical formula of a compound.
12. Balance simple chemical equations, given the reactants and products.
13. Perform, with balanced chemical equations, calculations involving masses or moles of reactants or products, including the determination of a limiting reactant.
14. Calculate the theoretical yield and the percent yield of a given product in a chemical reaction, given the amount of the limiting reactant and the actual mass of the product obtained in the reaction.
15. Calculate the molarity of a given solution, and use the molarity of a solution to calculate the number of grams or moles of solute contained in a given volume of solution.

PROBLEM SOLVING STRATEGIES

2.1 Essentials of Atomic Structure

Remember that the mass number for an isotope is the sum of its protons and neutrons, and that the atomic number is just the number of protons (or electrons). Thus ^{19}F contains 9 protons and 9 electrons, since the atomic number of F is 9, and 10 neutrons since 19 - 9 = 10.

2.2 Atomic Mass and 2.3 The Mole

In this chapter we learn how to calculate masses of individual atoms and molecules, as well as the mass of a mole of substance. All of these calculations are based on the unified atomic mass unit, u, which is exactly 1/12 of the mass of a $^{12}_{6}C$ atom. Since an individual atom has a very small mass (of the order of 10^{-24} g), one normally treats a **mole** of atoms (6.022×10^{23} atoms); in this case the mass is called the **molar** atomic mass. Various **average** molar atomic masses are listed on the inside cover of the textbook. An **average** is needed when there is more than one naturally-occurring isotope.

Example 2.1

When one mole of carbon reacts with one mole of oxygen molecules, one mole of carbon dioxide is produced. Calculate the mass of $^{12}C^{16}O_2$ produced when one mole of ^{12}C is combined with one mole of $^{16}O_2$.

Solution 2.1

Since carbon in this example is not naturally occurring carbon, but pure ^{12}C, we need the mass of a mole of ^{12}C atoms. In Table 2.1 the mass of **one** ^{12}C atom is given as 12.00000 u. Thus the mass of **one mole** of ^{12}C atoms is 12.00000 g since the mass of one atom, in unified atomic mass units, is numerically equal to the mass of one mole of atoms, in grams. Similarly, one $^{16}O_2$ molecule has a mass of 2 times the mass of one ^{16}O atom or from Table 2.1, 2×15.9949 u = 31.9898 u. Thus one $^{16}O_2$ mole of molecules has a mass of 31.9898 g. Then the mass of one mole of $^{12}C^{16}O_2$ is 12.0000 + 31.9898 = 43.9898 g.

You should be able to calculate the average atomic mass for an element from its isotope masses and their percentage abundances, as done in Example 2.2 of the text, or alternatively, to deduce the abundances given the average mass and the isotopic masses for an element that consists of two isotopes -- as done in Example 2.3 of the text. In the absence of exact values for isotopic masses, you can use their mass numbers.

2.4 Stoichiometric Calculations: Compositions and Formulas

With the table of average atomic masses we can calculate mass percentage compositions of the various atom types making up a substance, once we know its formula. In general the calculation can be based on a molecular formula or on an empirical formula. Both formulas will give the same answer. One counts how many atoms of the desired kind are present in the formula, determines the mass of this number of atoms, and divides this mass by the total formula mass. Multiplication by 100% gives the final result.

Example 2.2

Calculate the percentage composition of hydrogen peroxide, H_2O_2.

Solution 2.2

We will base the calculation on the molecular formula H_2O_2, and use one mole. For H, there are 2 moles of atoms in 1 mole of H_2O_2. The mass of these 2 moles of H atoms is 2 x 1.008 g = 2.016 g. The total mass of one mole of H_2O_2 is 2 x 1.008 + 2 x 16.00 = 34.02 g.

$$\% \text{ hydrogen} = \frac{2.016 \text{ g}}{34.02 \text{ g}} \times 100\% = 5.93\%$$

Likewise, for oxygen, there are 2 moles of atoms in one mole of H_2O_2:

$$\% \text{ oxygen} = \frac{2 \times 16.00 \text{ g}}{34.02 \text{ g}} \times 100\% = 94.06\%$$

Since there are only two types of atoms in H_2O_2, the % oxygen could have been calculated from 100% - % hydrogen.

In many situations, one determines **experimentally** the mass percentages of the various atoms in a substance, and wants to determine an empirical formula. (Additional measurements are required to convert an **empirical** formula into a **molecular** formula.) Here is a set of steps to follow in determining an empirical formula.

1. Find the mass of a given type of atom in a known amount of the compound, either from percentage composition or other given experimental data. If the percentage composition is given, it is convenient to assume that one is dealing with 100 g of the compound.
2. Convert this mass to the number of moles of atoms by using the appropriate unit conversion factor.
3. After repeating steps 1 and 2 for all the different atoms in the compound, find the relative numbers of moles of the constituent atoms by dividing throughout by the smallest number of moles of atoms; one of the resultant numbers will be unity. These numbers are also the ratios of the numbers of each kind of atom in the compound.
4. If all the numbers are not whole numbers, multiply them all by the smallest integer that will convert them all to whole numbers, or at least to within ±0.05 of whole numbers.

Example 2.3

A compound of sodium, oxygen and sulfur is found to contain 19.3% Na, 53.8% O, and 26.9% S. What is its empirical formula?

Solution 2.3

In 100 g of the compound there are 19.3 g Na, 53.8 g O and 26.9 g S. Unit conversion factors to convert mass to moles for Na, O, and S can be formed from the data on the inside cover of the textbook.

$$\frac{1 \text{ mol Na}}{22.99 \text{ g Na}} = 1, \quad \frac{1 \text{ mol O}}{16.00 \text{ g O}} = 1, \quad \frac{1 \text{ mol S}}{32.06 \text{ g S}} = 1.$$

Thus the numbers of moles of atoms in a 100 g sample are

$$19.3 \text{ g Na} \times \frac{1 \text{ mol Na}}{22.99 \text{ g Na}} = 0.839 \text{ mol Na}$$

$$53.8 \text{ g O} \times \frac{1 \text{ mol O}}{6.00 \text{ g O}} = 3.36 \text{ mol O}$$

$$26.9 \text{ g S} \times \frac{1 \text{ mol S}}{32.06 \text{ g S}} = 0.839 \text{ mol S}$$

The relative numbers of moles of Na:O:S are

$$\frac{0.839}{0.839} : \frac{3.36}{0.839} : \frac{0.839}{0.839} = 1.00{:}4.00{:}1.00$$

Thus the empirical formula is NaO_4S.

Sometimes percentage composition data are not given in a problem. Let's see how other data can be used to obtain an empirical formula. The steps are basically the same as before, but we find the mass of a given type of atom using unit conversion factors.

Example 2.4

A compound of carbon, hydrogen, oxygen and sulfur is analyzed by two methods. In the first, a 10.00 g sample is burned in oxygen to give 20.90 g CO_2 and 4.28 g H_2O. In the second, a separate 5.00 g sample has all of its sulfur converted to 9.23 g barium sulfate, $BaSO_4$. The oxygen content of the compound was obtained by difference. What is the empirical formula of the compound?

Solution 2.4

We will base our calculation on the 10.0 g sample. First let us find how much carbon is in the compound. We need the unit conversion factor for carbon in CO_2.

$$\frac{12.01 \text{ g C}}{(12.01 + 2 \times 16.00) \text{g } CO_2} = \frac{12.01 \text{ g C}}{44.01 \text{ g } CO_2}$$

By multiplying this unit conversion factor by the amount of CO_2 produced from the 10.00 g sample, we obtain the number of grams of carbon present in the sample.

$$20.90 \text{ g } CO_2 \times \frac{12.01 \text{ g C}}{44.01 \text{ g } CO_2} = 5.70 \text{ g C}$$

This corresponds to

$$5.70 \text{ g C} \times \frac{1 \text{ mol C}}{12.01 \text{ g C}} = 0.475 \text{ mol C}$$

Next, for hydrogen, we need a new set of conversion factors to find how much hydrogen was contained in the sample.

$$4.28 \text{ g } H_2O \times \frac{2.016 \text{ g H}}{18.02 \text{ g } H_2O} = 0.479 \text{ g H, which is } 0.475 \text{ mol H.}$$

For the sulfur content, let us first calculate what mass of $BaSO_4$ would have been produced if the sample had been 10.00 g rather than 5.00 g, since our C and H results are based on a 10.00 g sample. This mass is 2 x 9.23 = 18.46 g $BaSO_4$. Next we need a unit conversion factor relating S and $BaSO_4$.

$$18.46 \text{ g } BaSO_4 \times \frac{32.06 \text{ g S}}{233.4 \text{ g } BaSO_4} = 2.537 \text{ g S,}$$

and

$$2.537 \text{ g S} \times \frac{1 \text{ mol}}{32.06 \text{ g S}} = 0.0791 \text{ mol S}$$

Finally the mass of oxygen in the 10.00 g sample is found by difference:

$$10.00 \text{ g} - (5.70 + 0.48 + 2.54)\text{g} = 1.28 \text{ g O,}$$

and the moles of O are $1.28 \text{ g O} \times \frac{1 \text{ mol}}{16.00 \text{ g}} = 0.080 \text{ mol O}$.

The relative numbers of moles of C:H:O:S are

$$\frac{0.475}{0.079} : \frac{0.475}{0.079} : \frac{0.080}{0.079} : \frac{0.079}{0.079} = 6.01:6.01:1.01:1$$

Converting these numbers to whole numbers, we find that the empirical formula is C_6H_6OS.

2.5 Stoichiometric Calculations: Reactions

Now we have the required background to use balanced chemical equations in chemical calculations. Remember there are 2 ways to interpret a balanced chemical equation. First, it is a statement of the number of **atoms** and/or **molecules** of reactants that are completely converted into product **atoms** and/or **molecules.** Second, it tells us how many **moles** of products result from complete reaction of the given number of **moles** of reactants. We have already seen how to calculate **masses** of individual atoms or molecules, or **masses** of moles of atoms or molecules.

Example 2.5

Silver carbonate, Ag_2CO_3, can be decomposed into silver, oxygen gas and carbon dioxide. How many grams of Ag_2CO_3 are required to produce 10.00 g of silver? How many grams of O_2 are produced at the same time?

Solution 2.5

First we need to write a balanced equation. In an **unbalanced** form we have

$$Ag_2CO_3 \rightarrow Ag + O_2 + CO_2$$

By inspection we note that there are only 3 O atoms on the left but 4 on the right. Therefore we must **at least** insert a coefficient of 2 in front of Ag_2CO_3. This then forces us to write a 2 in front of CO_2 on the right, to balance the number of carbons. We now see that both carbon and oxygen atoms are balanced, and we need only to multiply Ag by 4 on the right to balance silver atoms.

$$2Ag_2CO_3 \rightarrow 4Ag + O_2 + 2CO_2$$

The equation tells us that 2 **moles** of Ag_2CO_3 decompose to 4 **moles** of Ag, 1 **mole** of O_2 and 1 **mole** of CO_2. The path to finding how many **grams** of Ag_2CO_3 are produced is

$$\text{g Ag} \rightarrow \text{moles Ag} \rightarrow \text{moles } Ag_2CO_3 \rightarrow \text{g } Ag_2CO_3$$

In general, we cannot convert from grams of one material to grams of another in a reaction without going through moles, since the only connection we have between materials is via moles from the balanced equation.

We need 3 unit conversion factors:

$\frac{1 \text{ mol Ag}}{107.9 \text{ g Ag}}$ to convert from g Ag to moles Ag;

$\frac{2 \text{ mol } Ag_2CO_3}{4 \text{ mol Ag}}$ to convert from moles Ag to moles Ag_2CO_3;

$\frac{275.8 \text{ g } Ag_2CO_3}{1 \text{ mol } Ag_2CO_3}$ to convert from moles Ag_2CO_3 to g Ag_2CO_3

$$\text{Thus mass } Ag_2CO_3 = 10.00 \text{ g Ag} \times \frac{1 \text{ mol Ag}}{107.9 \text{ g Ag}} \times \frac{2 \text{ mol } Ag_2CO_3}{4 \text{ mol Ag}} \times \frac{275.8 \text{ g } Ag_2CO_3}{1 \text{ mol } Ag_2CO_3} = 12.78 \text{ g } Ag_2CO_3$$

The second part of the problem is similar to the first; thus

$$\text{mass } O_2 = 10.00 \text{ g Ag} \times \frac{1 \text{ mol Ag}}{107.9 \text{ g Ag}} \times \frac{1 \text{ mol } O_2}{4 \text{ mol Ag}} \times \frac{32.00 \text{ g } O_2}{1 \text{ mol } O_2} = 0.7414 \text{ g } O_2$$

Limiting Reactant

Sometimes, the numbers of moles of reactants present at the start of a reaction are not in the ratios demanded by the balanced chemical equation. That is, some of the reactants are present in excess, and will not be converted to products. Here is a systematic approach to the solution of such problems that involve a **limiting reactant**.

1. Balance the chemical equation.
2. If **masses** of reactants are given, convert them to moles.
3. Deduce the amount (moles or grams) of one product that would be obtained by assuming in turn that each reactant is fully consumed. The one reactant that produces the **smallest** amount of product is the limiting reactant.
4. The maximum number of moles of a product that can be formed is the number of moles of that product formed from the limiting reactant, using the balanced equation. Convert to grams of product, if required.
5. If needed, the number of moles of excess reactants can be calculated.

$$\begin{pmatrix}\text{excess moles} \\ \text{of a reactant}\end{pmatrix} = \begin{pmatrix}\text{total moles of} \\ \text{reactant at start}\end{pmatrix} - \begin{pmatrix}\text{moles reactant needed to} \\ \text{react with limiting reactant}\end{pmatrix}$$

Example 2.6

Zinc oxide, ZnO, can be produced by heating zinc sulfide, ZnS, in the presence of O_2. Sulfur dioxide, SO_2 is also produced. How many kg of ZnO can be obtained from 1000 kg ZnS and 500.0 kg O_2? How much of one of the reactants remains?

Solution 2.6

The balanced equation is

$$2ZnS + 3O_2 \rightarrow 2ZnO + 2SO_2$$

If ZnS is fully consumed, the amount of ZnO produced is

$$1000 \text{ kg ZnS} \times \frac{10^3 \text{ g}}{1 \text{ kg}} \times \frac{1 \text{ mol ZnS}}{97.43 \text{ g ZnS}} \times \frac{2 \text{ mol ZnO}}{2 \text{ mol ZnS}} \times \frac{81.37 \text{ g ZnO}}{1 \text{ mol ZnO}}$$

$$= 8.352 \times 10^5 \text{ g ZnO}$$

Alternatively, if O_2 is fully consumed, the ZnO produced is

$$500.0 \text{ kg } O_2 \times \frac{10^3 \text{ g}}{1 \text{ kg}} \times \frac{1 \text{ mol } O_2}{32.00 \text{ g } O_2} \times \frac{2 \text{ mol ZnO}}{3 \text{ mol } O_2} \times \frac{81.37 \text{ g ZnO}}{1 \text{ mol ZnO}}$$

$$= 8.476 \times 10^5 \text{ g ZnO}$$

Thus the O_2 is in excess and ZnS is limiting, since less ZnO is obtained in the ZnS calculation. To determine how much O_2 remains, we calculate the mass of O_2 needed to produce 8.352×10^5 g ZnO and subtract this amount from the O_2 available:

$$\text{mass } O_2 \text{ left} = 5.000 \times 10^5 \text{ g } O_2 - (8.352 \times 10^5 \text{ g ZnO} \times \frac{1 \text{ mol ZnO}}{81.37 \text{ g ZnO}}$$

$$\times \frac{3 \text{ mol } O_2}{2 \text{ mol ZnO}} \times \frac{32.00 \text{ g } O_2}{1 \text{ mol } O_2}) = 7.32 \times 10^3 \text{ g } O_2$$

2.6 Molar Concentrations

The concentration unit used most frequently in **solution** chemistry is **molarity,** and is defined as the number of moles of solute in exactly 1 L solution. To use it, you need to be able to convert **mass** of solute into moles. Also, if the volume of the solution is not exactly 1 L, you must calculate how many moles of solute would be present if there were exactly 1 L of the solution. Let's see an example.

Example 2.7

1.03 g of silver nitrate, $AgNO_3$, are dissolved in sufficient water to make 0.200 L solution. What is the molarity of the solution? How many mL of this solution would be needed to prepare 2.00 L of a 1.00×10^{-3} M solution of $AgNO_3$?

Solution 2.7

We need to find how many moles of $AgNO_3$ are present.

$$1.03\ g\ AgNO_3 \times \frac{1\ mol\ AgNO_3}{169.9\ g\ AgNO_3} = 6.06 \times 10^{-3}\ mol\ AgNO_3$$

To find the molarity, we divide the number of moles of $AgNO_3$ by the volume of solution in liters.

$$\frac{6.06 \times 10^{-3}\ mol\ AgNO_3}{0.200\ L} = 3.03 \times 10^{-2}\ M$$

The second part of the problem requires a dilution. We know that 2.00 L of the desired 1.00×10^{-3} M solution of $AgNO_3$ will contain

$$\frac{1.00 \times 10^{-3}\ mol\ AgNO_3}{1\ L} \times 2.00\ L = 2.00 \times 10^{-3}\ mol\ AgNO_3$$

We convert moles $AgNO_3$ to volume of solution using a unit factor based upon the concentration:

$$2.00 \times 10^{-3}\ mol\ AgNO_3 \times \frac{1\ L}{3.03 \times 10^{-2}\ mol\ AgNO_3} \times \frac{1000\ mL}{1\ L} = 66.0\ mL$$

Thus, if 66.0 mL of the 3.03×10^{-2} M solution were placed in an empty 2 L volumetric flask, and diluted to the 2.00 L mark with distilled water, we would have the desired solution.

SELF TEST

A table of atomic masses may be consulted. The Avogadro constant = 6.022×10^{23}.

1. An isotope of carbon has mass number 13 and atomic number 6. How many protons, neutrons and electrons does this isotope contain?
   (a) 6 protons, 6 neutrons and 6 electrons
   (b) 13 protons, 13 neutrons and 6 electrons
   (c) 6 protons, 7 neutrons and 6 electrons
   (d) 6 protons, 13 neutrons and 6 electrons
   (e) 13 protons, 7 neutrons and 7 electrons
2. An element with average atomic mass 101.0 consists of two isotopes: Isotope A with mass number 100.0 and isotope B with mass number 104.0. What is the relative abundance of isotope A?
   (a) 33% (b) 50% (c) 66% (d) 25% (e) 75%
3. How many molecules are there in 2.5 moles of $C_6H_{12}O_6$?
   (a) 6.02×10^{24} (b) 1.5×10^{24} (c) 6.03×10^{23} (d) 2.5×10^{23}
   (e) 9.1×10^{23}
4. A compound with a molar mass of 30.04 g mol^{-1} contains nitrogen and hydrogen only. The percentage by mass of nitrogen is 93.3%. What is the molecular formula of this compound?
   (a) NH (b) NH_3 (c) HN_3 (d) N_2H_2 (e) N_2H_4

5. How many moles of NCl_3 are present in an 800.0 g sample of NCl_3?
(a) 8.00 (b) 6.64 (c) 76.5 (d) 12.5 (e) 60.2

6. A 5.82 gram silver coin is dissolved in nitric acid. When NaCl is added, all the silver is precipitated as AgCl. The AgCl precipitate has a mass of 7.20 grams. What is the percentage of silver in the coin?
(a) 93.1 (b) 99.4 (c) 76.5 (d) 86.5 (e) 53.2

7. A substance has the following percentage composition: S = 46.27%, Fe = 53.73%. What is the empirical formula of the substance?
(a) FeS_2 (b) Fe_2S (c) Fe_3S_2 (d) Fe_2S_3 (e) FeS

8. When the following equation is balanced (using the smallest integers possible): $C_2H_2 + O_2 \rightarrow CO_2 + H_2O$, what will be the coefficient in front of O_2?
(a) 1 (b) 2 (c) 3 (d) 4 (e) 5

9. The formation of hydrogen sulfide, H_2S, from its elements occurs at elevated temperatures. If 2.000 kg H_2 is reacted with 1.000 kg S, what is the limiting reactant?
(a) H_2 (b) S (c) H_2S (d) there is no limiting reactant

10. Using the data of question 9, how many grams of H_2S can be produced?
(a) 3000 (b) 1062 (c) 1.092 (d) 337.9×10^2 (e) 1000

11. The burning of propane, C_3H_8, is described by $C_3H_8(g) + 5O_2(g) \rightarrow 3CO_2(g) + 4H_2O(g)$. How many grams of propane must be completely burned to produce 7.2 grams of water?
(a) 2.2 (b) 4.4 (c) 8.8 (d) 17.6 (e) 70.4

12. How many grams of acetic acid, CH_3COOH, are required to prepare 500 mL of 0.40 M CH_3COOH?
a) 12 (b) 120 (c) 1.2 (d) 48 (e) 75

13. Aluminum reacts with hydrochloric acid, HCl(aq), according to the equation $2Al + 6HCl(aq) \rightarrow 2AlCl_3 + 3H_2(g)$. If 300 mL of 12.0 M hydrochloric acid reacted with an excess of aluminum, how many grams of hydrogen could be produced?
(a) 1.8 (b) 3.6 (c) 6.0 (d) 7.2 (e) 14.4

Answers to Self Test

1. c; 2. e; 3. b; 4. d; 5. b; 6. a; 7. d; 8. e; 9. b;
10. b; 11. b; 12. a; 13. b.

CHAPTER 3

THE ATMOSPHERE AND THE GAS LAWS

SUMMARY REVIEW

The gaseous **atmosphere,** the liquid **hydrosphere** and the solid **lithosphere** together comprise the earth's **crust,** in which more than half the atoms are oxygen atoms. Next in abundance, on an atom basis, are silicon and hydrogen (~ 15% each), aluminum (5%), sodium (2%), iron, calcium, magnesium, potassium (1 to 2% each) and titanium (0.2%). All of the remaining elements represent together only 0.5% of the total number of atoms, and among these the most abundant are Cl, Mn, C, S, Ar, N, Rb, Sr and F.

The atmosphere is divided into two layers. One layer extends from the surface to about 80 km and is largely composed of a uniform mixture of N_2 and O_2. The other extends out as far as 10000 km and is composed, in ascending order, of layers of molecular nitrogen, atomic oxygen, helium, and atomic hydrogen.

Beneath the earth's crust is a fluid **mantle,** composed mainly of iron and calcium and magnesium silicates, which surrounds a solid **core** of principally nickel and iron.

When heated with O_2 the majority of elements form **oxides** in **exothermic** reactions in which heat is liberated. When sufficient heat is given out to sustain an **oxidation** reaction, it is referred to as burning or **combustion.** Only the noble gases and the metals gold and platinum do not react directly with O_2. Many compounds are converted to the oxides of their constituent elements when burnt in oxygen.

Reactions that occur with the absorption of heat are **endothermic** reactions.

Any reaction in which oxygen is removed, either partially or completely, from a compound is called a **reduction** reaction. An oxidation reaction is always accompanied by a reduction.

Hydrogen is the most abundant element in the universe. Interstellar space is sparsely filled, mainly with hydrogen atoms. The stars are mostly composed of hydrogen, as is our sun. Compounds containing hydrogen are very common in the earth's crust. Industrially, H_2 is obtained from the mixture $H_2(g)$ and $CO(g)$ **(water gas)** that results from passing steam over coke (carbon). It also results from the reaction of reactive metals, such as Na or Ca, with water, or from the reaction of less reactive metals, such as Fe, with steam. In the manufacture of chlorine and sodium hydroxide by the electrolysis of NaCl(aq), hydrogen is an important by-product. **Synthesis gas** results when a mixture of methane and steam is passed over a heated nickel catalyst:

$$CH_4 + H_2O \rightarrow CO + 3H_2$$

H_2 reacts with many elements to give **hydrides,** such as NaH, CaH_2, CH_4, NH_3 and H_2O. It is a good **reducing agent.**

Nitrogen fixation is any process in which N_2 from the air is converted to nitrogen compounds. Certain bacteria cause such reactions in plants. Industrially, the manufacture of ammonia, NH_3, from H_2 and N_2, in the **Haber process** is the starting point for the synthesis of all nitrogen compounds.

A gaseous sample is characterized by its **mass, m, volume, V, pressure, P,** and **temperature, T.**

A **barometer** is an instrument for measuring pressure, such as atmospheric pressure. The Torricellian (mercury) barometer consists of a mercury column in an inverted closed tube which exactly balances the pressure of the external atmosphere. A height of 760 mm of Hg is defined as a standard pressure of 1 atmosphere.

$$1 \text{ atm} = 760 \text{ mm Hg} = 760 \text{ torr} = 101.33 \text{ kPa}$$

All gases and mixture of gases, including air, have a very similar physical behavior. They all approximately obey the following laws:

Boyle's law, PV = const, for a given mass of gas at constant temperature, or

$$P_1V_1 = P_2V_2 \text{ (const T)}$$

Charles's law, V/T = const, for a given mass of gas at constant pressure, or

$$V_1/T_1 = V_2/T_2 \text{ (const P)}$$

where T is the temperature in **kelvins.** The **absolute** zero of temperature (-273.15°C) is the temperature at which the volume would be zero. It is the basis of the **Kelvin temperature scale.** T in K and t in °C are related by the expression

$$T = t + 273.15$$

Standard temperature and pressure (STP) is defined as a temperature of 0°C (273.15 K) and a pressure of 1 atmosphere (760 mm Hg or 101.33 kPa).

The combination of Boyle's law and Charles's law gives the **combined gas law,** PV/T = const, for a given mass of gas, or

$$P_1V_1/T_1 = P_2V_2/T_2$$

Most gases and mixtures of gases obey this law to a good approximation.

A theoretical basis for the gas laws is given by the **kinetic molecular theory of gases,** which assumes:

1. gases are composed of molecules that are far apart relative to their size;
2. the molecules are in a constant state of random motion and continue to travel in straight lines unless they collide with each other, or with the walls of a container;
3. the molecules exert no intermolecular forces on each other, or on the walls of a container -- all collisions are elastic.

On this basis, the pressure results from the continuous bombardment of the walls of a container by rapidly moving molecules. Doubling the number of collisions by halving the volume doubles the pressure (Boyle's law). The average kinetic energy of the molecules ($mv^2/2$) is proportional to the absolute temperature. At 0 K, the kinetic energy and velocity are both zero; all the molecules are at rest and no lower temperature is possible.

In practice, all gases condense to liquids and form solids at some temperature above 0 K, which indicates that attractive forces **(intermolecular or van der Waals forces)** exist between all molecules.

At constant V and T, the pressure of a gas is proportional to the number of molecules and, thus, to the number of moles, n.

$$P/n = \text{const (V and T constant)}$$

Combining this with the combined gas law gives the **ideal gas law**

$$PV = nRT$$

where **R is the universal gas constant.** With P in atm, V in L and T in K, R = 0.08206 atm L K^{-1} mol^{-1}.

An important use of the ideal gas equation is to calculate the number of moles of a gas, n, from its volume, pressure and temperature. For a gaseous compound, if the mass of gas is also known, this gives an important way of obtaining its **molar mass.**

At constant T and P, the volume of a gas is proportional to the number of moles. Thus, at constant T and P, the volumes of the gaseous reactants and products of a reaction are in the ratio of the small whole numbers which are given by the balanced equation for the reaction. This is the **law of combining volumes.**

Dalton's law of partial pressures states that in a mixture of gases, the partial pressure of each constituent gas is the same as it would exert alone in the same volume at the same temperature. The total pressure is the sum of the partial pressures.

Diffusion is the process by which gases mix with each other and **effusion** is the process by which a gas escapes from a container through a small opening. The rate of diffusion of one gas into another, or its rate of effusion, is proportional to the average velocity of its molecules. The relative rates of diffusion or effusion of two gases is given by Graham's law,

$$r_1/r_2 = \sqrt{m_2/m_1} = \sqrt{M_2/M_1}$$

where M_2 and M_1 are the molar masses of the gases. The average velocities of gaseous molecules are very high at room temperature. However, as a consequence of many molecular collisions per second, molecules move by very tortuous paths. The average distance traveled between collisions is called the **mean free path.**

REVIEW QUESTIONS

1. What are the principal layers into which the atmosphere can be divided?
2. What are the main components of air?
3. Why is hydrogen found only in the outermost layer of the atmosphere and why as atoms rather than molecules?

4. How is the ozone layer formed? What is its importance?
5. What are the first and second most abundant elements found in the earth's crust?
6. How could a sample of oxygen be obtained in the laboratory? From air?
7. What is a compound containing oxygen and another element called?
8. What is an oxidation reaction? What is a reduction reaction?
9. Given that iron filings burn in oxygen to give an iron oxide of formula Fe_2O_3, write a balanced equation for the reaction.
10. How many grams of Fe_2O_3 would result from the combustion of 5.58 g of iron?
11. What is water gas, and how is it produced industrially?
12. What is synthesis gas, and how is it produced industrially?
13. What other methods may be used to produce hydrogen?
14. What is a compound formed between hydrogen and another element called?
15. What is meant by nitrogen fixation?
16. How is ammonia, NH_3, produced industrially?
17. Why are small amounts of methane, CH_4, found in the atmosphere?
18. What are the four fundamental properties used to describe a sample of a gas?
19. Draw a diagram of a mercury barometer and describe how it measures the pressure of the atmosphere.
20. Why is a pressure of 1 atmosphere often expressed as 760 mm of mercury?
21. State Boyle's law. State Charles's law.
22. What is meant by the absolute zero of temperature, and what is its value in °C?
23. Define standard temperature and pressure (STP).
24. State the combined gas law.
25. What is the relationship between the pressure of a given amount of gas and its temperature at constant volume?
26. In the kinetic molecular theory of gases, what three fundamental properties are attributed to gas molecules?
27. How does the kinetic molecular theory account for the pressure of a gas?
28. How does the kinetic molecular theory lead to the idea of an absolute zero of temperature?
29. What evidence is there that weak attractions exist between molecules?
30. Why is the pressure of a gas at constant temperature and volume proportional to the number of moles of gas?
31. State and explain Gay Lussac's law of combining volumes, with particular reference to the reaction between hydrogen and oxygen to give gaseous water.
32. Calculate the volume of 1 mole of an ideal gas at STP.
33. State and explain Dalton's law of partial pressures.
34. What is the relationship between the average velocities of the molecules of two gases at the same temperature and their molecular masses?
35. What law governs the rates of diffusion or effusion of gases?
36. How can effusion be used practically to separate a mixture of two gases?
37. Calculate the relative rates of effusion expected for H_2 and D_2.
38. Why for gases are the rates of effusion and diffusion small compared to their average velocities?

Answers to Selected Review Questions

9. $4Fe(s) + 3O_2 \rightarrow 2Fe_2O_3(s)$ 10. 7.99 g 32. 22.4 L
37. Ratio of rates = 1.41

OBJECTIVES

Be able to:

1. Name the main constituents of air and the earth's crust.
2. Describe some physical properties of oxygen and how oxygen is prepared industrially and in the laboratory.
3. Give the names, formulas, and descriptions of the products of reactions involving oxygen and common elements.
4. Define the terms endothermic, exothermic, oxidation, reduction.
5. Describe some physical properties of nitrogen and how nitrogen is prepared industrially.
6. Give the names and formulas of the products of high temperature reactions involving nitrogen with hydrogen, oxygen and magnesium.
7. Describe some physical properties of hydrogen and how hydrogen is prepared industrially and in the laboratory.
8. Give the names and formulas of the products of reactions involving hydrogen with oxygen, nitrogen, carbon, chlorine, and common metals.
9. Relate mass, volume, pressure and temperature of a gas by Boyle's law, Charles's law, and the combined gas law.
10. Define the terms Kelvin temperature scale and the conditions of standard temperature and pressure (STP).
11. State the assumptions of the kinetic theory of gases, and account for the gas laws with these assumptions.
12. Recognize the consequences of the existence of intermolecular forces in matter.
13. Define the term ideal gas and use the ideal gas law to interrelate in problems molecular mass, pressure, volume, temperature, density.
14. State Avogadro's law and Dalton's law of partial pressures.
15. Define the terms diffusion and effusion and use Graham's law to relate molecular mass and rate of effusion.

PROBLEM SOLVING STRATEGIES

Chapter 3 begins with some chemistry of oxygen and hydrogen. Be sure to learn the formulas of the oxides and hydrides, and the balanced equations by which they are prepared. We will see the oxides and hydrides again in later chapters.

3.5 Physical Properties of Gases and The Gas Laws

A gas sample is usually described in terms of four of its properties: mass m, volume V, pressure P and temperature T. The behavior of gases has been studied extensively, and the gas laws, or relationships between these properties, have been formulated. Since all these properties depend on one another, the dependence of **one** property of a gas on **one** other property can be investigated only if the values of the remaining **two** properties are kept constant.

Pressure and Volume: Boyle's Law

Boyle's law relates P and V of a given gas, **with m and T held fixed.** One form of the law is PV = constant. Usually pressures are expressed in **atmospheres,** rather than the SI unit, pascals. Unit conversion factors are needed to relate these two units or to express pressure in other units. These factors are derived from the relation

$$1 \text{ atm} = 101.33 \text{ kPa} = 760 \text{ mm Hg} = 760 \text{ torr}$$

Example 3.1

A sample of nitrogen occupies a volume of 740 mL at a pressure of 0.950 atm. What will be the volume, in liters, of the same sample of gas if the pressure is reduced to 620 torr at the same temperature?

Solution 3.1

Since mass and temperature are held constant, we recognize this as a Boyle's law problem.

$$P_1V_1 = \text{constant} = P_2V_2$$

Dividing both sides by P_2, we obtain

$$V_2 = \frac{P_1}{P_2} \times V_1$$

Since the volume V_2 needs to be expressed in liters, we convert V_1 to liters.

$$V_1 = 740 \text{ mL} \times \frac{1 \text{ L}}{10^3 \text{ mL}} = 0.740 \text{ L}$$

P_1 and P_2 are not in the same units. Let's choose atm as the desired pressure unit. Then

$$P_2 = 620 \text{ torr} \times \frac{1 \text{ atm}}{760 \text{ torr}} = 0.816 \text{ atm}$$

Finally

$$V_2 = \frac{0.950 \text{ atm}}{0.816 \text{ atm}} \times 0.740 \text{ L} = 0.862 \text{ L}$$

Is this answer reasonable? Human error can occur in manipulating even the simplest mathematical expressions, and mistakes in arithmetic can happen. Let's qualitatively check the answer. Boyle's law states that P and V are **inversely** related; if the **final pressure,** P_2, is **less than the initial pressure,** P_1, then the **final volume,** V_2, **must be greater than** the **initial volume,** V_1. In other words, V_1 must be multiplied by a pressure **ratio** greater than 1, so that V_2 will be larger than V_1. The above calculations shows $V_2 > V_1$, as expected. If we had calculated $V_2 < V_1$ for this problem, there must have been a mistake somewhere.

Temperature and Volume: Charles's Law

Charles's law relates V and T, **with m and P constant.** The law states that volume is **directly** proportional to the temperature under these conditions. The **temperature must be expressed on the Kelvin scale.** Errors with Charles's law calculations are usually due to inserting the wrong temperature (or a temperature in °C) into the expression. Let's see an example.

Example 3.2

A 1.00 L sample of oxygen is warmed from 25°C to 100°C at constant pressure. What volume, in liters, is occupied by the gas?

Solution 3.2

This is an example of Charles's law, with mass and pressure constant. Since the gas is **warmed,** we know the final volume will be **larger** than the initial volume. Thus, even if we have forgotten the mathematical form of Charles's law, we know

$V_2 = \frac{T_2}{T_1} \times V_1$, since the ratio T_2/T_1 greater than 1 gives $V_2 > V_1$

$T_1 = 25 + 273 = 298 \text{ K}$

$T_2 = 100 + 273 = 373 \text{ K}$

$V_2 = \frac{373 \text{ K}}{298 \text{ K}} \times 1.00 \text{ L} = 1.25 \text{ L}$

Combined Gas Law

The combined gas law allows us to calculate the effect of a change in **two** of the properties P, V, T on the value of the third, **with mass held constant.**

$$\frac{P_1 V_1}{T_1} = \frac{P_2 V_2}{T_2}$$

If you can work Boyle's law and Charles's law problems, this relationship should offer you no difficulties.

Example 3.3

A quantity of methane gas, originally at STP, has both its volume and pressure doubled. Will the final temperature be greater or less than the original temperature?

Solution 3.3

Starting with $\frac{P_1V_1}{T_1} = \frac{P_2V_2}{T_2}$, by rearrangement we get

$$\frac{T_2}{T_1} = \frac{P_2V_2}{P_1V_1} = \frac{2P_1 \times 2V_1}{P_1V_1} = 4.$$

Thus the final Kelvin temperature, T_2, is four times larger than the original temperature. Note that we did not make use of the fact that original conditions were STP, that is, 1 atm and 273 K. If the problem had asked for the value of T_2, then we could have calculated

$$T_2 = 4 \times 273 \text{ K} = 1.09 \times 10^3 \text{ K}$$

3.7 The Ideal Gas Law

The ideal gas law relates **all four** of the properties mass, pressure, volume, and temperature (always Kelvin scale). However the **mass** of gas must first be converted into n, the number of moles of the gas. The ideal gas law PV = nRT contains the gas constant R. If P is expressed in atm and V in L, then $R = 0.0821 \text{ L atm mol}^{-1} \text{ K}^{-1}$; if P has units of kPa and V is given in dm^3, then $R = 8.31 \text{ kPa dm}^3 \text{ mol}^{-1} \text{ K}^{-1}$. The simplest ideal gas law problems give R and any three of P, V, T, n and ask for the remaining variable.

Example 3.4

How many moles of ammonia are contained in a 500 mL vessel at 25°C and 730 mm Hg?

Solution 3.4

$$PV = nRT \qquad n = \frac{PV}{RT} \qquad R = 0.0821 \text{ L atm mol}^{-1} \text{ K}^{-1}$$

V must be in L, P in atm, and T in kelvins

$$V = 500 \text{ mL} \times \frac{1 \text{ L}}{10^3 \text{ mL}} = 0.500 \text{ L}$$

$$P = 730 \text{ mm Hg} \times \frac{1 \text{ atm}}{760 \text{ mm Hg}} = 0.961 \text{ atm}$$

$$T = 25 + 273 = 298 \text{ K}$$

$$n = \frac{0.961 \text{ atm} \times 0.500 \text{ L}}{0.0821 \text{ L atm mol}^{-1} \text{ K}^{-1} \times 298 \text{ K}} = 1.96 \times 10^{-2} \text{ mol}$$

The ideal gas law can also be used to calculate a molar mass M, since

$$n = \text{number of moles} = \frac{\text{mass of gas}}{\text{molar mass of gas}} = \frac{m}{M}$$

Thus we write $PV = \frac{m}{M} RT$. Of course m and M must be expressed in the same mass units.

Gas Density

Another useful relationship can be obtained by rearranging the above equation.

$$PVM = mRT$$

$$PM = \frac{m}{V} RT = dRT$$

where d is the density of the gas. Measurement of d, P, and T allows M to be calculated. By the way, a gas density measurement, coupled with an empirical formula calculation (Chapter 2), allows you to write a **molecular formula** for a compound.

Example 3.5

Nitrous oxide is a gaseous compound of nitrogen and oxygen atoms in a 2:1 ratio. At 25°C the density of a nitrous oxide sample was 1.73 g L^{-1} at a pressure of 730 torr. What is the molecular formula of nitrous oxide?

Solution 3.5

With gas density given, you should realize that a rearranged ideal gas law expression is needed.

$$PV = nRT = \frac{m}{M} RT \qquad M = \frac{mRT}{VP} = \frac{dRT}{P}$$

Conversion factors are required to change torr to atm and °C to K.

$$P = 730 \text{ torr} \times \frac{1 \text{ atm}}{760 \text{ torr}} = 0.961 \text{ atm}$$

$$T = 25 + 273 = 298 \text{ K}$$

$$M = \frac{1.73 \text{ g L}^{-1} \times 0.0821 \text{ L atm mol}^{-1} \text{ K}^{-1} \times 298 \text{ K}}{0.961 \text{ atm}}$$

$$= 44.0 \text{ g mol}^{-1}$$

With the empirical formula given as N_2O, the empirical formula mass is 2 x 14.01 + 16.00 = 44.02. Thus the empirical formula is identical to the molecular formula in this case. Nitrous oxide is N_2O.

Molar Volume of a Gas: Avogadro's Law

When gaseous reactants are converted to gaseous products in a balanced chemical reaction, we know the relative numbers of **moles** of substances involved. If there is a common constant T and P for each gas, then from PV = nRT for each gas it follows that the **volumes** of the gases are in the same ratios. In these cases we can work directly with **volumes,** and do not have to first convert to moles.

Example 3.6

What volume of ammonia can be produced from 3.00 L of nitrogen and sufficient hydrogen at 500°C? Assume each gas is at the same pressure.

Solution 3.6

First we need a balanced equation

$$N_2 + 3H_2 \rightarrow 2NH_3$$

Since P and T are the same for each gas, this is the same as 1 volume of N_2 producing 2 volumes of NH_3. Thus

$$3.00 \text{ L } N_2 \times \frac{2 \text{ vol } NH_3}{1 \text{ vol } N_2} = 6.00 \text{ L } NH_3 \text{ produced}$$

Of course this is the same result as could have been obtained by using **moles.**

$$\text{Moles } NH_3 = 2 \times \text{moles } N_2 \text{ from balanced equation.}$$

$$\frac{PV_{NH_3}}{RT} = \frac{2PV_{N_2}}{RT}$$

$$V_{NH_3} = 2V_{N_2} = 2 \times 3.00 \text{ L} = 6.00 \text{ L}$$

Dalton's Law of Partial Pressures

When a mixture of nonreacting gases is present, each gas occupies the total volume of the container and exerts a **partial pressure** on the container walls, identical to what it would exert if it alone were present. The sum of the partial pressures equals the total pressure of the mixture. With T and V constant, each partial pressure is directly proportional to the number of moles of that type of gas (Dalton's law).

Example 3.7

Air is approximately 21% O_2 and 79% N_2 by volume. What is this composition expressed in terms of % by mass?

Solution 3.7

For convenience, asume that we start with 100 L of air; then 100 L of air contains 21 L of O_2 (V_1), and 79 L of N_2 (V_2), and the number of moles of each gas (at constant T and P) is proportional to its volume, and we can write

$$\frac{V_1}{V_2} = \frac{21 \text{ L}}{79 \text{ L}} = \frac{n_1}{n_2} = \frac{m_1/M_1}{m_2/M_2}$$

where m_1 and m_2 are the masses of O_2 and N_2 and M_1 and M_2 are their molar masses (32.0 g mol^{-1} and 28.0 g mol^{-1}, respectively). Thus

$$\frac{m_1}{m_2} = \frac{V_1M_1}{V_2M_2} = \frac{21 \text{ L} \times 32.0 \text{ g mol}^{-1}}{79 \text{ L} \times 28.0 \text{ g mol}^{-1}} = 0.30$$

Since $m_1 = 0.30\ m_2$,

$$\% \text{ by mass } O_2 = \frac{m_1}{m_1 + m_2} \times 100\% = \frac{0.30\ m_2}{0.30\ m_2 + m_2} \times 100\% = 23\%$$

and % by mass of N_2 is (100 - 23) = 77%.

Example 3.8

Assume that 100 g of air is mixed with 100 g H_2, and the mixture ignited in a 1000 L vessel, in a reaction in which only the O_2 and H_2 react. All the water produced by the reaction, both gaseous and liquid, is subsequently removed. What are the partial pressures of the gases that remain and the total pressure, measured at 298 K?

Solution 3.8

The balanced equation for burning hydrogen is $2H_2 + O_2 \rightarrow 2H_2O$. Before reaction the number of moles of gases present are:

$$\text{mol } N_2 = 100 \text{ g air} \times \frac{77 \text{ g } N_2}{100 \text{ g air}} \times \frac{1 \text{ mol } O_2}{28.00 \text{ g } N_2} = 2.75 \text{ mol}$$

$$\text{mol } O_2 = 100 \text{ g air} \times \frac{23 \text{ g } O_2}{100 \text{ g air}} \times \frac{1 \text{ mol } O_2}{32.00 \text{ g } O_2} = 0.72 \text{ mol}$$

$$\text{mol } H_2 = 100 \text{ g } H_2 \times \frac{1 \text{ mol } H_2}{2.016 \text{ g } H_2} = 49.60 \text{ mol}$$

Since 1 mole of O_2 combines with 2 moles of H_2, we see that O_2 is the limiting reactant.

$$\text{mol } H_2 \text{ reacted} = 0.72 \text{ mol } O_2 \times \frac{2 \text{ mol } H_2}{1 \text{ mol } O_2} = 1.44 \text{ mol}$$

After H_2O is removed, there remains 2.75 mol N_2 and 49.60 - 1.44 = 48.16 mol H_2,

$$P_{N_2} = \frac{n_{N_2}RT}{V} = \frac{2.75 \text{ mol} \times 0.0821 \text{ L atm mol}^{-1} \text{ K}^{-1} \times 298 \text{ K}}{1000 \text{ L}} = 0.067 \text{ atm}$$

$$P_{H_2} = \frac{n_{H_2}RT}{V} = \frac{48.16 \text{ mol} \times 0.0821 \text{ L atm mol}^{-1} \text{ K}^{-1} \times 298 \text{ K}}{1000 \text{ L}} = 1.18 \text{ atm}$$

and total pressure exerted by the mixture is 1.25 atm.

3.8 Diffusion and Effusion

Two different gases at the same temperature effuse through a small hole at different rates, according to Graham's law. Let's see how we can use such a relationship.

Example 3.9

Thomas Graham's 1846 study of the rates of effusion of two different oxides of carbon reported that oxide A required 1.254 times as long as oxide B for equal volumes to effuse through a small hole. One of the oxides was CO_2. What was the other oxide? Was CO_2 oxide A or oxide B?

Solution 3.9

From Graham's law, $\frac{\text{rate 1}}{\text{rate 2}} = \sqrt{\frac{M_2}{M_1}}$.

We know that the **lighter** molecule has the **faster effusion rate,** and thus a given volume of the **lighter** type of molecules takes a **shorter** time to effuse. In other words, the ratio of **time** required for a given volume of effuse is just the **inverse** of the ratio of **effusion rates.**

Then $1.254 = \sqrt{\frac{M(\text{heavy oxide})}{M(\text{light oxide})}}$. Squaring and cross-multiplying, we get

$$1.573 \times M(\text{light oxide}) = M(\text{heavy oxide})$$

Let us assume CO_2 is the **heavier** oxide; it has a molar mass of 44.01 g. Then

$$M(\text{lighter oxide}) = \frac{44.01 \text{ g}}{1.573} = 27.98 \text{ g}$$

The molar mass of CO is 12.01 + 16.00 = 28.01. Thus it is reasonable to conclude that the lighter oxide is CO, and that CO_2 is oxide A.

SELF TEST

Part I True or False

1. Nitrogen is a very unreactive element.
2. Oxygen is a very unreactive element.
3. In the laboratory hydrogen can be generated by reacting magnesium with steam.
4. In industry, hydrogen is produced most cheaply by the electrolysis of water.
5. Anhydrous copper sulfate is white.
6. Hydrocarbons have the empirical formula $C_n(H_2O)_m$.
7. Boyle's law states that the volume of a gas is directly proportional to its pressure, if both mass and temperature are held constant.

8. When one mole of a gas, at constant pressure, has its temperature reduced from 500°C to 250°C, its volume is halved.
9. If the volume of a given quantity of gas is constant, heating the gas must increase its pressure.
10. Gas molecules are in constant random motion.
11. At low temperatures and high pressures the ideal gas law does not give a very exact description of the behavior of gases.

Part II Multiple Choice

12. Which element, either uncombined or combined with other elements, is the most abundant on the earth's crust?
(a) iron (b) helium (c) silicon (d) oxygen (e) nitrogen
13. If a 25.0 mL sample of air, initially at 23°C and 1.04 atm pressure, is forced under a pressure of 2.00 atm to occupy a volume of 15.0 mL, what will be the new temperature of the gas?
(a) -16°C (b) 27°C (c) 69°C (d) 96°C (e) 676°C
14. Which of the following oxides are solids at room temperature: CO, CO_2, CuO, SO_2, HgO?
(a) CuO and SO_2 (b) CuO and HgO (c) HgO and SO_2 (d) CO and HgO
(e) CO_2 and CuO
15. The ratio of the rate of effusion of N_2 to that of O_2 at the same temperature and pressure is
(a) 0.87 (b) 0.94 (c) 1.00 (d) 1.07 (e) 1.14
16. The number of molecules of fluorine, F_2, in 2.68 liters of fluorine gas at STP is:
(a) $\frac{2.68}{22.4} \times 6.02 \times 10^{23}$ (b) $\frac{22.4}{2.68} \times 6.02 \times 10^{23}$ (c) $\frac{2.68}{19.0} \times 6.02 \times 10^{23}$
(d) $\frac{2.68}{38.0} \times 6.02 \times 10^{23}$ (e) $2.68 \times 6.02 \times 10^{23}$
17. The volume in liters that 0.200 moles of oxygen could occupy at 393°C and 3.33 atm is
(a) 488 (b) 82.0 (c) 8.20 (d) 3.28 (e) 0.305
18. A gaseous compound of fluorine, carbon and oxygen contains 68.7% F, 21.7%C and the remainder O. A 2.06 g sample occupied a volume of 1100 mL at 174°C and 315 torr. What is the molecular formula of the compound?
(a) $C_6O_6F_{12}$ (b) COF (c) $C_{12}O_4F_{24}$ (d) C_3OF_6 (e) $C_4O_4F_4$
19. How many liters of CO_2, measured at STP, will be produced when 3.6 g H_2O are produced by the following reaction?
$C_3H_8(g) + 5O_2(g) \rightarrow 3CO_2(g) + 4H_2O(l)$
(a) 1.12 (b) 2.24 (c) 3.36 (d) 4.48 (e) 22.4
20. A 400 mL vessel, containing N_2 at 300 torr, was connected to an 800 mL vessel originally filled with 600 torr of O_2, so that both vessels were filled with a uniform mixture of gases. The temperature remained constant. What was the partial pressure of O_2, in torr, in the mixture?
(a) 300 (b) 400 (c) 480 (d) 720 (e) 900

Answers to Self Test

1. T; 2. F; 3. T; 4. F; 5. T; 6. F; 7. F; 8. F; 9. T; 10. T;
11. T; 12. d; 13. c; 14. b; 15. d; 16. a; 17. d; 18. d; 19. c; 20. b.

CHAPTER 4

THE PERIODIC TABLE AND CHEMICAL BONDS

SUMMARY REVIEW

In the **periodic table,** the elements are arranged in order of increasing atomic number, Z, so that elements with similar properties come in the same vertical column **(group),** of which there are eight main ones (numbered 1 to 8). The seven horizontal rows are **periods** and are numbered 1 to 7.

	1	2											3	4	5	6	7	8
1	H																	He
2	Li	Be											B	C	N	O	F	Ne
3	Na	Mg											Al	Si	P	S	Cl	Ar
4	K	Ca	Sc	Ti	V	Cr	Mn	Fe	Co	Ni	Cu	Zn	Ga	Ge	As	Se	Br	Kr
5	Rb	Sr					TRANSITION METALS						In	Sn	Sb	Te	I	Xe
6	Cs	Ba											Tl	Pb	Bi	Po	At	Rn
7	Fr	Ra																
		6					LANTHANIDES											
		7					ACTINIDES											

The other principal division of the elements is into **metals** and **nonmetals** -- by a diagonal line in the vicinity of Groups 3 to 5. Elements to the left of the line (including all the transition metals, lanthanides and actinides) are all **metals;** those to the right are **nonmetals,** except for a few elements near the dividing line which are **semimetals** (metalloids).

In contrast to metals, nonmetals are insulators rather than conductors of heat and electricity, are not shiny or reflective, are generally brittle rather than malleable and ductile, and do not show the thermionic and photoelectric effects.

Elements in the same group have related properties and the same common valence. The group 1 elements (except H) have a common valence of 1 and are called the **alkali metals.** They react with H_2 to give solid hydrides of formula MH, with Cl_2 to give solid chlorides of formula MCl, and with water to give MOH(aq) and H_2(g). Group 2 elements are called the **alkaline earth metals,** with a common valence of 2. They form solid hydrides and chlorides of formulas MH_2 and MCl_2, respectively. Although not as reactive as the alkali metals, they react with water or steam to give hydroxides of formula $M(OH)_2$. The group 6 elements have a common valence of 2, are nonmetals, and form hydrides with the general formula XH_2, all of which are gases, except H_2O, which is a liquid. The group 7 elements are the **halogens,** with a common valence of 1. They react with H_2(g) to give gaseous hydrides of formula HX, all of which are very soluble in water to give acidic solutions. With metals, they give solid halides. The group 8 elements are the **noble gases** and all of them are monatomic. Only Kr, Xe, and Rn react to give compounds, and then only directly with fluorine. Within a group all the elements have the **same** valence. The valence of an element is equal to the number of hydrogen atoms or halogens atoms that combine with one atom of the element. The trend in valences across groups 1-8 is 1,2,3,4,3,2,1,0. In general for a binary compound of empirical formula A_yB_z,

$$y \cdot (\text{valence of A}) = z \cdot (\text{valence of B})$$

Electrons in atoms are arranged in successive layers, or **shells;** the number of electrons in the outermost **(valence)** shell of an atom determines its **valence.**

The nucleus plus all of the completed electron shells constitute the **core** of an atom. The **core charge** equals the nuclear charge, Z+, minus the number of inner shell electrons, which for the main group elements equals the group number, e.g., +4 for group 4 atoms. Direct evidence for the shell structure comes from the measurement of **ionization energies,** which are the energies required to remove electrons from atoms in the gas phase.

$$M(g) + \frac{\text{ionization}}{\text{energy}} \rightarrow M^+(g) + e^-$$

The **first ionization energies** usually increase across any period (as core charge increases) and decrease down any group. Within a group the core charge is constant; the trend of ionization energies is consistent with an increase in the distance of the outermost electron from the nucleus in going from one period to the next, i.e., with an increasing number of completed inner electron shells, giving a valence shell that is increasingly distant from the nucleus.

The **electron affinity** of an atom is the energy change when a neutral atom in the gas phase captures an electron:

$$X(g) + e^- \rightarrow X^-(g) + \frac{\text{electron}}{\text{affinity}}$$

Atomic size is measured by the **covalent radius** of an atom (half the distance between identical atoms forming a covalent bond). Covalent radii are approximately additive and are used to predict bond lengths. They increase in descending any group and decrease from left to right along any period.

Only valence shell electrons are involved in bonding and these electrons are shown in Lewis symbols as dots. For example:

Li·   ·Be·   ·B·   ·C·   :N·   :O: :F: :Ne:

Atoms form bonds by gaining or losing or by sharing electrons. For example, $Na \rightarrow e^- + Na^+$ (with the same configuration as Ne), and $Cl + e^- \rightarrow Cl^-$ (with the same electron configuration as Ar). Thus in Na^+Cl^-, each ion has a noble gas configuration and oppositely charged ions are held together by electrostatic attraction, i.e., by an **ionic bond.** Alternatively, atoms can complete their valence shells by **sharing** electron pairs, as in the :Cl:Cl: molecules, where each atom contributes one electron to the Cl-Cl **electron pair,** or **covalent bond.**

Except for hydrogen, stable valence shells commonly consist of **eight electrons,** which is the basis of the **Lewis octet rule.** Thus, fluorine can complete its valence shell by forming the F^- ion, or by forming one covalent bond, as in HF or F_2; oxygen can form O^{2-} or two covalent bonds, as in H_2O or Cl_2O, and nitrogen can form N^{3-} or three covalent bonds, as in NH_3 or NCl_3, and so on. **Double bonds,** in which two electron pairs are shared between two atoms, and **triple bonds,** in which three electron pairs are shared, are also possible.

Diagrams showing all the valence electrons associated with an atom, or all the atoms in a molecule or ion, are **Lewis diagrams** or **structures.** Bonding electron pairs are designated either by : or by a bond line —. Electron pairs associated with atoms but not taking part in bonding are called **unshared pairs, nonbonding pairs,** or **lone pairs,** and are designated:.

Hydrogen is unique in that its valence shell is complete when it contains two electrons; it can form only one covalent bond. Alternatively, loss of an electron from H·, gives a proton, H^+, and gain of an electron gives the hydride ion, $:H^-$.

There are exceptions to the octet rule. For example, Be with two valence electrons normally forms two covalent bonds, while B with three valence electrons normally forms three covalent bonds. For example, reaction with fluorine gives BeF_2 and BF_3 respectively, with incomplete valence shells. However, such molecules can accept additional electron pairs to complete their octets.

Lewis diagrams are completed by assigning a **formal charge** to each atom of the molecule or polyatomic ion. The formal charge is equal to the number of electrons an atom possesses in the free state minus the average number it has in the molecule.

Molecular shape is determined by the relative positions of the atoms in a compound. **Ionic bonds** are **nondirectional** and the structures of **ionic crystals** are determined by the tendency of a cation to surround itself with as many anions as possible, and vice-versa. Atoms form **covalent bonds** in specific directions; these types of bonds are **directional,** so that the geometry of the atoms in covalent molecules and covalently bonded crystals depends on the directions of the covalent bonds formed by each atom. According to the **Valence Shell Electron Pair Repulsion** (VSEPR) **Model,** the electron pairs in the valence shell of an atom A arrange themselves to keep as far apart from each other as possible. Thus, two electron pairs have a linear arrangement; three electron pairs have a triangular arrangement, and four electron pairs have a tetrahedral arrangement. The geometric shape of a covalent molecule can be deduced from its Lewis structure, written in the form AX_nE_m, where there are **n** bonding (shared) pairs of electrons, and **m** nonbonding (lone) pairs of electrons in the valence shell of a central atom A bonded to n X ligands. In an AX_nE_m molecule, the geometric shape is given by the relative positions of the central atom A and those of the X ligands:

Molecular Type	Number and arrangement of electron pairs in valence shell of A		Number and arrangement of bonding electron pairs in valence shell of A	
AX_2	2	linear	2	linear
AX_3	3	equilateral triangle	3	equilateral triangle
AX_2E	3	equilateral triangle	2	angular
AX_4	4	tetrahedral	4	tetrahedral
AX_3E	4	tetrahedral	3	trigonal pyramidal
AX_2E_2	4	tetrahedral	2	angular
AXE_3	4	tetrahedral	1	linear

In general, AX_2 molecules are linear, and AX_3 molecules are triangular planar, irrespective of whether the bond to X is a single, double, or triple bond.

REVIEW QUESTIONS

1. What is the general definition of a chemical bond?
2. On what basis are elements ordered to form the periodic table?
3. What are the vertical columns of atoms in the periodic table called? What are the horizontal rows called?
4. What would be the atomic number of the next alkali metal after francium (Fr), if such an atom existed?
5. Draw the general form of the periodic table, showing the correct number of elements in each of the seven periods.
6. What special names are given to the elements in Groups 1, 2, 6, 7 and 8?
7. What special names are given to:
   (a) the blocks of ten elements that lie between Groups 2 and 3 in periods 4, 5 and 6,
   (b) the additional 14 elements in period 6, and
   (c) the additional 14 elements in period 7?
8. Where in the periodic table is the approximate dividing line between the metals and nonmetals?
9. Give three examples of semimetals (metalloids).
10. List six characteristic properties which could be used to distinguish a metal from a nonmetal.
11. What are the empirical formulas of the hydrides of the elements of the second period (Li to F)?
12. What is meant by the valence of an atom?
13. How is the valence of an atom related to its position in the periodic table?
14. What general formula relates the integers y and z in a compound A_yB_z to the valences of A and B?
15. What is meant by a shell of electrons?
16. For the first twenty elements, write out the arrangements of electrons in shells for each atom.
17. In any group of the periodic table, what common feature of the shell structure is the same for every element?
18. What is the valence shell of an atom?
19. What is meant by the core of an atom and its core charge?
20. What are the core charges of B, S, and Br?
21. Define the ionization energy of an atom.

22. Describe how the first ionization energies generally vary across any period of the periodic table.
23. Explain why the ionization energies vary across any period.
24. Explain why the ionization energies of each alkali metal is much smaller than that of the noble gas that precedes it in the periodic table.
25. Write electron dot symbols for each of the elements of the second period.
26. How is the covalent (atomic) radius of carbon obtained?
27. How and why do the covalent radii vary as one moves from left to right across any period of the periodic table?
28. How do the covalent radii vary in going from top to bottom of any group?
29. Define the term ionic bond and give three examples of ionic compounds.
30. Draw the Lewis structures of NaF, $CaCl_2$ and $BaSO_4$.
31. Define the term covalent bond and give three examples of diatomic molecules containing covalent bonds.
32. Draw Lewis structures for H_2O, NH_3 and HF.
33. Draw Lewis structures for CCl_4, F_2CCF_2 and H_3CCCCH_3.
34. What is meant by the Lewis octet rule?
35. In which of the following is the octet rule not obeyed?
$AlCl_3$, ClF, OF_2, OH^-, $BeCl_4^{2-}$
36. What is an unshared, nonbonding, or lone-pair of electrons?
37. How many lone pairs of electrons are there on each oxygen in each of the following? H_2O, H_3O^+, HOOH
38. Assign formal charges to each of the atoms in NH_4^+, BF_4^- and NO_2^+.
39. What is meant by a single bond, a double bond, and a triple bond? Give examples of molecules containing each.
40. What difference is there between ionic bonds and covalent bonds in terms of their directionality?
41. What factors determine the arrangements of ions in ionic crystals?
42. What factors determine the arrangement of the covalent bonds to the central atom A of a covalent molecule?
43. How is the geometric arrangement of bonds to a central atom A of a covalent molecule related to its Lewis structure?
44. What are the predicted shapes of each of the following molecules?
$BeCl_2$, BCl_3, CCl_4, NCl_3, OCl_2, HCl
45. What are the predicted shapes of each of the following?
CS_2, ClCN, H_2CO, ClNO, N_3^-

Answers to Selected Review Questions

30. Na^+ :F:$^-$ Ca^{2+} [:Cl:$^-$]$_2$ Ba^{2+} $^-$:O—S^{2+}—O:$^-$ (with :O:$^-$ above and :O:$^-$ below S)

32. H:O:H :N:H (H above N, H below N) H:F:

33. :Cl—C—Cl: (with :Cl: above and :Cl: below C) :F and :F bonded to C = C bonded to :F: and :F: H—C—C≡C—C—H (with H above and below each terminal C)

35. $AlCl_3$

37. 2,1,2 respectively
38. +1 on N in NH_4^+, -1 on B in BF_4^-, +1 on N in NO_2^+; otherwise zero.
44. linear, triangular, tetrahedral, trigonal pyramid, angular, linear.
45. linear, linear, triangular planar, angular, linear.

OBJECTIVES

Be able to:

1. Sketch the form of the periodic table, give the positions of the first 20 elements, and draw the boundary between metals and nonmetals.
2. Define the terms group, period, transition element, metalloid.
3. List the characteristic physical properties of metals and nonmetals.
4. Classify the elements into families and name the members of the families of groups 1, 2, 6, 7, 8.
5. Define the term valence, and state the values for the first 20 elements; give the formulas for the hydrides, fluorides, chlorides, and oxides of these elements.
6. Describe the shell model of the atom, and give the shell structures for the atoms of the first 20 elements.
7. Show how and why ionization energy and size of atoms vary with position in the periodic table, both across a period and down a group.
8. State the octet rule.
9. Define the terms ionic bond and covalent bond.
10. Give the charges of the common ions of the main group elements and deduce the formulas and Lewis structures (electron dot structures) for simple ionic compounds.
11. Determine Lewis structures for simple covalent compounds formed by electron pair sharing, and distinguish between shared and unshared electron pairs.
12. List several compounds which are exceptions to the octet rule.
13. Define the term formal charge and calculate formal charges on atoms in molecules and polyatomic ions.
14. State the arrangements of 2, 3 and 4 electron pairs on the valence shell of an atom.
15. Give examples of AX_4, AX_3E, AX_2E_2, AXE_3, AX_3, AX_2E and AX_2 molecules; draw their Lewis structures; draw and describe their shapes.
16. Give examples of molecules containing double and triple bonds; draw their Lewis structures; draw and describe their shapes.

PROBLEM SOLVING STRATEGIES

4.1 The Periodic Table

The periodic table lists elements with similar chemical properties in the same group. For example, all the group 1 alkali metals M react with water to give hydrogen gas and a solution of a metal hydroxide. The alkali metals also react with group 7 halogens to form the halides MX. Thus you can write at least 20 examples of MX once you know the alkali metals and the halogens. Similarly, when you know to which family any element of the first 20 belongs, and its valence, it is relatively easy to write the empirical formulas of simple compounds containing only two elements. Two steps are required. First, in assigning valences for the first 20 elements, those elements that are in groups 1-4 have common valences that are the same as the group number, while elements in the groups 5-8 have normal valences equal to 8 minus the group number. Second, the total combining power of all the atoms of one element must equal the total combining power of all the atoms of the other element in that compound. Each total combining power is the product of the valence of one atom times the number of atoms of that type in the compound.

Example 4.1

Predict the empirical formula of the compound formed from magnesium and nitrogen.

Solution 4.1

Magnesium is in group 2, so its valence is 2. Nitrogen is in group 5, and has a valence of 8 - 5 = 3. Let us write the compound as Mg_yN_z. Then, since the total combining power of Mg must be the same as that for N, we write

$$2y = 3z$$

By inspection we obtain y = 3, z = 2 as the simplest solution involving integers. Thus the empirical formula is Mg_3N_2.

4.2 The Shell Model

The periodic variations in the physical and chemical properties of the elements are related to the structure of their atoms. In the shell model of the atom, the number of electrons in the outer (or valence) shell is of utmost importance. For example, all alkali metal atoms have one electron in their valence shell; oxygen family members all possess 6 valence electrons. The shell model accounts for the trends in ionization energy and size of atoms. Remember two points. First, for the main group elements as we go from left to right across a period in the periodic table, just one shell is filling with electrons, and the core charge (atomic number minus number of inner shell electrons) is increasing. The greater attraction between the increasing core charge and the added electrons makes it more difficult to remove the most loosely bound electron. The **ionization energy increases as the core charge increases.** As well, the greater core charge means that the valence electrons are pulled closer to the core, and **atomic size decreases as core charge increases.** Second, as we move down a group in the table, the core charge stays constant, while the electron being added is farther away from the nucleus. Thus **ionization energy decreases and size increases down a group**.

Example 4.2

Arrange the following atoms in order of decreasing ionization energy: O, Na, F, Li.

Solution 4.2

First determine which shells, described by n, contain the **valence** electrons.

For Li, O and F, n = 2; for Na, n = 3. Next, determine the core charges by subtracting the number of inner electrons from the atomic number, or by remembering that core charge is equal to group number.

$$\left.\begin{array}{lll} \text{Li} & 3-2 & = 1 \\ \text{O} & 8-2 & = 6 \\ \text{F} & 9-2 & = 7 \end{array}\right\} n = 2$$

$$\begin{array}{lll} \text{Na} & 11-10 & = 1 \quad n = 3 \end{array}$$

Then consider all the atoms with **lowest n valence shell** electrons, and arrange them in order of **decreasing** core charge. This arrangement gives the relative ordering of ionization energies in that shell (with some minor exceptions). Thus we find F > O > Li for ionization energies. Finally we consider atoms with larger n valence shells. In this example, only Na remains. Since both Li and Na have a core charge of 1, but the valence electron of sodium is in a shell farther away from the core charge, the ionization energy of Na is less than that of Li. Thus the complete order is F > O > Li > Na.

Example 4.3

Using the atoms given in Example 4.2, arrange them in order of increasing size.

Solution 4.3

As before, determine which shells the valence electrons are in, and what are the core charges. Within a given shell, **size increases** as we go **from right to left.** Thus F < O < Li. Finally consider the atoms with valence electrons in larger n shells. Here we only have Na (n = 3) to consider. Since the core charges for Li and Na are the same, but the n = 3 shell is farther from the core charge than the n = 2 shell, Na is larger than Li. The final order of sizes is F < O < Li < Na. You might wish to compare these predictions with Figure 4.10 in the textbook.

4.3 Chemical Bonds and Lewis Structures

Ionic Bonds

Compound formation involves either the **transfer** or the **sharing** of valence electrons; in many cases there are **eight electrons** (an **octet**) surrounding each ion, or surrounding each atom in a molecule. First let's consider **ionic bonding**, in which oppositely charged ions, formed by loss or gain of electrons, attract each other. We will represent the valence electrons and the bonding situation by using Lewis (electron dot) diagrams.

Example 4.4

Draw the Lewis structures for K and F, and for the ionic compound KF.

Solution 4.4

From their positions in the periodic table we know that potassium has one valence electron, and fluorine has seven. Thus we write, for the **atoms,**

$$K\cdot \quad \text{and} \quad :\ddot{\underset{\cdot\cdot}{F}}\cdot$$

In the compound KF, K donates an electron to F and becomes K^+, potassium **ion.** Although there are 8 electrons in the n = 3 shell of K , there are none in its valence shell (n = 4), which is defined as the outermost shell of the neutral atom that contains electrons. Thus we write its Lewis structure simply as (K^+). A fluorine atom accepts the electron formerly belonging to potassium, and becomes F^- ion. It is represented as

$$(:\ddot{\underset{\cdot\cdot}{F}}:^-)$$

The ionic compound KF, potassium fluoride, is represented as **(K^+)** $(:\ddot{\underset{\cdot\cdot}{F}}:^-)$.

Note there are 8 electrons around K^+ (the eight in the n = 3 shell), and 8 electrons about F^-. However there are **no** individual KF molecules in solid KF. Instead the solid consists of a giant 3-dimensional array of + and - ions.

Covalent Bonds

We next consider **covalent** bonds, formed by **sharing** of a pair or pairs of electrons. When 2 electrons are shared by 2 atoms, we call the electrons a **bonding pair.** We say that a **single bond** exists between the atoms. Lewis diagrams show how many electron pairs are involved in bonding, and how many **nonbonding** valence shell electron pairs are present. These diagrams will also allow us to predict molecular geometries and other properties. Here are rules for drawing Lewis diagrams for simple covalent molecules.

1. Determine the number of valence electrons for all the atoms in the molecule.
2. Arrange the atoms in the order in which they are bonded. For simple compounds this is no problem. For example, you know that O is the central atom in H_2O.
3. Place the valence electrons in pairs around the atoms so that each atom has 8 electrons (2 for hydrogen). You may find it useful to use different symbols for the electrons supplied by different atoms.

Example 4.5

Draw the Lewis structure for NF_3, nitrogen trifluoride. Nitrogen uses 3 electrons for bonding in this compound.

Solution 4.5

Nitrogen has 5 valence electrons; each fluorine has 7. Since nitrogen uses 3 electrons for bonding here, 3 single bonds can be formed if each fluorine atom supplies 1 electron for bonding. Thus

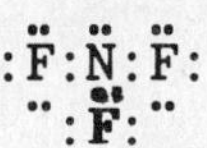

Note that now there is an octet of electrons around N, and an octet around each F. Usually a shared pair is represented by a dash, and we can write NF_3 as

$$:\ddot{\underset{..}{F}}-\ddot{N}-\ddot{\underset{..}{F}}:$$
$$|$$
$$:\underset{..}{F}:$$

Be sure to include all valence electrons, both shared and unshared, in a Lewis structure.

Polyatomic Ions and Formal Charge

Ionic compounds can be formed not only from **monatomic** ions, but may also include one or more **polyatomic** ions. In these cases, the bonding **within** the polyatomic ion is due to electron **sharing,** while **in the overall compound** the bonding is due to attraction between **ions.** When dealing with **polyatomic ions,** we would like to know where the net charge of the ion is located. An approximate answer to this question is found by calculating what is called the **formal charge** of an atom.

$$\begin{pmatrix}\text{formal}\\ \text{charge}\end{pmatrix} = \begin{pmatrix}\text{core}\\ \text{charge}\end{pmatrix} - \begin{pmatrix}\text{number of}\\ \text{unshared electrons}\end{pmatrix} - \frac{1}{2}\begin{pmatrix}\text{number of}\\ \text{shared electrons}\end{pmatrix}$$

Example 4.6

Sodium amide, $NaNH_2$, is an ionic solid consisting of positively charged sodium ions and negatively charged NH_2^- ions. Draw the Lewis structure for NH_2^-, including formal charges.

Solution 4.6

Each H has one valence electron, and N has 5. Since sodium atom donates an electron to form NH_2^-, there are altogether 8 electrons around N. We arrange these as follows

$$\overset{..}{\circ}\ddot{N}:H$$
$$..$$
$$H$$

Here the electron donated by Na has been represented by an open circle. To calculate formal charges, we take each atom in turn.

For N in NH_2^-, $\begin{pmatrix}\text{formal}\\ \text{charge}\end{pmatrix} = 5 - 4 - \frac{1}{2}(4) = 5 - 6 = -1$

For each H in NH_2^-, $\begin{pmatrix}\text{formal}\\ \text{charge}\end{pmatrix} = 1 - 0 - \frac{1}{2}(2) = 0$

Thus the Lewis structure for NH_2^- is $^-:\ddot{N}:H$

$$..$$
$$H$$

The -1 charge on nitrogen is the formal charge. The sum of the formal charges must equal the charge on the ion. Many **neutral** molecules also have formal charges associated with their Lewis structures. We will see one in Example 4.7 below and many more in Chapter 9.

Multiple Bonds

It is possible for two atoms to share more than one electron pair. In such cases there is multiple bonding. If **2 electrons pairs are shared** between two atoms, we speak of a **double bond;** 3 shared electron pairs constitute a **triple bond.**

Example 4.7

Draw the Lewis structure for CO, carbon monoxide.

Solution 4.7

Carbon has 4 valence electrons and oxygen has 6. We must arrange 10 electrons around 2 atomic centers, so that each atom has an octet. Thus we arrive at the structure

:C:::O: or :C≡O:

However, we are not finished yet. Let's see if there are formal charges to be assigned.

For carbon in CO, $\left(\begin{matrix}\text{formal}\\ \text{charge}\end{matrix}\right) = 4 - 2 - \frac{1}{2}(6) = 4 - 5 = -1$

For oxygen in CO, $\left(\begin{matrix}\text{formal}\\ \text{charge}\end{matrix}\right) = 6 - 2 - \frac{1}{2}(6) = 6 - 5 = +1$

Thus the complete Lewis structure for CO is $^{-}$:C≡O:$^{+}$

4.4 Molecular Shape and the VSEPR Model

The Valence Shell Electron Pair Repulsion (VSEPR) model is able to predict the geometries of a wide variety of covalent molecules. In this theory, electron pairs stay as far apart as possible from each other. However, not only is the arrangement of electron pairs important, but also the **number of bonding and nonbonding pairs.** Table 4.8 in the text lists the arrangements for 2, 3 and 4 electron pairs, and the shapes of typical molecules. You should distinguish very carefully between the **arrangement of electron pairs** and the **arrangement of bonds (bonding pairs).** The arrangement of **bonds** gives the geometry. Here is a set of rules to follow for determining molecular geometries:

1. Draw the Lewis structure for the molecule. (More complicated molecules will be treated in Chapter 9 of the text. In that chapter you will find a formal set of rules for drawing Lewis structures.)
2. Count the number of electron pairs around the central atom.
3. For 4 electron pairs around the central atom, determine if there are 0, 1, 2 or 3 nonbonding pairs (AX_4, AX_3E, AX_2E_2, AX_3E). AX_4 is tetrahedral; AX_3E is triangular pyramidal; AX_2E_2 is angular; AXE_3 is linear.
4. For 3 electron pairs around the central atom, determine if there are 0 or 1 nonbonding pairs (AX_3 or AX_2E). AX_3 is planar triangular and AX_2E is angular.
5. For 2 electron pairs around a central atom, where there are no nonbonding pairs (AX_2), the molecule is linear.

Example 4.8

Classify the following molecules as AX_4, AX_3E, AX_2E_2, AXE_3, AX_3, AX_2E or AX_2, and describe their shapes:

(a) H_2Se (b) PBr_3 (c) AsH_4^+ (d) SH^- (e) BeF_2

Solution 4.8

(a) Since selenium is in group 6, it contributes 6 valence electrons; each hydrogen contributes 1. The 8 valence electrons are arranged in pairs.

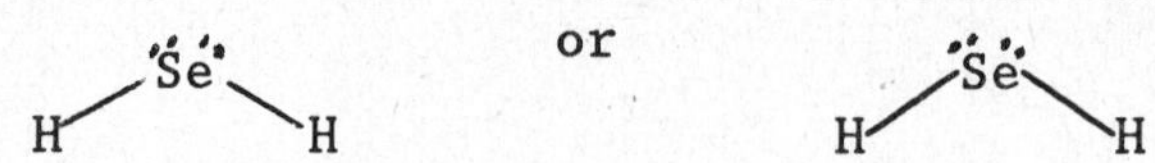

We recognize this molecule as an AX_2E_2 type; it is angular.

(b) Phosphorus has 5 valence electrons, and each bromine supplies 7, for a total of 26 electrons. If we form **single** Br-P bonds, there will be an octet around each Br and around P.

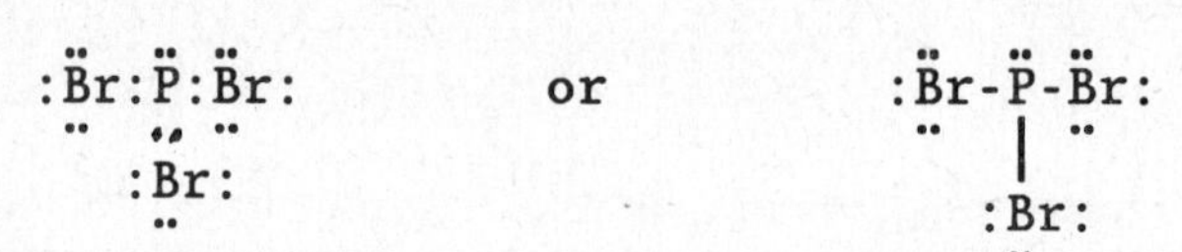

Note that a Lewis structure is not meant to show the geometry -- just use it as a 2-dimensional electron dot picture. From the Lewis structure we recognize PBr_3 to be an AX_3E molecule, which is triangular pyramidal. If we had not drawn the Lewis structure, we might have (incorrectly) assumed PBr_3 was AX_3, planar triangular.

(c) Arsenic (group 5) has 5 valence electrons; each hydrogen has 1; the **positive** ion has 5 + 4 - 1 = 8 valence electrons. We can immediately draw the Lewis structure as

```
        H
        |
   H———As⁺———H
        |
        H
```

The ion is of the type AX_4, and has a tetrahedral geometry.

(d) Any diatomic species is linear. We need to draw the Lewis structure to determine what classification the ion SH^- fits into. S has 6 valence electrons, H has 1, and in addition the ion is negatively charged. Thus there are 8 valence electrons around S.

$$^{-}:\ddot{\underset{..}{S}} - H$$

We recognize this as linear AXE_3.

(e) Beryllium has 2 valence electrons, and each fluorine supplies 7, for a total of 16 electrons. If we form single Be-F bonds, these will be an octet around each fluorine. Since fluorine is more electronegative than beryllium, we preferentially complete octets around fluorine.

$$:\ddot{F}\text{-Be-}\ddot{F}:$$

With this structure beryllium does not have an octet, but it does have a formal charge of zero. If we move nonbonding electron pairs from fluorine to form two **double** bonds from fluorine to beryllium, there will be octets around all three atoms. However, such a structure is discarded, since extra formal charges are created. The molecule is a linear AX_2 type.

Once you know the geometry of a molecule, you also know the bond angles. When there are **no nonbonding electrons around the central atom, and all the atoms X are identical,** the following holds. For AX_2, the bond angles are 180°; for AX_3, 120°; for AX_4, 109.5°. We will see in Chapter 9 that these bond angle predictions need slight modification when nonbonding electrons are present around the central atom.

Example 4.9

Match the molecules CH_4, NH_3 and BF_3 with the predicted bond angles 90°, 109.5°, 120°.

Solution 4.9

First determine the type of each of the molecules. These have already been presented in the text as examples. Review them if necessary. The types are: CH_4, AX_4; NH_3, AX_3E; BF_3, AX_3. The bond angles for AX_n molecules have just been presented in the previous paragraph (memorize them), so we have CH_4, 109.5° bond angles; NH_3, 109.5° bond angles; BF_3, 120° bond angles. You will learn in Chapter 9 that the AX_3E molecule, NH_3, has an HNH angle slightly less than the tetrahedral angle of 109.5°, which is expected only for AX_4 molecules.

The Shapes of Molecules Containing Double and Triple Bonds

To predict the geometry of a molecule containing one or more multiple bonds, after you have a correct Lewis structure, proceed by assuming that electrons in double and/or triple bonds can be treated as if they were a single bond as far as VSEPR rules are concerned. Thus,. for example, two bonding pairs forming a double bond are handled as if they were one bonding pair. Table 4.8 summarizes the shapes of molecules containing double and triple bonds. Let's take an example.

Example 4.10

What is the shape of the ClO_2^- ion?

Solution 4.10

We start with the Lewis structure. There are 7 valence electrons from chlorine, and 6 from each oxygen, and 1 more supplied by the negative charge, so 20 valence electrons must be arranged about the atoms. However, oxygen must obey the octet rule, and the formal charge should be as small as possible. There must be a formal charge, since we are considering an **ion.** We write at first single bonds from O to Cl.

$$^{-}:\ddot{\underset{..}{O}} - \ddot{\underset{..}{Cl}}^{+} - \ddot{\underset{..}{O}}:^{-}$$

We see there are 3 formal charges here. We can remove 2 of them by transferring a nonbonding pair on an oxygen into the bonding region, forming a **double** bond. When we recalculate formal charges, we find

$$\ddot{\underset{..}{O}} = \ddot{\underset{..}{Cl}} - \ddot{\underset{..}{O}}:^{-}$$

This is a reasonable single Lewis structure. (There is another one, which we will see in Chapter 9.) For geometry purposes we consider the double bond to be a single bond. So this is an AX_2E_2 molecule, and is angular. By the way, don't be surprised that chlorine is surrounded by 10 electrons. We will find many exceptions to the octet rule in later chapters.

SELF TEST

Part I True or False

1. The metallic elements are all solids at room temperature.
2. The nonmetals are located in the lower left of the periodic table.
3. Metals are good conductors of electricity.
4. The boron atom is larger than the B^{3+} ion.
5. All hydrides are covalent compounds.
6. The majority of the first 20 elements that are nonmetals exist in the gas state at room temperature.
7. In the compound NH_4Cl, both ionic and covalent bonds are present.
8. In the compound Cl_2, each chlorine atom is surrounded by 7 valence electrons.
9. H_2S is an example of a compound with 2 unshared pairs of valence electrons.
10. The compound BCl_3 is an exception to the octet rule.

Part II Multiple Choice

11. What family name is given to the elements of group 7 of the periodic table?
    (a) alkali metals (b) halogens (c) alkaline earths
    (d) noble gases (e) representative elements
12. A **horizontal** row in the periodic table is called
    (a) a group (b) a family (c) a period (d) a column (e) a list
13. The empirical formula for the compound formed between carbon and sulfur, assuming the valences listed in Table 4.1, is
    (a) CS_4 (b) CS_2 (c) C_2S (d) C_3S_2 (e) C_2S_3
14. The number of valence electrons for phosphorus is
    (a) 1 (b) 3 (c) 5 (d) 13 (e) 15
15. The usual valence expected for phosphorus is
    (a) 1 (b) 3 (c) 5 (d) 13 (e) 15
16. The empirical formula for the ionic compound formed between aluminum and fluorine is
    (a) Al_2F (b) Al_2F_3 (c) AlF_3 (d) Al_3F_2 (e) AlF_2

17. The element selenium is situated just below sulfur in the periodic table. The empirical formula of the compound formed between magnesium and selenium is
(a) $MgSe_2$ (b) $MgSe$ (c) Mg_2Se (d) Mg_2Se_3 (e) Mg_3Se_2

18. The element with the largest value for the first ionization energy is located in which portion of the periodic table?
(a) upper right (b) lower right (c) middle
(d) lower left (e) upper left

19. Which of the following is the correct order of ionization energies?
(a) Cl > Ge > Rb (b) Cl > Rb > Ge (c) Rb > Cl > Ge
(d) Ge > Rb > Cl (e) Rb > Ge > Cl

20. The species with smallest radius is
(a) N^{3-} (b) O^{2-} (c) F^- (d) Ne (e) Na^+

21. Each of the following is being compared to lithium. Which statement is **incorrect?**
(a) Fluorine has a smaller atomic radius and a larger ionization energy.
(b) Potassium has a larger atomic radius and a larger ionization energy.
(c) Helium has a smaller atomic radius and a larger ionization energy.
(d) Sodium has a larger atomic radius and a smaller ionization energy.
(e) Beryllium has a smaller atomic radius and a larger ionization energy.

22. Beryllium forms a high melting ionic oxide. The charge on the beryllium ion is
(a) -2 (b) -1 (c) 0 (d) +1 (e) +2

23. Which of the following is the correct Lewis structure for calcium oxide?
(a) (Ca) $(:\ddot{\underset{..}{O}}:^-)$ (b) $(Ca:^{2+})(:\ddot{\underset{..}{O}}:^{2-})$ (c) $(Ca^{2+})(O^{2-})$
(d) $(Ca^{2+})(:\ddot{\underset{..}{O}}:^-)_2$ (e) $(Ca^{2+})(:\ddot{\underset{..}{O}}:^{2-})$

24. Which of the following is the correct Lewis structure for ethane, C_2H_6?

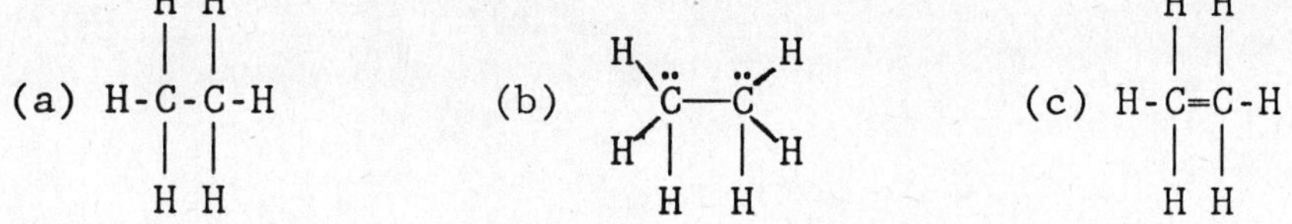

(a) 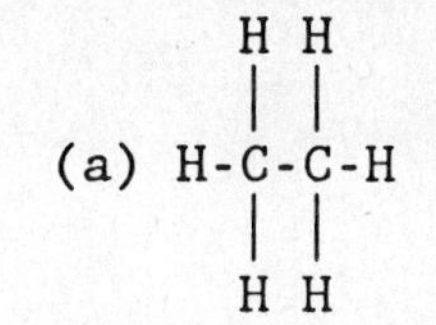

(b)

(c) 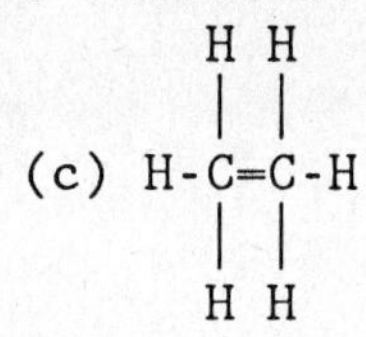

(d) 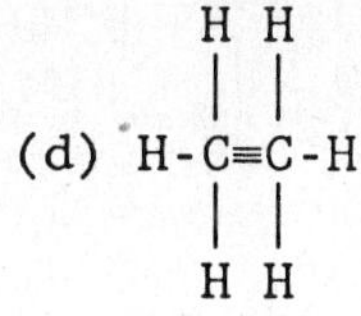

(e) $H_2\ddot{C} \equiv \ddot{C}H_2$ (each C also bonded to H below)

25. The formal charges on C and N of CN^- are, respectively,
(a) 0 and -1 (b) +1 and -2 (c) -1 and 0 (d) -2 and +1
(e) $\frac{-1}{2}$ and $\frac{-1}{2}$

26. In the ionic compound NH_4Cl, the +1 formal charge resides on which atom(s)?
a) N b) one H c) each H d) Cl

27. The shape of PH_3 is a) equilateral triangular planar b) triangular pyramidal c) angular d) linear e) tetrahedral

28. The shape of CH_3^+ is

a) equilateral triangular planar b) triangular pyramidal
c) angular d) linear e) tetrahedral

29. The shape of BeH_2 is a) equilateral triangular planar b) triangular pyramidal c) angular d) linear e) tetrahedral

30. The shape of H_2S is a) equilateral triangular planar b) triangular pyramidal c) angular d) linear e) tetrahedral

Answers to Self Test

1. F; 2. F; 3. T; 4. T; 5. F; 6. T; 7. T; 8. F; 9. T; 10. T; 11. b; 12. c; 13. b; 14. c; 15. b; 16. c; 17. b; 18. a; 19. a; 20. e; 21. b; 22. e; 23. e; 24. a; 25. c; 26. a; 27. b; 28. a; 29. d; 30. c.

CHAPTER 5

THE HALOGENS

SUMMARY REVIEW

Fluorine, F, chlorine, Cl, bromine, Br, iodine, I, and astatine, At, called the halogens (group 7), occur as diatomic molecules, and are nonmetals. Intermolecular forces increase with increasing atomic size; F_2 and Cl_2 are gases, Br_2 is a liquid, and I_2 is a solid. With **metals,** they react to give **ionic halides,** and with **nonmetals,** volatile **covalent halides** are formed. The hydrogen halides, HX, are all soluble gases that dissolve in water to give acidic solutions.

The **electronegativity of an atom in a molecule** measures its relative ability to attract the electrons of a covalent bond. Bonds between atoms of identical electronegativity, in which the bonding electrons are exactly equally shared between the atoms, are **pure covalent** or nonpolar bonds. When the two atoms have different electronegativities, the atom of higher electronegativity has a greater share of the bonding electrons than has that of lower electronegativity, and the bond is **polar covalent.** Electronegativity increases from left to right across any **period** and decreases in going down any **group;** metals on the left side of the periodic table have low electronegativities relative to the nonmetals on the right side. The greater the electronegativity difference between two atoms forming a bond, the greater its polarity. For a large electronegativity difference, the more electronegative atom essentially **captures** the bonding pair, giving an **ionic** bond, which is the limiting case of the polar bond.

In general, **ionic compounds** are formed between metals from the left side of the periodic table and nonmetals from the right side. **Covalent compounds** result from elements of similar electronegativity, especially nonmetals.

The nature of a chemical bond depends on the relative electronegativities of the atoms forming a bond. Bonds between a group 1 or a group 2 metal and a nonmetal are **ionic**; bonds between two different nonmetals are **polar covalent**, and bonds between two identical atoms are **nonpolar covalent**.

Evidence that the metallic halides are composed of ions comes from their behavior as ionic conductors in the molten state, and in aqueous solution. Substances that give conducting aqueous solutions are **electrolytes; nonelectrolytes** dissolve without the formation of ions and give nonconductive solutions.

Salts, such as NaCl, exist at high temperature as highly polar molecules, such as NaCl and $(NaCl)_2$. In **solid** NaCl, no individual NaCl molecules can be distinguished. Each ion is surrounded by as many ions of opposite charge as possible. In **NaCl** each Na^+ ion is surrounded by six Cl^- ions, and each Cl^- ion by six Na^+ ions. In **CsCl,** the coordination number of each ion is eight rather than six. Ionic crystals are composed of infinite arrays of positive and negative ions. Their high melting points reflect the large amount of energy needed to break ionic bonds so that the ions become mobile and can carry an electric current.

X-ray diffraction is used to measure interionic distances, from which **ionic radii** can be estimated. The type of ionic structure formed depends primarily on the ratio of anion to cation radius. Adding an electron to a neutral atom gives an anion that is larger; removing an electron gives a cation that is smaller.

In oxidation-reduction reactions, electrons are transferred between reactants. The reactant which **loses** electrons is **oxidized;** that which **gains** electrons is **reduced. Oxidizing agents** are **electron acceptors; reducing agents** are **electron donors.**

The oxidizing strengths of the halogens decrease in the series from fluorine to iodine. $F_2(g)$ oxidizes Cl^-, Br^- and I^- to the corresponding halogens; $Cl_2(g)$ oxidizes Br^- and I^-, while Br_2 oxidizes only I^-. Halogens are prepared by oxidation of halide ion using halogens, MnO_2, or by electrolysis.

Aqueous solutions of **two** or more soluble salts contain all the component ions. On concentration by evaporation, the first salt to separate is the combination of ions that gives the salt of lowest solubility. Mixing two salt solutions gives a **precipitate** if one of the possible combinations of ions is an **insoluble** salt.

There is no general theory that predicts the solubility of a salt with certainty but some **solubility rules** usefully summarize the facts:

	EXCEPTIONS	
SOLUBLE SALTS	INSOLUBLE	SPARINGLY SOLUBLE
All nitrates (NO_3^-)		
All perchlorates (ClO_4^-)		
Fluorides (F^-)	Group 2 metals, Pb^{2+}	
Chlorides (Cl^-)	Ag^+, Hg_2^{2+}	Pb^{2+}
Bromides (Br^-)	Ag^+, Hg_2^{2+}	Pb^{2+}
Iodides (I^-)	Ag^+, Hg_2^{2+}, Pb^{2+}	
Sulfates (SO_4^{2-})	Sr^{2+}, Ba^{2+}, Pb^{2+}	Ca^{2+}, Ag^+, Hg_2^{2+}
Acetates ($CH_3CO_2^-$)	Ag^+, Hg_2^{2+}	

	EXCEPTIONS	
INSOLUBLE SALTS	SOLUBLE	SPARINGLY SOLUBLE
Carbonates (CO_3^{2-})	Na^+, K^+, NH_4^+	
Phosphates (PO_4^{3-})	Na^+, K^+, NH_4^+	
Sulfides (S^{2-})	Na^+, K^+, NH_4^+, Mg^{2+} Ca^{2+}, Sr^{2+}, Ba^{2+}	
Hydroxides (OH^-)	Na^+, K^+, NH_4^+, Ba^{2+}	Ca^{2+}, Sr^{2+}

In a precipitation reaction, the ions of soluble salts remain in solution as spectator ions.

Acidic aqueous solutions of hydrogen halides, and aqueous solutions of all acids, contain the hydronium, H_3O^+, ion, which gives all acidic solutions their characteristic properties.

Bronsted acids are proton donors and **Bronsted bases are proton acceptors.** In aqueous solution, acids donate H^+ to H_2O to give H_3O^+; bases accept H^+ from H_2O to give OH^-, hydroxide ion. Acids and bases are **strong** if completely ionized in solution, **weak** if incompletely ionized:

$$HA(aq) + H_2O \rightleftarrows H_3O^+(aq) + A^-(aq) \qquad \text{acid}$$

$$B(aq) + H_2O \rightleftarrows BH^+(aq) + OH^-(aq) \qquad \text{base}$$

In solutions of weak acids or bases, all the species in solution are in **dynamic equilibrium** and quickly achieve apparent constant concentrations. For strong acids and bases, the position of equilibrium favors the formation of ions to such an extent that the detectable concentration of unionized acid, HA (or base B), is negligible. The common **strong** acids are HCl, HBr, HI, HNO_3, H_2SO_4 and $HClO_4$. Most other acids are **weak.** Common **strong** bases in water are soluble hydroxides, such as NaOH, KOH, and $Ba(OH)_2$, soluble oxides (containing the O^{2-} ion), hydrides (containing the H^- ion), and amides (containing the NH_2^- ion).

$$O_2^-(aq) + H_2O(l) \longrightarrow 2OH^-(aq)$$

$$H^-(aq) + H_2O(l) \longrightarrow H_2(g) + OH^-(aq)$$

$$NH_2^-(aq) + H_2O(l) \longrightarrow NH_3(aq) + OH^-(aq)$$

The commonest **weak** base is ammonia, $NH_3(aq)$,

$$NH_3(aq) + H_2O(l) \rightleftharpoons NH_4^+(aq) + OH^-(aq)$$

Water is **amphoteric** with both acid and base properties, which accounts for its small degree of **self-ionization** or **autoprotolysis:**

$$\underset{\text{acid}}{H_2O} + \underset{\text{base}}{H_2O} \rightleftarrows H_3O^+ + OH^-$$

Neutralization of an acid by a base is the reverse of this reaction and goes almost to completion:

$$\underset{\text{acid}}{H_3O^+} + \underset{\text{base}}{OH^-} \rightleftarrows 2H_2O$$

Other cations or anions present play no active part in the reaction but remain in solution and constitute a salt. A reaction in which a solution of known concentration of an acid is added to a solution of a base of unknown concentration (or vice-versa) until the resulting solution contains only salt, is an **acid-base titration.**

In general, the anion A^- of a weak acid HA is a weak base and the cation BH^+ of a weak base B is a weak acid. A^- is the conjugate base of the acid HA. HA and A^- constitute **a conjugate acid-base pair.** BH^+ is the conjugate acid of the base B, and BH^+ and B are a **conjugate acid-base pair.** The strengths of acid-base conjugate pairs are related. If HA is a strong acid, then A^- is a very weak base; if HA is weak then A^- is a relatively strong base, because appreciable concentrations of HA and A^- are present at equilibrium. Conversely, if HA is a very weak acid, then A^- has appreciable base strength.

H_3O^+ is the strongest acid species that can exist in aqueous solution and OH^- is the strongest base. All acids intrinsically stronger than H_3O^+ are quantitatively changed to H_3O^+; all bases intrinsically stronger than OH^- (such as O^{2-}, NH_2^- or H^-) are converted quantitatively in solution to OH^-.

REVIEW QUESTIONS

1. List the elements in group 7 of the periodic table in order of increasing atomic numbers.
2. Why do the halogens exist as diatomic molecules rather than as atoms?
3. Why are halogens in the elemental form not found in nature?
4. What is the common valence for all of the halogens?
5. List the empirical formulas of the compounds formed between F and the second period elements (Li to F).
6. Draw Lewis structures for HF, ClF, SCl_2, PCl_3 and CBr_4.
7. How is the shape of each of the above molecules described?
8. Define electronegativity. How does it vary with respect to positions in the periodic table:
   (a) in going from left to right across any period;
   (b) in going from top to bottom of any group?
9. What can be said about the electronegativities of metallic elements in general as opposed to those of nonmetals?
10. Which of the elements is the most electronegative? Which of the elements is the least electronegative?
11. An element has an electronegativity of 1.5. Is it a metal or a nonmetal?
12. What is a homonuclear diatomic molecule and what can be said about the distribution of its bonding electrons?
13. What is meant by the term nonpolar bond? What is meant by polar bond?
14. What can be said about the electronegativities of two atoms that form a polar bond?
15. What is meant by the term ionic bond? Between what types of elements are ionic bonds expected?
16. What are the characteristic properties of ionic compounds that differentiate them from covalent compounds?
17. What evidence is there that ionic compounds are composed of ions?
18. Draw Lewis structures for the following ions: Na^+, Ca^{2+}, F^-, Cl^-, O^{2-}, H^- and S^{2-}.
19. Arrange the following in order of increasing size:
   (a) Cl^-, Ar, K^+; (b) Li^+, Be^{2+}, Al^{3+}.
20. What is the nature of the forces that hold the ions together in an ionic crystal?
21. Why is it impossible to distinguish discrete molecules in an ionic solid, such as NaCl?
22. Why is the structure of sodium chloride different from that of cesium chloride?
23. Would it be possible to distinguish between a solution containing 1.42 g Na_2SO_4 and 1.49 g KCl in a liter of solution, and one containing 1.74 g K_2SO_4 and 1.17 g NaCl per liter?

24. Under what conditions does a precipitate form when two solutions of different salts are mixed?
25. In terms of their solubilities in water, what can be said about the hydroxides of the alkali metals and the hydroxides of the alkaline earth metals?
26. Which are the common insoluble chlorides, bromides, and iodides?
27. Which sulfates are insoluble, and which are sparingly soluble?
28. Which carbonates are soluble?
29. What test is used to detect the presence of halide ions in aqueous solution?
30. What would you expect to observe when solutions of potassium nitrate and sodium iodide are mixed?
31. Give a general definition of an oxidation-reduction reaction.
32. In terms of the transfer of electrons between reagents, define an oxidizing agent and a reducing agent.
33. In the reaction $Zn + Cl_2 \rightarrow ZnCl_2$, which reagent is **oxidized** and which is **reduced?** Which reagent is the **oxidizing agent** and which is the **reducing agent?**
34. What is the order of the relative strengths of the halogens as oxidizing agents?
35. What reaction occurs (if any):
    (a) when an aqueous solution of Cl_2 is added to a solution of NaI?
    (b) when an aqueous solution of Br_2 is added to a solution of KF?
36. Write an equation for the preparation of chlorine from sodium chloride. How is bromine prepared?
37. What are the common properties of aqueous solutions of HF, HCl, HBr and HI, that enable them to be classified as acids?
38. Draw the Lewis structure of ammonia, NH_3, and that of the hydronium ion, H_3O^+.
39. What is meant when we refer to two molecules as being isoelectronic?
40. What is the definition of a strong acid? A weak acid?
41. Classify aqueous solutions of each of the following as either strong acids or weak acids:
    $HClO_4$ HF CH_3CO_2H HCl HNO_3 H_2CO_3 H_3PO_4 H_2SO_4 HBr HI
42. Define a base in aqueous solution. Give examples of three compounds that behave as strong bases in aqueous solution and one that behaves as a weak base.
43. A weak acid, HA, ionizes in aqueous solution; what reactions occur that lead to the establishment of a dynamic equilibrium?
44. How is the formal charge assigned to an atom in a molecule or ion?
45. What is the difference between a compound that behaves as a strong base and one that behaves as a weak base, in aqueous solution?
46. Give three examples of sodium salts that behave as strong bases in aqueous solution.
47. Write an equation to represent the behavior of ammonia as a weak base in aqueous solution.
48. Why is water classified as an amphiprotic or amphoteric substance? Write an equation to represent the self-ionization of water.
49. What is a salt? How is a solution of a salt prepared from an aqueous solution of an acid and an aqueous solution of a base?

50. Write equations for suitable reactions by which each of the following salts could be prepared from hydrochloric acid:
(a) $MgCl_2$ (b) $CuCl_2$ (c) $CaCl_2$.
51. What conjugate acid-base pairs are formed when a weak acid, such as HF, ionizes in aqueous solution? What is the conjugate acid of the base ammonia, NH_3?
52. Write, in its simplest terms, the reaction the occurs when an acid is neutralized by a base in aqueous solution?
53. What is the strongest acidic species and the strongest basic species that can exist in aqueous solution?

Answers to Selected Questions

23. First solution contains 1.42 g/142 g mol^{-1} = 0.010 mol Na_2SO_4 per liter solution and 1.49 g/77.45 g mol^{-1} = 0.020 mol KCl per liter solution. Second solution contains 0.010 molar K_2SO_4 and 0.020 molar NaCl; i.e., both solutions contain 0.020 mol L^{-1} of Na^+, K^+ and Cl^-, and 0.010 mol L^{-1} of SO_4^{2-}. They are indistinguishable.

OBJECTIVES

Be able to:

1. List the halogens, describe their physical states and colors, tell in what combined forms they occur in nature, and write equations for their industrial and laboratory preparation.
2. Write the typical reactions of halogens with nonmetals and metals.
3. Draw Lewis structures and molecular geometries for covalent halides.
4. Define the term electronegativity, show how electronegativity varies in the periodic table, and use electronegativity differences to predict polarities of bonds and the signs of partial charges on atoms in molecules.
5. Classify bonds between pairs of atoms as ionic, polar covalent, or nonpolar covalent.
6. Describe the trends in ionic radii in the periodic table, and predict the relative sizes of ions or atoms having the same number of valence shell electrons.
7. Describe the 3-dimensional structure of solid sodium chloride and cesium chloride.
8. Define the terms oxidation, reduction, oxidizing agent, and reducing agent and identify the oxidizing agent, the reducing agent, the substance being oxidized, and the substance being reduced in an oxidation-reduction reaction.
9. Write the order of the halogens as oxidizing agents and the halide ions as reducing agents.
10. Classify common salts as soluble, sparingly soluble, or insoluble, and write net ionic reactions for precipitation reactions.
11. List the characteristic properties of acids and bases.
12. Define the terms Bronsted-Lowry acid and base and describe acid-base reactions in terms of proton transfer.

13. Define the terms strong and weak as related to acids and bases and recognize common strong and weak acids and bases.
14. Calculate unknown concentrations or volumes in neutralization reactions (titrations).
15. Recognize conjugate acid-base pairs in acid-base reactions and the relationships between the strength of an acid or base and its conjugate base or conjugate acid.
16. Classify a reaction as oxidation-reduction, acid-base or precipitation.

PROBLEM SOLVING STRATEGIES

Chapter 5 contains much material about the bonding and chemical reactions of the halogens and halogen-containing compounds. You should recall the two extremes in bonding situations introduced in Chapter 4: the **complete transfer** of an electron from one atom to another, leading to electrostatic attraction between oppositely charged **ions**, called an **ionic bond;** and **equal sharing** of an electron pair between two atoms in a molecule, leading to a **covalent bond.**

5.2 Electronegativity

In many examples of bonding, the two bonded atoms do **not** have an equal share of the electron pair. We use the concept of **electronegativity** to give a rough measure of the ability of an atom in a molecule to attract the bonding electrons. With electronegativity **differences,** we can predict the type of bonding situation likely to occur (pure covalent, polar covalent, ionic).

Example 5.1

Which of the following bonds is expected to be the most ionic? H-F, H-Cl, Li-H, Li-F, Li-Cl.

Solution 5.1

Electronegativities increase from left to right across the periodic table, and decrease from top to bottom of the table. One guiding principle is that the **larger** the electronegativity difference, the more ionic the bond is likely to be. As well, a useful rule of thumb is that compounds of group 1 or group 2 metals with nonmetals are usually ionic. Thus the five choices in the problem are quickly reduced to two, since the largest electronegativity difference will be between Li (group 1) and the halogen atoms F and Cl (group 7). Also, F is the most electronegative atom, so Li-F is the correct choice.

Example 5.2

Which of the following exhibits bonding closest to pure covalent? H_2S, SO_2, CO_2. Indicate the partial charges on the atoms of the molecule you have chosen.
[Electronegativities: H (2.2), S (2.4), C(2.5), O (3.5)]

Solution 5.2

All the above molecules are composed of nonmetallic atoms. They are all polar covalent. The **smallest** electronegativity difference is between H and S in H_2S. The sulfur atom in H_2S is slightly negative and the hydrogen atom is slightly positive. We write this as

$$\overset{\delta^+}{H} \diagup \overset{2\delta^-}{S} \diagdown \overset{\delta^+}{H}$$

In later chapters we will make further use of bond polarities to predict several molecular properties.

5.4 Oxidation-Reduction Reactions

Oxidation-reduction reactions involve **electron** transfer from one substance (the **reducing agent,** which gives up electrons and is oxidized) to another (the **oxidizing** agent, which takes on electrons, and is reduced). When trying to identify if a given reaction involves electron transfer, it is useful to try to write the reaction in two halves, an oxidation part and a reduction part. If this is possible, then you know it is an oxidation-reduction reaction, and at the same time you have identified what substances are being oxidized and reduced.

Example 5.3

Decide whether the following reaction is an oxidation-reduction reaction.

$$Mg(s) + Cl_2(g) \rightarrow MgCl_2(s)$$

If it is, name the oxidizing agent and reducing agent.

Solution 5.3

Let us try to write the reaction as the sum of two halves, in one of which electrons are given up and in the other electrons are taken on. We know that magnesium chloride, $MgCl_2$, is an ionic material, since magnesium is a group 2 metal and chlorine is a nonmetal. We also know the typical valence of magnesium is 2. Thus we write

$$Mg \rightarrow Mg^{2+} + 2e^-$$

The 2 electrons appear on the right side so that the equation is charge balanced. We note that this is an **oxidation** of Mg to Mg^{2+}. The 2 electrons generated in this **half** reaction must be accepted by some other reagent; in this case it is Cl_2.

$$2e^- + Cl_2 \rightarrow 2Cl^-$$

This represents a **reduction** of Cl_2 to $2Cl^-$.

The sum of the two half reactions gives us the desired reaction. In the overall reaction we do not see any electrons, since they have been generated by Mg, but consumed by Cl_2. Also note that the total number of electrons **produced** must be equal to the number of electrons **consumed.** In this example Mg is the reducing agent; it is the substance that provides the electrons. Mg is oxidized to Mg^{2+}. The reducing agent is always oxidized. The other reactant, Cl_2, is the oxidizing agent. Cl_2 is reduced by the reducing agent.

The Halogens as Oxidizing Agents

You should know some basic facts about the ability of the halogens to act as oxidizing agents.

Example 5.4

List which halogens, if any, are capable of oxidizing Cl^- to Cl_2.

Solution 5.4

The order of the halogens as oxidizing agents is $F_2 > Cl_2 > Br_2 > I_2$. Let us begin by writing two half reactions. We would like one of them to be

$$2Cl^- \rightarrow Cl_2 + 2e^-$$

which is an oxidation. We need to combine it with a reduction half reaction

$$X_2 + 2e^- \rightarrow 2X^-$$

where X_2 is a halogen.

Is fluorine, F_2, a suitable example of X_2? Yes, since F_2 is the strongest oxidant among the halogens.(It will oxidize Cl^-, Br^-, I^-.) How about Br_2? Will it oxidize Cl^- to Cl_2? The answer here is no, since **if** the reaction

$$Br_2 + 2Cl^- \rightarrow 2Br^- + Cl_2$$

did take place, we would produce Cl_2, which is a better oxidizing agent than Br_2. In that case the Cl_2 formed would react with the Br^-, giving back the starting materials. Using the same argument for I_2, we conclude that I_2 is not a suitable oxidizing agent for Cl^- conversion to Cl_2. So we have found that, among the halogens, only F_2 is capable of oxidizing Cl^- to Cl_2.

$$F_2 + 2Cl^- \rightarrow 2F^- + Cl_2$$

It is also helpful to remember that the order of **halide ions** as reducing agents is $I^- > Br^- > Cl^- > F^-$. In other words, since F_2 is such a good oxidizing agent, it is very easily reduced to F^-. However, F^- has very little tendency to go back to F_2; we say F^- is a poor reducing agent.

5.5 Precipitation Reactions

The second type of reaction discussed in the chapter is a **precipitation reaction**. Here solutions of ions, when mixed, form an insoluble solid, or precipitate -- no oxidation or reduction necessarily takes place.

Example 5.5

Give the net ionic equation for the reaction of aqueous $Pb(NO_3)_2$ with aqueous NaCl.

Solution 5.5

The two solutions, $Pb(NO_3)_2$ and NaCl, contain the ions Pb^{2+}, NO_3^-, Na^+ and Cl^-. When the solutions are mixed, we need to decide if any salts of low solubility will be formed so that precipitation results. (Solubility rules are summarized in Table 5.9.) Since nitrates are soluble, we do not expect a $NaNO_3$ precipitate. However, lead chloride, $PbCl_2$, is an insoluble chloride. The remaining ions, Na^+ and NO_3^-, are spectators. Thus the net ionic reaction is

$$Pb^{2+}(aq) + 2Cl^-(aq) \rightarrow PbCl_2(s)$$

If no possible insoluble salt can be formed in a solution, no precipitate will result.

5.6 Acid-Base Reactions

The last type of reaction discussed in the chapter is an **acid-base reaction.** The reaction of an acid with a base is often called a **neutralization.** One example is the reaction between the acid H_3O^+ and the base OH^- to produce water

$$H_3O^+(aq) + OH^-(aq) \rightleftharpoons 2H_2O(l)$$

For neutralization problems, it is useful to remember that, at the neutralization point,

$$\text{moles } H_3O^+ = \text{moles } OH^-$$

It makes no difference whether the acid or base is strong or weak. Since concentrations are usually expressed in units of molarity, or moles/liter, at the neutralization point,

$$\text{moles } H_3O^+ = \left(\begin{matrix}\text{concentration of} \\ H_3O^+ \text{ in moles } L^{-1}\end{matrix}\right) \times (\text{volume in L of } H_3O^+ \text{ required})$$

$$= \left(\begin{matrix}\text{concentration of} \\ OH^- \text{ in moles } L^{-1}\end{matrix}\right) \times (\text{volume in L of } OH^- \text{ required}) = \text{moles } OH^-$$

Example 5.6

What volume in mL of 0.100 M NaOH is required for complete neutralization of 2.00 mL of 0.050 M H_2SO_4?

Solution 5.6

Remember each mole of H_2SO_4 provides 2 moles of H_3O^+.

$$H_2SO_4 + 2H_2O \rightarrow SO_4^{2-} + 2H_3O^+$$

Thus moles H_3O^+ $= 2(\text{moles } H_2SO_4)$

$$= 2(0.050 \frac{\text{moles}}{\text{L}} H_2SO_4) \times 200 \text{ mL } H_2SO_4 \times \frac{1 \text{ L}}{1000 \text{ mL}}$$

$$= 0.020 \text{ moles } H_3O^+$$

For neutralization, then, 0.020 moles OH^- are needed. The volume associated with this amount can be deduced from the concentration:

$$0.020 \text{ moles } OH^- \times \frac{1 \text{ L solution}}{0.100 \text{ moles } OH^-} \times \frac{10^3 \text{ mL}}{1 \text{ L}} = 200 \text{ mL}$$

Thus 200 mL of 0.100 M NaOH are needed.

Conjugate Acid-Base Pairs

After an acid transfers a proton to a base, the species formerly an acid is now called the **conjugate base** of the acid. Likewise, the base that has accepted the proton is now called the **conjugate acid** of the base. To identify acid-base reactions, look for a pair of species, one on the left side of the equation and the other on the right, that differ only by loss or gain of a proton, H^+.

Example 5.7

Identify the acid-base pairs in the reaction

$$HI(g) + H_2O(l) \rightarrow H_3O^+(aq) + I^-(aq)$$

Solution 5.7

HI and I^- form an acid-base pair, since they differ only by a proton. The **acid** HI donates a proton to H_2O, and the **conjugate base** I^- is formed. Likewise H_2O and H_3O^+ are an acid-base pair. H_2O is acting as a **base** here, because it **accepts** a proton from the acid HI. H_3O^+ is the conjugate acid of the base H_2O.

Note in the above example that there was no arrow written from right to left. This arrow is omitted when one of the substances on the left side of the equation is a **strong** acid or a **strong base.** In the example, HI is a strong acid when placed in water. The stronger the acid, the weaker is the conjugate base. This means that I^- is too weak a base to accept a proton from H_3O^+. Thus HI and H_2O are **not** produced. Clearly it is important to know the common strong acids and bases. On the other hand, when neither of the reactants is a **strong** acid or base, the reverse reaction must be considered. For example

$$NH_3 + HCN \rightleftarrows NH_4^+ + CN^-$$

Here the acid HCN donates a proton to the base NH_3, forming NH_4^+, the conjugate acid of base NH_3, and CN^-, the conjugate base of HCN. But the reverse reaction also takes place, and NH_4^+ donates a proton to HCN, reforming the reactants. A situation exists called dynamic equilibrium, in which reactants continually form products, and products continually reform reactants.

SELF TEST

1. Which sequence represents the ordering of the halogens in the periodic table?
   (a) F, Br, Cl, I, At (b) Cl, Br, F, I, At (c) Cl, I, F, Br, At
   (d) F, Cl, I, Br, At (e) F, Cl, Br, I, At
2. Choose the correct Lewis diagram for F_2.
   (a) :F-F: (b) $\underset{\cdot\cdot}{\ddot{F}}=\underset{\cdot\cdot}{\ddot{F}}$ (c) $:\underset{\cdot\cdot}{\ddot{F}}=\underset{\cdot\cdot}{\ddot{F}}:$ (d) $:\ddot{F}-\ddot{F}:$ (e) $:\underset{\cdot\cdot}{\ddot{F}}-\underset{\cdot\cdot}{\ddot{F}}:$

3. When phosphorus is heated with excess chlorine gas, what product(s) is (are) formed?
(a) only PCl_3 (b) only PCl_5 (c) PCl_3 and PCl_5
(d) PCl_3, PCl_4 and PCl_5 (e) no reaction

4. Give the product of the reaction of magnesium with fluorine gas; name the reaction type.
(a) MgF, oxidation-reduction (b) MgF_2, precipitation
(c) MgF_3, precipitation (d) MgF_2, oxidation-reduction
(e) MgF, acid-base

5. Which one of the following statements concerning the electronegativities of the elements is **incorrect**?
(a) The electronegativity of an element is a measure of the charge on its most commonly found ion.
(b) The most electronegative element is fluorine (F).
(c) The least electronegative elements are cesium (Cs) and francium (Fr).
(d) Compounds of elements with large differences in electronegativity are ionic.
(e) Compounds of elements with roughly equal electronegativities are covalent.

6. Which is the correct order of electronegativity among the elements cesium, arsenic and chlorine?
(a) Cs > As > Cl (b) Cs > Cl > As (c) As > Cs > Cl
(d) As > Cl > Cs (e) Cl > As > Cs (f) Cl > Cs > As

7. How many of the following molecules have polar covalent (as opposed to nonpolar covalent or ionic) bonds? HBr Br_2 KBr CBr_4
(a) 0 (b) 1 (c) 2 (d) 3 (e) 4

8. Which of the following bonds would be the **least** polar?
(a) H-F (b) O-F (c) Cl-F (d) Ca-F (e) B-F
(Electronegativities: H 2.2, B 2.0, F 4.1, Cl 2.8, Ca 1.0, O 3.5)

9. How many **nearest** neighbors does an Na^+ ion have in solid NaCl?
(a) 4 (b) 6 (c) 8 (d) 10 (e) 12

10. Which of the statements is(are) correct about the sizes of halogen atoms and ions?
(a) $I^- > I > Br$ (b) $Cl > F > I$ (c) $F^- > Cl^- > I^-$
(d) $Br^- > Cl^- > F$ (e) $Cl > F > F^-$

11. Which is the correct order of size for the following atoms and ions: Rb^+, Sr^{2+}, Br?
(a) $Rb^+ > Sr^{2+} > Br$ (b) $Rb^+ > Br > Sr^{2+}$ (c) $Sr^{2+} > Rb^+ > Br$
(d) $Sr^{2+} > Br > Rb^+$ (e) $Br > Rb^+ > Sr^{2+}$ (f) $Br > Sr^{2+} > Rb^+$

12. Which of the following compounds is expected to conduct an electrical current when it is molten?
(a) Cl_2 (b) SCl_2 (c) PBr_3 (d) $MgCl_2$

13. Which of the following statements is(are) **true?**
(a) If a reagent loses electrons it undergoes oxidation.
(b) In any redox reaction one substance gains electrons and another substance loses electrons.
(c) The oxidizing agent is oxidized in an oxidation-reduction reaction.
(d) When a reagent is reduced it gains electrons.
(e) An oxidizing agent gains electrons.

14. For the reaction $I_2 + 2ClO_3^- \rightleftharpoons 2IO_3^- + Cl_2$, indicate which **two** of the following statements are false:

(a) In this reaction, I_2 is the oxidizing agent.
(b) ClO_3^- is acting as an oxidizing agent.
(c) IO_3^- is the product of oxidation of I_2 by ClO_3^-.
(d) Neither oxidation nor reduction occurs in this reaction.
(e) Cl_2 is the reduced product of ClO_3^-.

15. Iodine can be liberated as a result of
(a) oxidizing I^- by Cl^- (b) oxidizing I^- by Cl_2
(c) oxidizing Cl_2 by I_2 (d) oxidizing Cl_2 by I^-

16. Consider the reaction $Cl_2 + 2Br^- \rightarrow 2Cl^- + Br_2$
Which of the following statements is true?
(a) Cl_2 is reduced and Br^- is the oxidizing agent.
(b) Cl_2 is reduced and Br^- is the reducing agent.
(c) Cl_2 is oxidized and Br^- is the oxidizing agent.
(d) Cl_2 is oxidized and Br^- is the reducing agent.
(e) The reaction is not an oxidation-reduction process.

17. Which of the following reactions will proceed substantially from left to right as shown?
i) $Br_2 + 2I^- \rightarrow 2Br^- + I_2$ ii) $Br_2 + 2Cl^- \rightarrow 2Br^- + Cl_2$
iii) $I_2 + 2Cl^- \rightarrow 2I^- + Cl_2$
(a) Only i (b) Only ii (c) Only iii (d) Two of i, ii and iii
(e) All of i, ii and iii (f) None of i, ii and iii

18. How many of the following statements is(are) correct?
i) The reaction of fluorine with hydrogen is more violent than between H_2 and any other halogen.
ii) The order of oxidizing ability among halogens is $F_2 > Cl_2 > Br_2 > I_2$.
iii) The order of size among halogens is $F_2 > Cl_2 > Br_2 > I_2$.
iv) The most abundant halogen in the earth's crust is fluorine.
(a) 0 (b) 1 (c) 2 (d) 3 (e) 4

19. How many of the following statements is(are) correct?
i) The electronegativity of Br is greater than that of F.
ii) Most commercial iodine is currently produced from seaweed.
iii) All covalent molecular halides are high-melting solids.
iv) Among nonmetal chlorides, only CCl_4 fumes in moist air.
(a) 0 (b) 1 (c) 2 (d) 3 (e) 4

20. A substance is an acid if it:
(a) turns phenolphthalein red (b) dissociates to give hydrogen (hydronium) ions (c) turns litmus blue (d) will yield OH^- ions
(e) always forms an acid salt

21. Which of the following can be a Bronsted-Lowry acid in **aqueous** solution?
(a) MgO (b) OH^- (c) NH_3 (d) NH_4^+ (e) CO_3^{2-}

22. Which of the following is **not** a strong acid in water?
(a) HCl (b) H_2SO_4 (c) HI (d) HF (e) HNO_3

23. Which of the following molecules is least likely to act as an acid?
(a) H_2O (b) CH_4 (c) HCl (d) H_2S

24. Which of the following reactions is **not** an acid-base reaction?
(a) $CO_3^{2-} + H_3O^+ \rightarrow HCO_3^- + H_2O$ (b) $Ca + 2H_3O^+ \rightarrow Ca^{2+} + H_2 + 2H_2O$
(c) $HNO_3 + H_2O \rightarrow NO_3^- + H_3O^+$ (d) $H_3O^+ + OH^- \rightarrow 2H_2O$

25. Which of the following acids is(are) **weak?**
HF HCl HBr H_2CO_3 Acetic acid
(a) All except HCl (b) All except HF and HCl (c) All except HF

(d) Only HCl and HBr (e) Only H_2CO_3 and acetic acid (f) All except HCl and HBr

26. Which of the following is a salt?
(a) KNO_3 (b) $HC_2H_3O_2$ (c) $Ca(OH)_2$ (d) $H_2C_2O_4$ (e) NH_4OH

27. 20 mL of a 0.10 M NaOH solution completely neutralized 25 mL of an HCl solution. The molarity of the HCl solution was:
(a) 0.80 M (b) 0.080 M (c) 0.125 M (d) 0.0125 M (e) 0.050 M

28. For the reaction $HPO_4^{2-}(aq) + H_2O(l) \rightleftharpoons H_2PO_4^-(aq) + OH^-(aq)$
(a) HPO_4^{2-} is an acid and OH^- its conjugate base
(b) H_2O is an acid and OH^- its conjugate base
(c) HPO_4^{2-} is an acid and $H_2PO_4^-$ its conjugate base
(d) H_2O is an acid and HPO_4^{2-} its conjugate base
(e) there are no conjugate acid-base pairs.

29. Which of the following statements is(are) **true?**
i) The conjugate base of HF is F^-
ii) The conjugate acid of H_2O is OH^-
iii) The conjugate base of H_3O^+ is H_2O
(a) Only i (b) Only ii (c) Only iii (d) Two of i, ii and iii
(e) All of i, ii and iii (f) None of i, ii and iii.

30. The strongest acid that can exist in dilute aqueous solution is
(a) H_2SO_4 (b) HCl (c) $HClO_4$ (d) H_2O (e) H_3O^+

31. Which one of the following statements is correct?
(a) F^- is a strong base, and the conjugate base of the weak acid HF.
(b) NH_4^+ is the conjugate base of the weak acid NH_3.
(c) Acetate ion is the conjugate base of the strong acid acetic acid.
(d) ClO_4^- is a weak base and is the conjugate base of the strong acid $HClO_4$.
(e) $Cl^-$is a strong base, and is the conjugate base of the strong acid HCl.

32. In which of the following reactions will a precipitate form?
i) $KI(aq) + Na_2SO_4(aq) \rightarrow$
ii) $KI(aq) + Mg(NO_3)_2(aq) \rightarrow$
iii) $AgNO_3(aq) + NaBr(aq) \rightarrow$
(a) None (b) i only (c) ii only (d) iii only (e) two of i,ii,iii
(f) All of them.

Answers to Self Test

1. e; 2. e; 3. c; 4. d; 5. a; 6. e; 7. c; 8. b; 9. b; 10. a, d; 11. e; 12. d; 13. a,b,e; 14. a,d; 15. b; 16. b; 17. a; 18. c; 19. a; 20. b; 21. d; 22. d; 23. b; 24. b; 25. f; 26. a; 27. b; 28. b; 29. d; 30. e; 31. d; 32. d.

CHAPTER 6

CARBON, ENERGY AND THERMOCHEMISTRY

SUMMARY REVIEW

A few simple compounds of carbon, including CO, CO_2, CS_2 and carbonic acid, are classified as **inorganic compounds**; the remainder are called **organic compounds**. Compounds containing only carbon and hydrogen are **hydrocarbons;** other organic compounds may be regarded as derived from hydrocarbons by replacement of hydrogens by other atoms or groups of atoms (functional groups).

Carbon **as the element occurs as several different forms or allotropes** which include **diamond, graphite** and also microcrystalline forms of graphite, such as **charcoal, carbon black** and **coke** -- all are used as reducing agents. **Carbon monoxide,** CO, results from burning carbon in a limited supply of air. It is very toxic. Mixtures with H_2 are important fuels and are used in synthesis of organic compounds. **Water gas** (CO + H_2) results from passing steam over coke, and **synthesis gas** (CO + $3H_2$) is produced by the catalytic oxidation of methane with steam. Carbon monoxide is readily oxidized to **carbon dioxide,** CO_2, by burning in air or reaction with steam at high temperature. CO reduces many metal oxides to metal and reacts catalytically with H_2 at high temperature to give methanol, CH_3OH. **Carbon dioxide** results from burning any carbon compound in excess oxygen, from decomposition of metal carbonates by heat or the action of dilute acid, and as a byproduct of fermentation. Aqueous solutions contain small amounts of diprotic **carbonic acid,** $(HO)_2C{=}O$, and its anions, **hydrogen carbonate** ion, HCO_3^-, and **carbonate ion,** CO_3^{2-}. On heating, carbon combines with sulfur to give **carbon disulfide,** CS_2; CS_2 reacts with Cl_2 to give a mixture of **tetrachloromethane,** CCl_4 , and S_2Cl_2 . **Hydrogen cyanide,** HCN, results from heating a mixture of methane and ammonia. It resembles the hydrogen halides; it is a covalent substance and gives an acidic aqueous solution containing HCN(aq), **hydrocyanic acid.** The salts of hydrocyanic acid are the **cyanides,** containing the CN^- anion. HCN and cyanides are very toxic.

With metals, carbon forms **carbides** with group 1 and 2 metals. Compounds such as Na_2C_2 and CaC_2 are formed; they contain the carbide ion, $^-{:}C{\equiv}C{:}^-$, the strongly basic anion of the very weak acid ethyne (acetylene), $H{-}C{\equiv}C{-}H$. The very hard silicon carbide (carborundum), SiC, is a covalent carbide with the diamond structure, in which alternate C atoms are replaced by Si atoms.

Carbon forms a very large number of compounds with hydrogen, called **hydrocarbons. Alkanes** have the general formula C_nH_{2n+2}, where n is an integer, and contain only single bonds. The simplest alkane is methane, CH_4, the major component of natural gas. Alkanes with up to 4 carbon atoms are gases, those from C_5 to C_{15} are liquids, and the higher alkanes are solids. The C-H bonds are weakly polar and alkanes do not behave as acids, and the carbon atoms have no unshared pairs of electrons to which a hydrogen ion can be added, so that alkanes have no basic properties. The lack of acid and base properties and the strong C-H and C-C bonds make alkanes very unreactive, except at high temperatures, when, for example, complete combustion occurs with oxygen to give carbon dioxide and water,

$$2C_4H_{10}(g) + 13O_2(g) \longrightarrow 8CO_2(g) + 10H_2O(g)$$

Compounds such as **ethene**, C_2H_4, containing a C=C bond, are formed when alkanes are strongly heated. $H_2C{=}CH_2$ has AX_3 geometry at each C atom and is a planar molecule. It polymerizes to give $(CH_2)_n$, polyethylene (polyethene). Compounds such as **ethyne**, C_2H_2, contain a C≡C bond. H-C≡C-H has AX_2 geometry at each C atom and is linear. It results from the high temperature decomposition (cracking) of ethane,

$$C_2H_6(g) \xrightarrow{\text{heat}} C_2H_2(g) + 2H_2(g)$$

or from the reaction of carbide ion, $^-$:C≡C:$^-$, with water,

$$C_2^{2-} + 2H_2O \longrightarrow C_2H_2(g) + 2OH^-$$

Ethene and ethyne (acetylene) are the first members of the series of hydrocarbons called alkenes and alkynes, with the general formulas C_nH_{2n} and C_nH_{2n-2}, respectively.

Benzene, $C_6H_6(l)$, is the simplest **arene** and is obtained when coal is heated to high temperature in the absence of air, or when alkanes are heated in the presence of suitable catalysts. Compounds such as benzene are also called **aromatic** hydrocarbons. The Lewis structure of benzene contains the six carbon atoms arranged in a planar ring with alternating C-C and C=C double bonds (Kekulé structure). The H atoms of benzene may be replaced by other groups to give compounds such as methylbenzene (toluene).

```
          H                         CH3
          |                          |
          C                          C
        //  \                      //  \
  H—C         C—H          H—C         C—H
     ||         |               ||         |
  H—C         C—H          H—C         C—H
        \   //                    \   //
          C                          C
          |                          |
          H                          H
      benzene                     toluene
```

An enormous number of **organic** substances result from replacing the H atoms of hydrocarbons by other groups. Simple examples include -OH, to give **alcohols**, -C(=O)OH, to give **carboxylic acids**, and $-NH_2$, to give **amines**.

Stored energy that can be released as light or heat in chemical reactions is **chemical energy**. It constitutes most of the world's energy reserves and it occurs in substances such as natural gas, oil and coal. The energy changes that accompany chemical reactions result largely from the making and breaking of chemical bonds; **bond formation is exothermic** and **bond breaking is endothermic** -- an overall reaction can either absorb heat (endothermic) or give out heat (exothermic). The quantitative study of heat changes in chemical reactions is **thermochemistry**, a branch of **thermodynamics**.

Energy is the capacity to do work or transfer heat. **Work, w**, is given by force times the displacement of an object that it causes (**w** = F d), and in SI is measured in newton meters (joules). Thus, work and energy are both measured in **joules**: $1 \text{ J} = 1 \text{ N m} = 1 \text{ kg m s}^{-2} \times \text{m} = 1 \text{ kg m}^2 \text{ s}^{-2}$. **Heat** is energy that is transferred between objects as the result of a temperature difference, and flows spontaneously from a hot body to a cooler body until both have the same temperature. In other words, molecules with a relatively high average kinetic energy transfer energy to molecules with a lower average kinetic energy, until the molecules of both bodies have the same average kinetic energy.

The **first law of thermodynamics** states that **energy can neither be created nor destroyed but only changed from one form to another**. In other words, the total energy of a system isolated from its surroundings remains constant. During a process, if the amount of **heat added to the system** is **q**, and the amount of **work done on the system** is **w**, the change in the **internal energy** of the system is

$$\Delta E = E_{final} - E_{initial} = q + w$$

where $E_{initial}$ is the initial internal energy of the system and E_{final} is the final internal energy. ΔE is independent of the path by which the change from initial state to final state is achieved; q and w may be different for different pathways but the sum of q and w is always the same, independent of the pathway.

When a system expands by an amount ΔV, the work **done on the system** against an external pressure P, is $-P\Delta V$, and two situations are of practical importance: (i) For a system at **constant volume**, e.g., for a reaction carried out in a closed vessel such as a bomb calorimeter, $\Delta V = 0$, and $\Delta E = q_V$, i.e., **the change in internal energy is equal to the amount of heat that flows into the system.** (ii) For a system at **constant pressure**, e.g., for a reaction carried out in a vessel open to the atmosphere, if we define a **state function** $H = E + PV$, where H is called the **enthalpy**, the change in enthalpy is given by $\Delta H = \Delta E + \Delta(PV)$, and at constant pressure $\Delta H = \Delta E + P\Delta V$, so that $q_p = \Delta E + P\Delta V = \Delta H$, i.e., **at constant pressure the heat flow into the system is equal to the enthalpy change ΔH.**

Commonly, reactions are carried out at constant pressure, e.g., atmospheric pressure, or it can easily be arranged that the reactants are initially at a given pressure and the products end up at the same pressure. For a reaction the heat absorbed by the system in going from reactants to products at the same constant pressure is called the **enthalpy change**, ΔH. It is **negative** for **exothermic reactions** and **positive** for **endothermic reactions**. The enthalpy change, ΔH, is the difference between the total enthalpy of the products and the total enthalpy of the reactants

$$\Delta H = \sum H_{products} - \sum H_{reactants}$$

Reactions are carried out under various conditions but a pressure of 1 atmosphere and a temperature of 25°C are defined as **standard conditions**. The enthalpy change for a reaction in which both reactants and products are at standard conditions is given by the symbol $\Delta H°$.

The **molar heat capacity** of a substance is the amount of heat needed to raise the temperature of 1 mole by 1 K. **Calorimetry** describes the experimental measurement of the heat changes accompanying chemical reactions; thermochemical changes are measured in a calorimeter, designed so that there is negligible exchange of heat with the surroundings during a reaction. The heat given off, or absorbed, by the reactants during a chemical reation is equal to that gained, or lost, by the calorimeter and its contents.

Hess's law states that the ΔH for a reaction depends only on its initial and final states and **not** on how it is carried out; ΔH is exactly the same whether a reaction goes by one step or occurs in many steps. Thus, ΔH for a reaction can be calculated indirectly by summing the ΔH values for any other series of reactions that added together give the same balanced equation.

A standard way of calculating enthalpy changes for reactions is to use tables of **standard enthalpies of formation,** ΔH_f°, the standard enthalpy changes for the reactions where 1 mole of a substance is formed from its elements in their standard states. For any reaction

$$\Delta H^\circ = \sum n_p(\Delta H_f^\circ)_p - \sum n_r(\Delta H_f^\circ)_r$$

where **p** refers to products, **r** to reactants, and **n** refers to the number of moles of each.

The energy required to break a bond in a gaseous molecule is its **bond energy.** For $CH_4(g)$, the **average CH bond energy** is one-quarter of ΔH for the reaction

$$CH_4(g) \rightarrow C(g) + 4H(g)$$

and similarly for all molecules with only one type of bond. For molecules with more than one type of bond, bond energies can be evaluated assuming that the individual values are the same in different molecules; this is true to a first approximation. A table of such average bond energies may be used to calculate approximate ΔH° values for gas phase reactions:

$$\Delta H = \sum BE \text{ (bonds broken)} - \sum BE \text{ (bonds formed)}$$
$$= \sum BE \text{ (reactants)} - \sum BE \text{ (products)}$$

Values so obtained are usually within a few percent of those obtained experimentally.

In consideration of chemical fuels, the prime economic factor is cost per kJ of heat evolved; sometimes, as in the case of rocket fuels, energy content per unit mass is important. The energy needs of animals are supplied by oxidation reactions; one of the most important for humans is the oxidation of glucose. This occurs in the body in a series of slow steps to give $CO_2(g)$ and water; nevertheless, the overall enthalpy change is exactly the same as if glucose were burnt outside the body (provided reactants and products start out and end up at the body temperature). The energy contents of foods are obtained experimentally by burning them in a calorimeter. Because they are often mixtures, rather than pure substances, the energy content is often expressed in units of **kJ gram^{-1}** or **Calories gram^{-1}** (where 1 Calorie = 1 kilocalorie = 4.184 kJ).

REVIEW QUESTIONS

1. Write the formulas and draw Lewis structures for the following substances:
   carbon monoxide, carbon dioxide, carbon disulfide, carbonic acid, hydrogen carbonate ion, carbonate ion, hydrogen cyanide, cyanide ion
2. What makes charcoal a useful substance for purifying others?
3. How is coke manufactured?
4. Write an equation for the reduction of cupric oxide to copper metal, using coke.
5. Write equations representing the reactions that give (a) water gas, (b) synthesis gas, as the products.
6. Write equations for the reaction of carbon monoxide with O_2, H_2O, and H_2, respectively.
7. Why is CO_2 a linear molecule?
8. How is CO_2 conveniently prepared in the laboratory?
9. What simple test is often used to detect CO_2 as a product of a reaction?
10. Why is carbonic acid described as a diprotic acid? What salts are formed between carbonic acid and potassium hydroxide?
11. How is HCN(g) prepared from sodium cyanide?
12. What names are used to distinguish HCN(g) from HCN(aq)?
13. How is calcium carbide prepared industrially?
14. Write an equation representing the reaction of calcium carbide with water to give acetylene.
15. What evidence is there that the carbide ion behaves as a strong base in aqueous solution?
16. Draw Lewis structures for the carbide ion and for acetylene.
17. How is the structure of silicon carbide related to that of diamond?
18. Give a possible reason why there is no structure for silicon carbide that is analogous to that for graphite.
19. Write the general formulas for (a) an alkane, (b) an alkene, and (c) an alkyne, containing n carbon atoms.
20. What is the general formula for a cycloalkane with n carbon atoms?
21. What is the difference between a **saturated** and an **unsaturated** hydrocarbon?
22. Name the hydrocarbons with the formulas: CH_4 C_2H_6 C_3H_8 C_4H_{10}
23. Draw Lewis structures for C_3H_8, C_3H_6, and C_3H_4.
24. Write the structural formulas for benzene and methylbenzene.
25. Write structural formulas for ethanol, ethylamine, and acetic (ethanoic) acid.
26. What is the amount of work done on a system at constant pressure, P, when its volume increases by ΔV?
27. What equation relates the enthalpy change, ΔH, to the change in the internal energy of a system, ΔE, at constant pressure?
28. Explain why for most systems the value of $\Delta H°$ is not too different from that of $\Delta E°$.
29. What quantity is the heat absorbed by a system equal to: (a) at constant pressure; (b) at constant volume?
30. In addition to heat, name two other kinds of energy.
31. What is the molar heat capacity of a substance?
32. If 1.508 kJ of heat is added to 18.02 g of water initially at 23°C, what will be the final temperature?

33. How is the enthalpy change, ΔH, for a reaction defined?
34. If the value of ΔH for a reaction is -200 kJ, is the reaction exothermic or endothermic?
35. What is calorimetry? Describe a simple calorimeter and list the important properties it should have if accurate thermochemical measurements are to be obtained using it.
36. Would you expect the enthalpy of neutralization of HCl(aq) by NaOH(aq) to be the same as the enthalpy of neutralization of HBr(aq) by KOH(aq), assuming the same concentrations and conditions? Explain. What if HBr were to be replaced by HF(aq)?
37. Draw a diagram of a bomb calorimeter.
38. In thermochemistry, what is meant by the term standard conditions? Under standard conditions, how is the enthalpy change for a reaction designated?
39. If ΔH for the reaction $2A + B \rightarrow C$, has a value of ΔH_1°, what is the value for the reaction $6A + 3B \rightarrow 3C$?
40. Water can be formed by burning $H_2(g)$ and $O_2(g)$, for which ΔH° is -285.8 kJ mol^{-1}. What is ΔH° for the overall reaction where $H_2(g)$ and $O_2(g)$ are first dissociated into atoms, and then the atoms are combined to give water? What principle does your answer exemplify?
41. State Hess's law. In a practical sense, why is Hess's law useful?
42. Define enthalpy of formation, ΔH_f°, with specific reference to the enthalpy of formation of $SO_2(g)$.
43. How is the standard enthalpy of a reaction related to the standard heats of formation of the participating reactants and products?
44. Explain why the standard heat of formation of $Cl_2(g)$ is zero, while that of $I_2(g)$ is not zero.
45. Ethyne burns in oxygen to give CO_2 and H_2O. If ethyne is less stable than the elements from which it is formed, and CO_2 and H_2O are more stable, would you expect the combustion of ethyne to be exothermic or endothermic? Explain.
46. What energy measures the strength of the bond between two atoms? How could this energy be measured for a diatomic molecule?
47. Define the average bond energy for the SF bond in $SF_6(g)$.
48. Would you expect the average PCl bond energy in $PCl_3(g)$ to be the same as that in $PCl_5(g)$?
49. The average CH bond energy in $CH_4(g)$ is 415.8 kJ mol^{-1}. What energy is required to completely dissociate one mole of methane gas into atoms? The standard enthalpy of formation of methane, $\Delta H_f^\circ(CH_4, g)$, is -74.5 kJ mol^{-1}. How is this quantity related to your answer?
50. List some advantages and disadvantages of the use of hydrogen as a fuel.
51. Write the balanced equation for the oxidation of glucose, $C_6H_{12}O_6(s)$ to $CO_2(g)$ and liquid water. What is the standard enthalpy change for this reaction when it occurs slowly in the body?

<u>Answers to Selected Review Questions</u>

4. $CuO(s) + C(s) \longrightarrow Cu(s) + CO(g)$ 19. C_3H_8, C_3H_6, C_3H_4 20. C_nH_{2n}

```
        H H H        H H H       H
        | | |        | | |       |
23.   H-C-C-C-H    H-C-C=C-H   H-C-C≡C-H     32.  44°C     39.  3 ΔH1
        | | |        |           |
        H H H        H           H
```

40. -285.8 kJ mol^{-1}

OBJECTIVES

Be able to:

1. Describe the structure of diamond and graphite; list the properties of these two allotropes and the other elemental forms of carbon.
2. Give the physical and chemical properties of CO, CO_2, H_2CO, CS_2, CCl_4, HCN, and carbides.
3. Describe the geometry around each carbon atom in alkanes, alkenes,and alkynes.
4. Give balanced chemical equations for the reactions of alkanes, alkenes, and alkynes with oxygen, and show how ethene and ethyne can be prepared by high temperature thermal decomposition of ethane.
5. Define the terms heat, work, internal energy, system, surroundings, state function.
6. State the first law of thermodynamics in terms of heat and work.
7. Define the term enthalpy and state how an enthalpy change is related to heat.
8. Define the term molar heat capacity and use molar heat capacities to calculate the amount of heat evolved in a calorimeter.
9. Define the term standard state of an element or compound.
10. State the convention used for the value of the enthalpy of the most stable form of an element in its standard state.
11. State Hess's law; manipulate chemical reactions and their enthalpy changes to calculate the enthalpy change for a desired reaction.
12. Define the term formation reaction.
13. Define the term combustion reaction and give the products and their phases for combustion of hydrocarbons.
14. Calculate the enthalpy changes accompanying various types of chemical reactions, using enthalpies of formation and/or enthalpies of combustion.
15. Calculate the energy required to break and form chemical bonds and approximate enthalpy changes from bond energy data.
16. Appreciate the role of thermochemical data in considerations of fuels and foods.

PROBLEM SOLVING STRATEGIES

6.4 Thermochemistry

In thermochemistry we are interested in energy changes, usually in the form of heat, accompanying chemical reactions. When treating any numerical problem in thermochemistry, it is always necessary to write a balanced chemical equation corresponding to the change. As well, the precise meaning of **formation reaction** and **combustion reaction** must be known. Sometimes the same reaction can be described in more than one way. For example

$$C(graphite) + O_2(g) \rightarrow CO_2(g)$$

is **both** the formation of $CO_2(g)$ from its elements in their standard states and, as well, the combustion of C(graphite). In addition, to do well with thermochemical calculations, the concept of **state function** must be understood. For enthalpy changes, this means that as long as one starts with a given set of reactants under given conditions, and ends with a given set of products under given conditions, any series of steps that connects the reactants to the products will give the correct ΔH. As you gain experience with enthalpy calculations, you should be able to choose the **simplest** set of steps.

The same basic procedures are followed in solving almost all thermodynamic problems:

(a) Write the **balanced** chemical equation for the required change for which the thermodynamic quantity is to be calculated. Note that in this equation the reactants and products must be in the correct phase (s), (l), (g) or (aq)) and if elements are present they must be in the correct allotropic form.

(b) Using the thermodynamic data given (or standard data from tables), write a series of balanced chemical equations (or draw a set of steps) that added together will give the balanced equation for the required change. Each of these equations will have associated with it the appropriate thermodynamic quantity. Make sure that each of these quantities has the correct sign (+ or -). Remember that the enthalpy of formation of an **element in its most stable form** is zero. The object is to add together the series of balanced equations to give the required balanced equation, so make sure that the reactants and products are initially on the appropriate sides of each equation that is to be added. This may require multiplying throughout by an appropriate numerical factor, in which case any associated thermodynamic quantity is also multiplied by the same factor, or an equation reactants → products may have to be reversed, in which case the associated thermodynamic quantities change sign.

(c) In calculating enthalpy changes, one can often use the equation

$$\Delta H° = \sum n_p(\Delta H_f°)_p - \sum n_r(\Delta H_f°)_r$$

if the given data involves standard heats of formation.

(d) Solve for the unknown.

Reaction Enthalpies from Enthalpies of Formation

Hess's law states that the sum of standard enthalpies of formation of the products in a chemical reaction minus the sum of standard enthalpies of formation of the reactants equals the standard enthalpy change for a reaction.

$$\Delta H° = \sum n_p\ (\Delta H_f°)_p - \sum n_r\ (\Delta H_f°)_r$$

Example 6.4 in the text illustrates a typical application of this equation. Let's see another one.

Example 6.1

Calculate the enthalpy change at 25°C and 1 atm for $2HBr(g) \rightarrow H_2(g) + Br_2(g)$.

The standard enthalpies of formation are -36.23 kJ mol^{-1} and +30.96 kJ mol^{-1}, for HBr(g) and $Br_2(g)$.

Example 6.1

(a) The balanced equation is already given in the statement of the problem. Note that $Br_2(g)$, and not $Br_2(l)$, is one of the products.

(b) and (c) Using Hess's law, we write

$$\Delta H° = \Delta H_f°(H_2,g) + \Delta H_f°(Br_2,g) - 2\Delta H_f°(HBr,g)$$

(d) Since $H_2(g)$ is the most stable form of the element hydrogen at 25°C, it has an enthalpy of formation of zero. However, the most stable form of the element bromine at 25°C is $Br_2(l)$. Gaseous bromine, $Br_2(g)$, has an enthalpy of formation of +30.96 kJ mol^{-1}. Then

$$\Delta H° = 0 + (1 \text{ mole})(30.96 \text{ kJ mol}^{-1}) - (2 \text{ moles})(-36.23 \text{ kJ mol}^{-1})$$

$$\Delta H = 103.42 \text{ kJ}$$

Enthalpies of Formation from Enthalpies of Combustion

In a combustion reaction, $O_2(g)$ reacts with a compound to form products. If the compound is a hydrocarbon, the products at 25°C are $CO_2(g)$ and $H_2O(l)$. In the text's Example 6.6, a measured enthalpy of combustion is used to calculate an unknown enthalpy of formation. We can do the same type of calculation in reverse, and determine an enthalpy of combustion from several given enthalpies.

Example 6.2

Calculate $\Delta H°$ for the combustion of one mole of ethyne, $C_2H_2(g)$ at 25°C and 1 atm, given

(1) $C(graphite) + O_2(g) \rightarrow CO_2(g)$ $\qquad \Delta H_1 = -393.5$ kJ

(2) $H_2(g) + 1/2O_2(g) \rightarrow H_2O(l)$ $\qquad \Delta H_2 = -285.8$ kJ

(3) $2C(graphite) + H_2(g) \rightarrow C_2H_2(g)$ $\qquad \Delta H_3 = +226.8$ kJ

Solution 6.2

(a) Since the products of the combustion of a hydrocarbon at 25°C and 1 atm are $CO_2(g)$ and $H_2O(l)$, we can immediately write the balanced equation:

$$C_2H_2(g) + 5/2\ O_2(g) \rightarrow 2CO_2(g) + H_2O(l)$$

If you do not want to work with fractional stoichiometric coefficients, you can multiply the above equation by 2, and solve for the ΔH for the combustion of 2 moles of ethyne. At the end, you must then divide your answer by 2 to obtain the ΔH per mole of ethyne combusted.

(b) Since $CO_2(g)$ and $H_2O(l)$ are products in the balanced equation, but $C_2H_2(g)$ is a reactant, we need to combine the three given reactions algebraically so as to end with the overall reaction written in part (a). We see that the reactant C_2H_2 is needed for the C_2H_2 combustion reaction. Reaction 3 listed above has C_2H_2 as a product, and there are no other reactions involving C_2H_2. Thus, by writing the reverse of reaction 3 and reversing the sign of its enthalpy change we have

$$C_2H_2(g) \rightarrow 2C(graphite) + H_2(g) \qquad \Delta H_i = -\Delta H_3$$

But this reaction produces 2 moles of C(graphite). No C(graphite) appears in the combustion of C_2H_2 reaction we have written in part (a). So we must look for a reaction that uses up the undesired C(graphite) and produces the product $CO_2(g)$. So we write

$$2C(graphite) + O_2(g) \rightarrow 2CO_2(g) \qquad \Delta H_{ii} = 2\Delta H_1$$

Finally, the reaction that produced 2 moles of C(graphite) from C_2H_2 also yielded 1 mole of $H_2(g)$. This $H_2(g)$ must be converted to $H_2O(l)$, since that is the other desired product in the equation of part (a). Thus

$$H_2(g) + 1/2\ O_2(g) \rightarrow H_2O(l) \qquad \Delta H_{iii} = \Delta H_2$$

Now we have manipulated the given reactions to obtain the desired combustion reaction.

$$C_2H_2 \rightarrow 2C(graphite) + H_2(g)$$
$$2C(graphite) + 2O_2(g) \rightarrow 2CO_2(g)$$
$$H_2(g) + 1/2\ O_2(g) \rightarrow H_2O(l)$$
$$\overline{C_2H_2(g) + 5/2\ O_2(g) \rightarrow 2CO_2(g) + H_2O(l)}$$

(c) We must combine the enthalpy changes in exactly the same fashion as we manipulated the reactions. Thus

$$\Delta H°(\text{combustion of } C_2H_2) = \Delta H_i + \Delta H_{ii} + \Delta H_{iii} = -\Delta H_3 + 2\Delta H_1 + \Delta H_2$$

(d) From the given data, we get

$$\Delta H° = -(226.8\ kJ) + 2(-393.5\ kJ) + (-285.8\ kJ) = -1299.6\ kJ$$

Note that in this problem all three given reactions are **formation** reactions, and the problem could have been solved using

$$\Delta H° = \sum \Delta H_f°(\text{products}) - \sum \Delta H_f°(\text{reactants})$$

For the reaction

$$C_2H_2(g) + 5/2\ O_2(g) \rightarrow 2CO_2(g) + H_2O(l)$$

we get

$$\begin{aligned}\Delta H° &= 2\Delta H_f°(CO_2,g) + \Delta H_f°(H_2O,l) - \Delta H_f°(C_2H_2,g) - 5/2\Delta H_f°(O_2,g)\\ &= 2(-393.5\ kJ) + (-285.8\ kJ) - (226.8\ kJ) - 5/2(0)\\ &= -1299.6\ kJ\end{aligned}$$

Bond Energies

Sometimes we would like to have an estimate of the enthalpy change for a reaction, but all the enthalpy of formation data needed might not be available. In that case, we could use bond energy data, and follow the four steps outlined earlier.

Example 6.3

Use the bond energy data below to estimate ΔH for the conversion of methane,

$CH_4(g)$, and chlorine, $Cl_2(g)$, into chloroform, $CHCl_3(g)$, and hydrogen chloride, $HCl(g)$.

Bond	Bond Energy, kJ/mol
C-H	413
Cl-Cl	239
C-Cl	326
H-Cl	431

Solution 6.3

(a) Remember that bond energies refer to **gaseous** molecules and **gaseous** atoms. The required balanced equation is

$$CH_4(g) + 3Cl_2(g) \rightarrow CHCl_3(g) + 3HCl(g)$$

(b) Whenever a bond energy calculation is asked for, it is very useful to write **Lewis structures** for all the reactants and products, and to break and form bonds, as shown in the following. Don't forget that some hydrocarbons, for example, have double bonds or triple bonds.

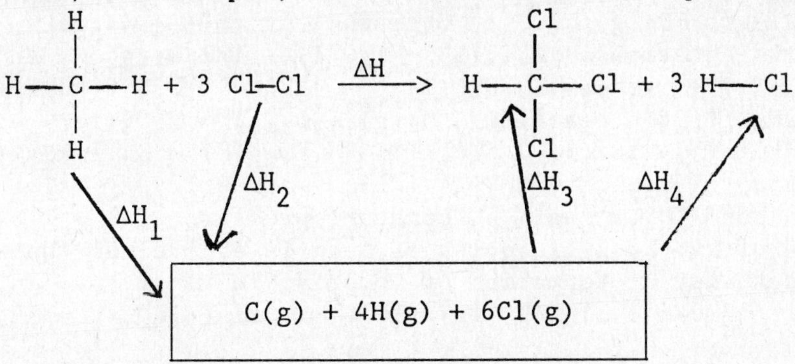

Here we have broken all the bonds in the reactants to give atoms and then recombined the atoms to form the required products, so that we can write an equation relating the ΔH's.

(c) $\Delta H = \Delta H_1 + \Delta H_2 + \Delta H_3 + \Delta H_4$

$= 4B.E.(C\text{-}H) + 3B.E.(Cl\text{-}Cl) - B.E.(C\text{-}H) - 3B.E.(C\text{-}Cl) - 3B.E.(H\text{-}Cl)$

$= 3B.E.(C\text{-}H) + 3B.E.(Cl\text{-}Cl) - 3B.E.(C\text{-}Cl) - 3B.E.(H\text{-}Cl)$

An alternative method for this step is to use

$$\Delta H = \sum B.E.(\text{bonds broken}) - \sum B.E.(\text{bonds formed})$$

which yields the same expression.

(d) $\Delta H = 3(413\ kJ) + 3(239\ kJ) - 3(326\ kJ) - 3(431\ kJ) = -315\ kJ$

SELF TEST

1. Water gas is a mixture of
   a) H_2O and CO b) H_2O and CO_2 c) H_2 and H_2O d) H_2 and CO_2 e) H_2 and CO
2. Substantial conversion of aqueous NaCN to HCN occurs when one mixes a solution of the salt with
   a) strong acid b) strong base c) more water d) $H_2(g)$

3. A major use for ethyne is
a) for welding b) as a lubricating agent c) for asphalt
d) as a solvent e) for nuclear power

4. What is the range of carbon content in the mixture of hydrocarbons used for gasoline?
a) C_1 to C_4 b) C_6 to C_{10} c) C_{13} to C_{17}
d) C_{18} to C_{25} e) C_{26} and higher.

5. Given ΔH = -602 kJ for the reaction: $Mg(s) + 1/2O_2(g) \rightarrow MgO(s)$, what would you expect to happen if the reaction were allowed to proceed at constant pressure in such a way that no heat transfer could take place between the reaction mixture and the surroundings?
(a) no reaction could occur
(b) the temperature of the reaction mixture would increase
(c) the temperature of the reaction mixture would decrease
(d) insufficient information is given.

6. The enthalpies of formation of compounds assume that elements are initially in their most stable states at 25°C and 1 atm. Of the following species, the one which is not the most stable state of the element is:
(a) $N_2(g)$ (b) $O_3(g)$ (c) C(graphite) (d) $H_2(g)$ (e) $Br_2(l)$

7. Given the two equations: $H_2O_2(l) \rightarrow H_2O(l) + 1/2\ O_2(g)$ $\Delta H°$ = -98.3 kJ
$H_2(g) + 1/2O_2(g) \rightarrow H_2O(l)$ $\Delta H°$ = -285.8 kJ
determine $\Delta H°$ for the reaction: $H_2(g) + O_2(g) \rightarrow H_2O_2(l)$
(a) +187.5 kJ (b) -187.5 kJ (c) +384.1 kJ (d) -384.1 kJ (e) +98.3 kJ

8. When 1.00 gram of ammonia, $NH_3(g)$ (molar mass = 17.0 g mol^{-1}) is produced from $N_2(g)$ and $H_2(g)$ at a constant pressure of 1 atm and a constant temperature of 298 K, 2711 joules of heat is evolved to the surroundings. The molar enthalpy of formation of ammonia, in kJ is
(a) -(2.711)(17.0) (b) -(2.711)/(17.0) (c) (17.0)/(2.711)
(d) (2.711)/(17.0) (e) (17.0)/(-2.711)

9. The enthalpy of formation, $\Delta H_f°$, of carbon dioxide is -393.5 kJ/mole.

Assuming charcoal is pure graphitic carbon, calculate the enthalpy change (in kJ) at 298 K due to the combustion of 10 lb of charcoal (1.00 lb = 454 g). Molar mass of carbon = 12.0.
(a) $\frac{(454)\ (+393.5)}{(12)}$ (b) $\frac{(10)(454)(-393.5)}{(298)}$ (c) $\frac{(10)(-393.5)}{(12)(454)}$
(d) $\frac{(-393.5)(10)(454)}{(12)}$ (e) (393.5)(10)(454)(12)

10. What is $\Delta H°$ for the following reaction: $2Fe_2O_3(s) + 3C(graphite) \rightarrow 4Fe(s) + 3CO_2(g)$?

substance	$\Delta H_f°$ (kJ mol^{-1})
$CO_2(g)$	-393
$Fe_2O_3(s)$	-284

(a) -469 kJ (b) -427 kJ (c) +431 kJ (d) +469 kJ
(e) Insuffient information; need $\Delta H_f°$ of C and Fe

11. Calculate $\Delta H°$ for the following reaction
$2LiOH(s) + CO_2(g) \rightarrow Li_2CO_3(s) + H_2O(g)$

Substance	ΔH°_f (kJ mol^{-1})
$H_2O(l)$	-285.8
$H_2O(g)$	-241.8
$CO_2(g)$	-393.5
$Li_2CO_3(s)$	-1215.6
$LiOH(s)$	-487.2

(a) -133.5 (b) -576.7 (c) -1792.3 (d) -89.5 (e) +133.5

12. Calculate the standard enthalpy of formation of phosgene ($COCl_2(g)$), in kJ mol^{-1}, given the following.

$C(graphite) + O_2(g) \longrightarrow CO_2(g)$ $\Delta H^\circ = -393.5$ kJ mol^{-1}
$CO(g) + Cl_2(g) \rightarrow COCl_2(g)$ $\Delta H^\circ = -112.5$ kJ mol^{-1}
ΔH°combustion (CO,g) $= -283.0$ kJ mol^{-1}

(a) -789.0 (b) -223.0 (c) +564.0 (d) +789.0 (e) -564.0

13. One can often calculate the enthalpy change for a reaction before carrying out the reaction because
(a) all heats of reactions have already been measured and tabulated
(b) enthalpies of formation for all known compounds have been measured
(c) the given reaction, and its enthalpy change, may be related algebraically to other reactions already carried out
(d) since enthalpy change depends on amount of substance reacting, one can always adjust the amount of materials so as to observe an enthalpy change of any value.

14. The bond energy of carbon monoxide, CO, is the change in enthalpy for the reaction:
(a) $CO(g) \rightarrow C(graphite) + 1/2\ O_2(g)$ (b) $CO(g) \rightarrow C(g) + O(g)$
(c) $1/2C(graphite) + 1/2CO_2(g) \rightarrow CO(g)$
(d) $C(graphite) + 1/2\ O_2(g) \rightarrow CO(g)$

15. Instead of calculating the **average** bond energy of a C-H bond in $CH_4(g)$ (methane), what is the energy needed to break only the first C-H bond (in kJ/mole)?

Compound or Atom	ΔH°_f (kJ mol^{-1})
$CH_4(g)$	-74.9
$CH_3(g)$	+142
$H(g)$	+218

(a) 142 (b) 218 (c) 285 (d) 435 (e) 870

16. Given the following average bond energies:

	kJ mol^{-1}
C-H	413
Cl-Cl	239
C-Cl	326
H-Cl	431

calculate ΔH, in kJ, for the reaction $CH_4(g) + 4Cl_2 \rightarrow CCl_4(g) + 4HCl(g)$
(a) -105 (b) -420 (c) +105 (d) +420 (e) +352

17. From the thermochemical equations:

$1/2N_2(g) \rightarrow N(g)$ $\Delta H = +470.5$ kJ
$1/2O_2(g) \rightarrow O(g)$ $\Delta H = +247.0$ kJ
$1/2N_2(g) + 1/2O_2(g) \rightarrow NO(g)$ $\Delta H = +90.3$ kJ

one estimates the bond energy of NO to be (in kJ mol^{-1})
(a) +807.9 (b) +627.2 (c) -807.8 (d) +1254.4 (e) -90.3

18. The absorption of 15 kJ of heat by a system which does 49 kJ of work on its environment brings about a change in its internal energy, ΔE, equal to:
(a) -34 kJ (b) +64 kJ (c) +34 kJ (d) -64 kJ

19. Which **one** of the following statements is correct?
(a) The standard enthalpy of formation of an element is negative
(b) ΔH is always more negative than ΔE
(c) q and w are state functions
(d) q + w is a state function
(e) $\Delta P = \Delta V$ for an ideal gas

20. Of the following reactions at constant temperature and constant pressure, the one for which $\Delta H = \Delta E$ is:
(a) $H_2(g) + 1/2O_2(g) \rightarrow H_2O(l)$ (b) $F_2(g) + H_2(g) \rightarrow 2HF(g)$
(c) $N_2(g) + 3H_2(g) \rightarrow 2NH_3(g)$ (d) $2KClO_3(s) \rightarrow 2KCl(s) + 3O_2(g)$

21. In the balanced equation for the combustion of one mole of n-octane (C_8H_{18},l) at 25°C, the coefficient of oxygen is: (a) 12.5 (b) 15 (c) 25 (d) 34 (e) 50

Answers to Self Test

1. e; 2. a; 3. a; 4. b; 5. b; 6. b; 7. b; 8. a; 9. d; 10. d; 11. d; 12. b; 13. c; 14. b; 15. d; 16. b; 17. b; 18. a; 19. d; 20. b; 21. a.

CHAPTER 7

QUANTUM THEORY AND THE ELECTRONIC STRUCTURE OF ATOMS AND MOLECULES

SUMMARY REVIEW

Light and other forms of electromagnetic radiation are described either as waves or as streams of particles (photons) traveling with a velocity c = 3.00×10^8 m s^{-1}. A wave is characterized by a **wavelength, λ**, a **frequency, ν**, and an **amplitude, A**; **λ** and **ν** are related by $\lambda\nu$ = **c**, and A^2 measures the intensity of the wave. Wave theory alone accounts for phenomena such as interference and diffraction that cannot be explained by the photon theory; photon theory alone accounts for the photoelectric effect and photochemical reactions. In the photon theory, each light photon has an energy **E = hν**, where **h = 6.63 x 10^{-34} J s, Planck's constant.**

In photochemical reactions, a reaction is started by a photon of light breaking a specific bond in a molecule. This creates highly reactive intermediates (radicals) that react with more reactant molecules, to give more intermediates, and so on. However, no reaction occurs unless the photons used have an energy in excess of the energy of the initial bond that is to be broken.

Emission spectra from atoms do not contain a complete spectrum of frequencies but only a limited number (spectral lines) that are characteristic for a particular atom. This makes it possible to use emission spectra to identify the atoms of which a substance is composed. Emission spectral lines originate from an electron dropping from one energy level of an atom to a lower one, with the difference in energy appearing as a photon of light. Thus, each atom has only a certain number of allowed energy levels; its energy is said to be **quantized.**

Quantization of energy levels occurs in all **microscopic** systems (atoms and molecules). In **macroscopic** systems the energy appears to vary continuously only because the separations between the allowed energy levels are too small to be observed.

Atomic hydrogen has an emission spectrum consisting of several **series** of lines in the ultraviolet (UV), visible, and infra-red (IR) regions. Each permitted energy level of the single electron in hydrogen is associated with a **quantum number n,** which can have only the integral values 1, 2, 3, 4, ... etc. The ground state of hydrogen corresponds to n = 1, and n = 2, 3, 4, ... etc., correspond to the allowed excited states. The emission spectrum originates from transitions to the n = 1 energy level from excited states with n = 2, 3, 4,..., respectively; to the n = 2 level from states with n = 3, 4, 5, ..., and so on. The single set of energy levels designated by the **n** quantum numbers have energies given by E_n = $-1312/n^2$ kJ mol^{-1}, and accounts for all the spectral lines. For any specific line, for Avogadro's number N of electrons undergoing the given transition, the energy of N photons is

$$E = Nh\nu = 1312 \left(\frac{1}{n_1^2} - \frac{1}{n_2^2}\right) \text{ kJ mol}^{-1}$$

where $n_2 > n_1$.

In the **Bohr model** for the hydrogen atom the electron moves around the proton only in one of a set of circular orbits, each associated with a quantum number **n** that can have only the integral values 1, 2, 3, 4, ... Bohr showed that the energy of a particular orbit depends on the value of n and is given by $E_n = -2\pi me^4/n^2h^2$, where m is the electron mass, e its charge and h is Planck's constant. Numerically this gives $E_n = -1312/n^2$ kJ mol^{-1}, identical with what was deduced experimentally. The Bohr model was, however, unable to be applied to the spectra of atoms with more than one electron.

Experimentally, the energy levels occupied by the electrons in an atom can be determined using **photoelectron spectroscopy.** Such experiments show that in many-electron atoms the electrons are arranged in **shells** and **subshells.** Each shell is identified with a particular value of the quantum number n, but is found to be divided into n subshells which differ from one another slightly in energy.

shell	the n subshells	capacity for electrons	total electrons $= 2n^2$
n = 1	1s	2	2
n = 2	2s 2p	2, 6	8
n = 3	3s 3p 3d	2, 6, 10	18
n = 4	4s 4p 4d 4f	2, 6, 10, 14	32

Electron configurations for the elements are obtained by systematically filling the subshells (energy levels) with electrons. For the first 36 elements, the order of energy levels is 1s<2s<2p<3s<3p<4s<3d<4p. The **main group** elements of groups 1 and 2 have s outer electrons and are called the **s-block.** The main group elements of groups 3 to 8 have s and p outer electrons and are called the **p-block.** The **transition metal (d-block) elements** have electrons in outer d shells, and the **inner transition metal (f-block) elements** (lanthanides and actinides) have outer f electrons.

de Broglie postulated that electrons and other atomic particles have **wave properties,** such the wavelength, λ, of a body of mass **m** moving with a velocity **v** is given by $\lambda = h/mv$, and **Heisenberg** showed that if there is an uncertainty $\Delta p = \Delta(mv)$ in the **momentum** of an electron and an uncertainty Δx in its **position,** then $\Delta x \cdot \Delta p > h$, with the consequence that if p is known more accurately then x becomes more uncertain. This means that electrons cannot be described as moving in atoms in defined orbits with known velocities; such **classical descriptions** have to be replaced by **quantum mechanical descriptions,** where only the **probability** of locating an electron at some point in space can be known.

The **Schrödinger wave equation** describes an electron as a **standing wave (orbital).** By analogy with other vibrating systems (such as a violin string or a drumhead), only certain orbitals are allowed to describe the behavior of the electron **in a hydrogen atom,** and each of these orbitals is characterized by **three** quantum numbers, n, ℓ, and m. The **n** quantum number has the possible values 1,2,3,4,..., and defines the **energy level** of the orbital; the ℓ quantum number has the possible values 0,1,2,3,...(n-1), for a given value of **n**, and defines the **shape** of the orbital, and the **m** quantum number has the possible values $0,\pm1,\pm2,\pm3,\ldots,\pm\ell$, and specifies the **relative orientation in space** of an orbital.

For an orbital, the total number of **nodes** is n-1, of which ℓ are **planar** nodes and n-1-ℓ are **spherical** nodes. At any node, the value of the amplitude of the standing wave is zero. Orbitals with ℓ=0 are **ns orbitals** with n-1 spherical nodes; orbitals with ℓ=1 are **np orbitals** with 1 planar node and n-2 spherical nodes, and orbitals with ℓ=2 are **nd orbitals** with 2 planar nodes and n-3 spherical nodes. For **ns orbitals**, ℓ=0, and **m=0**, and there is **one** ns orbital; for **np orbitals**, ℓ=1, and m=+1, 0, and -1, and there are **three** np orbitals, designated np_x, np_y, and np_z; and for **nd orbitals**, ℓ=2, and m=+2,+1,0,-1, or -2 and there are **five** nd orbitals that differ only in their relative orientations in space.

In the **Schrödinger equation,** the **amplitude** of the standing wave is given the symbol ψ (psi) and ψ is called the **wavefunction.** For all **s orbitals,** the variation in all directions from the hydrogen nucleus is the same and s orbitals are **spherical.** For a **1s orbital,** ψ has its maximum value at the nucleus, and decreases with increasing distance from the nucleus, while for a **2s orbital** with one **spherical** node, ψ has its maximum value at the nucleus, decreases to zero at the node, then becomes negative with increasing distance from the nucleus until a minimum value is achieved , and then increases again to an infinitesimally small value. For each of the three **2p orbitals,** ψ is zero at the nodal plane that passes through the nucleus and can be approximately described as a sphere sliced in half by this nodal plane. In one hemisphere, ψ increases with increasing distance out from the nucleus to a maximum positive value and then decreases to a infinitesimally small positive value, and in the other hemisphere it has a negative value which decreases with increasing distance out from the nucleus to a minimum value and then increases to an infinitesimally small negative value. For the $2p_x$, $2p_y$, and $2p_z$ orbitals, ψ has its maximum and minimum values along the x,y, and z axes, respectively, and 2p orbitals are not truly spherical.

ψ has no physical meaning, but ψ^2 at any point gives the **probability** of finding the electron at that point. ψ^2 varies in a similar way to ψ, but is everywhere positive (or zero at a node). The electron is conveniently thought of as spread out in the form of a negative charge cloud surrounding the nucleus; ψ^2 at any point represents the **electron density** at that point.

A **1s orbital** has a spherical charge cloud and the density decreases with increasing distance from the nucleus. A **2s orbital** also has a spherical charge cloud, but the electron density decreases with increasing distance from the nucleus to **zero** at the spherical node, then increases again and passes through a maximum before decreasing to an infinitesimally small value.

A **2p orbital** has two charge clouds, each with the shape of a flattened sphere. Each has zero density at the nuclear nodal plane, which for a $2p_z$ orbital is the xy plane. For a $2p_z$ orbital, the electron density increases along the z axis towards the centre of each cloud, reaches a maximum value, and then decreases to an infinitesimally small value. The charge cloud distributions for the $2p_x$ and $2p_y$ orbitals are similar. For a $2p_x$ orbital the nodal plane is the yz plane and the two regions of maximum density are along the x-axis, and for a $2p_y$ orbital, the xz plane is the nodal plane and the two maxima are along the y axis, above and below this plane. When a **boundary surface** is drawn to enclose 90% of the total electron density, an s orbital has a spherical boundary surface and each of the p orbitals has a boundary surface that can be approximately described as a two flattened spheres, one on each side of a nuclear nodal plane.

For **polyelectronic atoms,** the orbitals (standing wave functions) have approximately the same form as the hydrogen atom orbitals, and are described by the same three quantum numbers, n, ℓ, and m. The maximum number of electrons that any orbital can accommodate is **two**, which is a consequence of the **quantization of electron spin.** The **Pauli exclusion principle** states that an orbital can accommodate no more than two electrons, provided that they have **opposite** spins.

Box (orbital) diagrams are used to show ground (and excited) state electron configurations in detail. Conventionally, these show the orbitals as boxes arranged in order of increasing energy, which for polyelectronic atoms is $1s<2s<2p_x=2p_y=2p_z<3s<3p_x=3p_y=3p_z<4s$, etc. Orbital boxes are filled with electrons **in order of increasing energy,** according to the **Pauli exclusion principle,** and **Hund's rule,** to give the ground state electron configuration of any atom or monatomic ion. **Hund's rule** states that in a ground state electron configuration, electrons in the same energy level (e.g., a set of three 2p orbitals) occupy as far as is possible separate orbitals and have the same spin (which minimizes their repulsion energy).

Valence electrons in polyelectronic atoms experience an **effective nuclear charge** of Z-S, where **S** is the **screening constant. S** is different for electrons in different subshells; for example, **S** is greater for np electrons than for ns electrons, because an s electron has a greater electron density close to the nucleus than does a p electron. As a consequence, an np energy level is slightly higher than an ns energy level, and an np electron has a smaller ionization energy than does an ns electron. This accounts for the reversal, between group 2 and group 3 of any period, in the expected increase in ionization energy from left to right with increasing core charge. Thus, IE(B) < IE(Be), in period 2, and IE(Al) < IE(Mg), in period 3. Since it is easier to remove an electron from a filled orbital than from a singly occupied orbital, IE(O) < IE(N), in period 2, and IE(S) < IE(P), in period 3.

Covalent (electron pair) bond formation occurs when the electron densities (orbitals) of unpaired electrons on two separate atoms overlap each other to form a **molecular orbital.** Provided the electrons have opposite spins, this gives an increase in electron density between the two atoms, which holds the two atoms together by electrostatic attraction and constitutes the **covalent bond.** Two electrons with opposite spins constitute a **bonding**

electron pair and occupy a **bonding molecular orbital.** The number of unpaired electrons in the valence shell of an atom determines the number of bonds that it can form (its valence). In the **localized orbital model** each bond between **two** nuclei is represented by a **localized bonding orbital** and any nonboding electron pair is assumed to remain associated with only one nucleus in a **nonbonding orbital,** which is assumed to be an atomic orbital of the atom with which it is associated. According to the localized orbital model, an atomic orbital on one atom can be combined with a suitable orbital of another atom to give a bonding orbital, provided each atomic orbital is initially occupied by only one electron.

While the **ground state** electron configurations of Li, N, O, F, and Ne, in **period 2** are consistent with their observed valences, those of Be, B, and C correspond to the **excited states** in which an electron from a 2s orbital is promoted to an empty 2p orbital to give the **valence state.**

REVIEW QUESTIONS

1. What is the nature of an electromagnetic wave?
2. How are the wavelength and the frequency of an electromagnetic wave related to the speed of light?
3. Arrange the following forms of electromagnetic radiation in order of increasing wavelength (decreasing frequency): visible light, ultraviolet radiation, infra-red radiation, X-rays, radiowaves.
4. Explain what is meant by constructive interference and destructive interference between two waves.
5. How is the energy of a photon related to the frequency of the light?
6. In the photoelectric effect, why are no electrons emitted if the light used has less than some characteristic frequency, ν_0, however great the intensity?
7. In the photoelectric effect, why does the kinetic energy of the emitted electrons depend linearly on the frequency of the light used?
8. What is a photochemical reaction?
9. How does the minimum frequency of a photon necessary to break a chemical bond depend on the bond strength?
10. Why do the emission spectra of atoms consist only of a limited number of spectral frequencies?
11. What is meant by the ground state of an atom, and what is an excited state?
12. If an electron in an atom falls from an energy level with energy E_2 kJ mol^{-1} to an energy level with energy E_1 kJ mol^{-1}, what is the wavelength of the resulting emitted photon?
13. What is the energy of a photon emitted from a hydrogen atom when its electron moves from the energy level with $n = 2$ to the ground state ($n = 1$)?
14. What assumption was made by Niels Bohr about the path followed by the electron in a hydrogen atom?
15. What was the major flaw in Bohr's theory?
16. What is the momentum p of a body of mass m moving with velocity v?
17. What is the law of conservation of momentum?

18. What is the relationship between the wavelength of a photon and its momentum?
19. How was it shown experimentally that moving atomic particles, such as electrons, have an associated wavelength?
20. What is the de Broglie relationship?
21. What is the Heisenberg uncertainty principle?
22. Why is it impossible to state that an electron in an atom moves around the nucleus in a precisely defined orbit?
23. What information can be obtained from photoelectron spectra?
24. How does the photoelectron spectrum of an atom such as argon show the existence of shells and subshells?
25. What is the Schrödinger equation?
26. What experimental evidence supports the idea that moving electrons have wave properties?
27. Why are the vibrations of a violin string said to be quantized?
28. Each of the allowed vibrational modes of a violin string of length L is associated with a quantum number n. What is the value of n for (a) the fundamental, and (b) the second overtone?
29. What is a node?
30. For a particular mode of vibration of a string with a quantum number n, what is the number of nodes?
31. For the vibrational modes of a circular drumhead, what is the nature of the two types of nodes?
32. How many nodes are there for the mode of vibration of a drumhead designated by a quantum number n?
33. For a drumhead with n=3, what are the possible numbers of linear nodes?
34. Draw diagrams to illustrate each of the possible modes of vibration of a circular drumhead with (a) n=2, and (b) n=3.
35. What is a standing wave?
36. What name is used to describe a standing wave?
37. What types of nodes are found in the standing waves for the single electron in a hydrogen atom?
38. For n=4, what are the possible combinations of planar nodes and spherical nodes?
39. What quantum number gives the total possible numbers of independent orientations of the orbitals described by a quantum number n?
40. What are the possible values of ℓ and m for n=4?
41. What quantum numbers arise in solving the Schrödinger equation for the hydrogen atom?
42. In the solutions of the Schrödinger equation for the hydrogen atom, what are the possible values of the ℓ and m quantum numbers for a given value of n?
43. How are orbitals corresponding to ℓ = 0,1,2, and 3, respectively, designated?
44. What information about an orbital is given by the description 3d?
45. What symmetry is possessed by all s orbitals?
46. How does a 2s orbital differ from a 1s orbital?
47. For an orbital, what quantity is designated by the symbol ψ?
48. Draw sketches to illustrate how the value of ψ varies with increasing distance from the nucleus for 1s, 2s, and 3s hydrogen orbitals, respectively.
49. In what way do $2p_x$, $2p_y$, and $2p_z$ orbitals differ?.

50. Draw a sketch to illustrate how the value of ψ varies along the x axis of a $2p_x$ orbital and indicate the position of the planar node.
51. Where are the planar nodes situated in the $2p_y$ and $2p_z$ orbitals?
52. Describe the approximate shape of each of the orbitals designated 1s, 2s, $2p_x$, 3s, and $3p_x$.
53. How is the wavefunction squared, ψ^2, interpreted?
54. What is represented by the value of ψ^2 at any given point in space?
55. Draw approximate electron density maps to show how the electron density varies in any plane through the hydrogen nucleus in (a) a 1s orbital, and (b) a 2s orbital.
56. Draw an approximate electron density map to represent the electron density of a $2p_x$ orbital in the xy or xz planes.
57. Where are the maximum regions of electron density for the $2p_x$, $2p_y$, and $2p_z$ orbitals, respectively?
58. How is the overall shape and size of an orbital represented?
59. Draw orbital boundary surfaces to illustrate each of the 1s, 2s, and 2p orbitals.
60. What orbital represents the **ground state** of the hydrogen atom?
61. What orbitals represent the first excited state of the hydrogen atom?
62. For a polyelectronic atom, what is an orbital?
63. In what important way will an orbital of a polyelectronic atom differ from a hydrogen orbital?
64. What quantum numbers designate the energy levels of orbitals of polyelectronic atoms?
65. Why is a 2p orbital at a higher energy level than a 2s orbital?
66. State the Pauli exclusion principle.
67. State Hund's rule.
68. What is meant by the effective nuclear charge of an atom?
69. What is meant by the screening constant, **S**?
70. Why does S have different values for electrons in the 2s and 2p orbitals of atoms with more than one electron?
71. Why is the ionization energy of a 2s electron greater than that of a 2p electron?
72. Give an example of a degenerate set of orbitals.
73. What is the total number of electrons in a shell designated by the quantum number n.
74. How many subshells are there, and how are they designated, for $n = 3$?
75. In what order do the energies of ns, np, nd, and nf orbitals increase.
76. Draw an orbital box diagram to illustrate (a) the ground state, and (b) the first excited state of (i) a Be atom, and (ii) a C atom.
77. Draw a diagram to illustrate how the electrons in two 1s atomic orbitals of H atoms can be combined to give a molecular orbital.
78. What is a bonding orbital?
79 What orbitals may be combined to give an approximate description of the electron pair bond in (a) HF, (b) HCl, and (c) Cl_2?
80. What orbitals are occupied by the nonbonding electron pairs in each of the molecules in question 79?
81. Using orbitals on nitrogen, how is the bonding in ammonia, NH_3, described, and what is the predicted HNH bond angle?
82. Draw orbital box diagrams to illustrate (a) the ground states, and (b) the valence states of (i) Si, (ii) Al, and (iii) P.

Answers to Selected Review Questions

2. $\lambda\nu = c$. 5. $E = h\nu$. 16. $p = mv$. 18. $\lambda = h/p$. 27. Only vibrations for which the length of the string is divided into an integral or half-integral number of wavelengths are allowed. 28. (a) n = 1; (b) n = 3. 29. Any point in space where the wavefunction has the value zero, $\psi = 0$; any point at which the probability of finding the electron is zero, $\psi^2 = 0$. 30. n-1. 31. The nodes are either circular or linear. 32. n-1. 33. 0, 1, or 2. 37. The nodes are either spherical or planar. 38. 3 spherical nodes and 0 planar nodes (4s); 2 spherical nodes and 1 planar node (4p); 1 spherical node and 2 planar nodes (4d), and 3 planar nodes (4f). 39. The quantum number m with the values 0, ±1, ±2, ±3, ... ±(n-1). 40. ℓ = 0, 1, 2, or 3; m = 0; m = 0, ±1; m = 0, ±1, ±2, and m = 0, ±1, ±2, ±3. 41. n, ℓ, and m. 43. s,p,d, and f. 45. All s orbitals have a spherically symmetrical distribution of electron density. 47. The standing wave. 51. In the xz and xy planes, respectively. 60. 1s. 61. 2s, $2p_x$, $2p_y$, or $2p_z$, all of which have the same energy. 64. n and ℓ. 72. a set of p, d, or f orbitals. 73. $2n^2$. 74. Three; 3s, 3p, and 3d.

OBJECTIVES

Be able to:

1. Name the various regions of the electromagnetic spectrum and relate frequency, wavelength and speed of light.
2. Perform calculations relating frequency and wavelength, and the energy of a photon of light and its frequency.
3. Use the particle description of light to account for the photoelectric effect and to explain the occurrence of photochemical reactions.
4. Explain the line spectra observed for excited hydrogen atoms in terms of differences in quantized energy levels, and calculate wavelengths and frequencies of emitted light from a knowledge of the energy levels involved.
5. Describe the Bohr model of the hydrogen atom.
6. Show how ionization energy results for atoms with more than one electron provide evidence for the existence of subshells of differing energy when n is greater than one.
7. Write ground state electron configurations for the first 36 elements.
8. Describe experiments illustrating the wave nature of particles, and relate the wavelength of an electron to its momentum.
9. State the Heisenberg uncertainty principle and tell how quantum mechanics describes the position of a particle.
10. Give the interpretation associated with ψ^2 (the square of the amplitude of the wave described by the Schrödinger equation) for an electron.
11. Describe the shapes of s and p orbitals, and how ψ^2 varies with distance from the nucleus; draw boundary surfaces that enclose 90% of the electron density for the electron in a hydrogen atom in an s and a p orbital.
12. Use the Pauli exclusion principle and Hund's rule in writing orbital (box) diagrams for the ground states of atoms and ions.
13. Use the concept of effective nuclear charge to account for the fact that the energy level of a 2s electron in a polyelectronic atom is lower than that of a 2p electron.

14. Account for the covalent bonding resulting from a shared pair of electrons between two atoms in terms of attractions between the nuclei and the increased electron density in the internuclear region.
15. Describe a covalent single bond by a pair of electrons occupying a bonding localized molecular orbital.
16. Rationalize the valences of atoms from groups 2, 3 and 4 by promoting a paired electron to a higher orbital, forming a valence state of the atom.

PROBLEM SOLVING STRATEGIES

7.1 Light and Electromagnetic Waves

Some experiments with light are best interpreted using a wave picture. The frequnecy ν, wavelength λ, and speed of light c are related by $c = \nu\lambda$. There are several different units used for λ and ν in the textbook. You should be able to form unit conversion factors relating all of them, and use them to calculate λ and ν.

Example 7.1

What is the wavelength, in meters and nanometers, of light of frequency 7.0×10^{14} Hz?

Solution 7.1

Since $c = \nu\lambda$, $\lambda = c/\nu$. We know $c = 3.0 \times 10^{8}$ m s^{-1} and 1 Hz = 1 s^{-1}. Then we can form

$$1 = \frac{1 \text{ s}^{-1}}{1 \text{ Hz}}$$

So we are now ready to substitute ν and c into the λ expression.

$$\lambda = \frac{c}{\nu} = \frac{3.0 \times 10^{8} \text{ m s}^{-1}}{7.0 \times 10^{14} \text{ Hz} \times \frac{1 \text{ s}^{-1}}{1 \text{ Hz}}} = 4.3 \times 10^{-7} \text{ m}$$

To express the wavelength in nanometers we only need the unit conversion factor relating meters and nanometers.

$$\lambda = 4.3 \times 10^{-7} \text{ m} \times \frac{10^{9} \text{ nanometers}}{1 \text{ m}}$$

$$\lambda = 4.3 \times 10^{2} \text{ nanometers}$$

The Electromagnetic Spectrum

You should remember that **visible** light corresponds to wavelengths between 400 and 750 nm. **Infra-red** light has **longer** wavelengths than visible light and **ultraviolet** light has **shorter** wavelengths than those in the visible region.

See Figure 7.4 in the text for the complete range of electromagnetic radiation. For Example 7.1 above, we conclude that the light is **visible** light.

The Photoelectric Effect

Other experiments involving light are explained with the assumption that light consists of a stream of **particles,** called **photons.** Each photon has an energy E which is proportional to its frequency ν via the relationship $E = h\nu$, where h is called Planck's constant. For a **mole** of photons, all having the same frequency, the total energy is N times the energy of **one** photon, where N is the Avogadro constant. Note that since $\lambda = c/\nu$, all these photons have the **same** wavelength. In the **photoelectric effect,** if the energy of a photon striking a metal surface is great enough, an electron at the surface of the metal can be ejected. Be careful that you distinguish between calculations that involve **one** photon and **one** photoelectron versus those that treat a **mole** of photons and photoelectrons.

Example 7.2

About 480 kJ are required to remove one mole of electrons from the surface of platinum metal. What is the minimum frequency of light, in Hz, required to observe the photoelectric effect in platinum? To what wavelength, in nm, does this frequency correspond? Is this light described as infra-red, visible, or ultraviolet?

Solution 7.2

The problem gives data for the energy, NE, required for the production of 1 **mole** of photoelectrons; thus we use $NE = Nh\nu$. The frequency ν that we calculate will be the **minimum** frequency needed, since with $E < h\nu$, **no** photoelectron is produced.

$$\text{Energy per photoelectron} = \frac{NE}{N} = \frac{480 \text{ kJ mol}^{-1} \times \frac{10^3 \text{ J}}{\text{kJ}}}{\frac{6.022 \times 10^{23} \text{ electrons}}{1 \text{ mol}}}$$

$$= 7.97 \times 10^{-19} \frac{\text{J}}{\text{electron}}$$

For one electron to be ejected, the frequency ν must be

$$\nu = \frac{E}{h} = \frac{7.97 \times 10^{-19} \text{ J}}{6.63 \times 10^{-34} \text{ J s}} \times \frac{1 \text{ Hz}}{1 \text{ s}^{-1}}$$

$$= 1.20 \times 10^{15} \text{ Hz}$$

The corresponding wavelength is obtained from

$$\lambda = \frac{c}{\nu} = \frac{3.0 \times 10^{8} \text{ m s}^{-1} \times \frac{10^{9} \text{ mn}}{1 \text{ m}}}{1.20 \times 10^{15} \text{ Hz} \times \frac{1 \text{ s}^{-1}}{1 \text{ Hz}}}$$

$$= 250 \text{ nm}$$

This is ultraviolet light.

Light can also be used to cause photochemical reactions. A typical problem is discussed in the text's Example 7.3.

7.2 Atomic Spectra

Bohr suggested early in this century that the electron of the hydrogen atom could exist only in certain energy levels, and that emission of light from excited hydrogen atoms was due to an electron falling from a higher energy level to some lower one. The energy of the photon of light emitted is equal to the difference in energy between the two levels involved in the transition. In other words, the electron of hydrogen atom loses energy as it proceeds from one level to another; the energy appears as a photon of light. Since the energy levels are quantized, only certain photon energies can be observed; we observe a **line** spectrum. In a large sample of excited hydrogen atoms, many transitions are possible. All those that end in a given level form what is called a **series** of lines. For example, the Lyman series involved transitions like 2→1, 3→1, 4→1, 5→1 and so on, where 1 is the value of the quantum number n in the lower level. All the electronic energy levels of the hydrogen atom are described with the quantum number n, and all of the energy levels are negative. Those with smaller values of n are more negative, since the energy is proportional to $-1/n^2$. To get a positive value for the difference in energy between the two levels, you must subtract the lower energy level from the higher one. For emission line spectra, the higher level is the initial level involved in the transition. So the photon emitted for the hydrogen atom emission has

$$h\nu = E_{initial} - E_{final}.$$

Once the initial (higher) and final (lower) levels are known, the frequency ν or the wavelength λ of the spectral line can be calculated.

Example 7.3

Calculate the wavelength in nanometers for the longest wavelength line in the Lyman series for atomic hydrogen.

Solution 7.3

Remember $E = h\nu = hc/\lambda$. Since we need the longest wavelength line, we want the smallest photon frequency. The photon's energy will be smallest when the difference between the two levels involved is smallest, with the restriction that the quantum number n must be 1 for the final level of the Lyman series. Thus we see that the initial level must be 2. (If it were n = 3 or n = 4, then the transitions 3→1 or 4→1 would have greater photon energies associated with them than the chosen 2→1.)

$$E_{photon} = \left(E_{initial} - E_{final}\right)_{hydrogen\ atom}$$

For **one atom** undergoing the 2→1 transition,

$$E_{photon} = -1312\left(\frac{1}{2^2} - \frac{1}{1^2}\right) \frac{kJ}{mol} \times \frac{10^3\ J}{kJ} \times \frac{1\ mol}{6.022 \times 10^{23}\ atoms}$$

$$= 1.63 \times 10^{-18}\ J$$

$$\lambda = hc/E = \frac{6.63 \times 10^{-34}\ J\ s \times 3.00 \times 10^8\ m\ s^{-1} \times \frac{10^9\ nm}{m}}{1.63 \times 10^{-18}\ J}$$

$$\lambda = 1.22 \times 10^2\ nm$$

Note that the photon energy E refers to **one** photon, and **not** to a mole of photons.

7.3 Electron Configurations

A modification to the shell model of the atom is necessary, since experiments show that not all electrons described by a particular value of n have the same energy. We assume there are **subshells,** of differing energies, corresponding to the same n. We describe in what shells and subshells the electrons of an atom are when we give the **electron configuration** of the atom. Here is a set of steps to follow in writing such configurations.

1. Determine the atomic number Z, which gives the total number of electrons associated with the neutral atom. Correct this value for any charge on the atom.
2. Recall that for a given value of n, there are n different subshells, labelled by s, p, d, f ... .(The n = 1 shell has only the 1s subshell. For n = 2, there is a 2s and a 2p subshell. For n = 3, there can be 3s, 3p and 3d subshells.) An s subshell can contain a maximum of 2 electrons; for a p subshell, the number is 6; for a d subshell, 10 electrons can be held.
3. We follow a recipe, or filling order, to give the correct number of electrons. This filling order, diagrammed in Figure 7.16 in the text, is 1s, 2s, 2p, 3s, 3p, 4s, 3d, 4p ... .
4. The electrons are placed in the subshells, first completely filling one subshell before starting to fill the next one in the filling order. (There are some minor exceptions to this, as Table 7.2 in the text shows. Ignore these exceptions for now.)

Example 7.4

Write the electron configuration for the ground state of bromine atom, Z = 35.

Solution 7.4

First write the filling order (which you should memorize or be able to replicate quickly from a drawing or diagram like Fig. 7.16).

1s,2s,2p,3s,3p,4s,3d,4p ...

Next realize that you must account for 35 electrons. Use the rule that tells how many electrons can be placed in s, p, d subshells, and add electrons, using the filling order, until all 35 electrons are used up. Then the electron configuration is

$$1s^2\ 2s^2\ 2p^6\ 3s^2\ 3p^6\ 4s^2\ 3d^{10}\ 4p^5$$

Many times an alternative form of the electron configuration is written, one that rearranges the order of writing the subshells, but not changing any subshell occupation. In this scheme one writes all the subshells of a lower n first, before writing any subshells of higher n. But the filling order is unchanged. Thus for bromine, we have

$$1s^2\ 2s^2\ 2p^6\ 3s^2\ 3p^6\ \underline{3d^{10}\ 4s^2}\ 4p^5$$

The underlined part of this configuration is in a different order, compared to our earlier configuration. Both of these configurations are correct. An electron configuration is not meant to convey exactly what is the subshell energy ordering.

7.4 Wave Properties of the Electron

Not only does light exhibit a wave nature, but so do electrons. The wavelength associated with an electron can be calculated from

$$\lambda = \frac{h}{mv}$$

where the product of mass times velocity, mv, is called the **momentum** of the electron. Be sure to take care with units here, as with all of the numerical problems in the chapter.

Example 7.5

Calculate the wavelength, in nm, and frequency, in kHz, of an electron moving at 1% of the speed of light. The mass of the electron is 9.11×10^{-31} kg.

Solution 7.5

The key to success in this problem is remembering the definition of the unit joule. We know that 1 joule = 1 kg m² s⁻². Then we can write the wavelength associated with a particle (an electron here) as

$$\lambda = \frac{h}{mv} = \frac{6.63 \times 10^{-34}\ \text{J s} \times \dfrac{1\ \text{kg m}^2\ \text{s}^{-2}}{1\ \text{J}} \times \dfrac{10^9\ \text{nm}}{1\ \text{m}}}{9.11 \times 10^{-31}\ \text{kg} \times \dfrac{1}{100} \times 3.00\ 10^8\ \text{m s}^{-1}}$$

$$\lambda = 0.242\ \text{nm}$$

The frequency is

$$\nu = \frac{c}{\lambda} = \frac{3.00 \times 10^{8}\ \text{m s}^{-1} \times \frac{1\ \text{Hz}}{1\ \text{s}^{-1}} \times \frac{1\ \text{kHz}}{10^{3}\ \text{Hz}}}{0.242\ \text{nm} \times \frac{10^{-9}\ \text{m}}{1\ \text{nm}}}$$

$$\nu = 1.24 \times 10^{15}\ \text{kHz}$$

7.5 Quantum Mechanics

Orbital Box Diagrams and Electron Configurations

Quantum mechanics provides us with a mathematical model of the structure of the atom. In this view, an electron in an atom with a fixed energy has a probability of being found at various distances from the nucleus. This probability is related to a mathematical function called an orbital. Also experiments have shown that electrons behave as though they are spinning about their own axis. Relative to a magnetic field, this spin can be in only two orientations, up or down. Only two electrons, of opposite spin, can belong to the same orbital. Orbitals have the same labels as subshells; thus we speak of one 1s orbital, one 2s orbital, three 2p orbitals, and so forth. With these ideas we can extend electron configurations to include electron spin. One usually represents an orbital by a box, and uses up or down arrows for spin. Follow these rules in writing orbital diagrams.

1. Write the electron configuration for the atom.
2. Draw boxes to represent all the orbitals listed in the electron configuration. For each p subshell, make 3 boxes and for each d subshell, make 5.
3. Put electrons into the orbitals, following the filling order used to construct the electron configuration. Each s orbital can contain 2 electrons of opposite spin. For the three p orbitals, each is assigned one electron of (say) spin up, before electrons with spin down are added to these orbitals. For d orbitals, each of the five gets one electron of the same spin before any pairing of spins takes place.

Example 7.6

Draw the orbital box diagram for the ground state of bromine atom, Z = 35.

Solution 7.6

From Example 7.4 we know the electron configuration to be $1s^2\ 2s^2\ 2p^6\ 3s^2\ 3p^6\ 4s^2\ 3d^{10}\ 4p^5$. Then we draw boxes to represent orbitals.

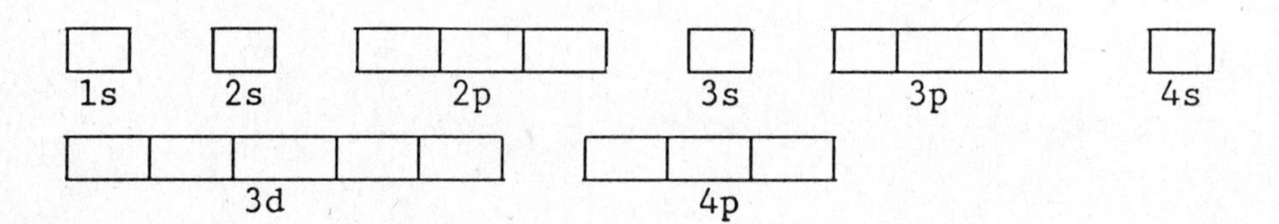

Finally we place electrons into the orbitals. For the first seven electrons, we have

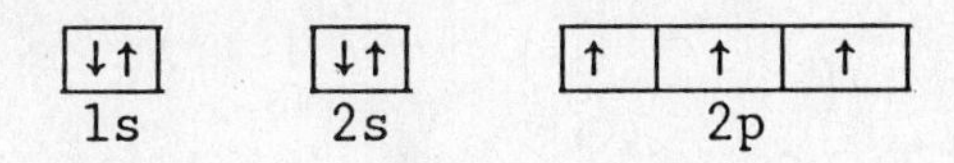

The next 3 electrons are placed in the 2p orbitals with spins down. After we have added a total of 25 electrons, the incomplete orbital diagram looks like

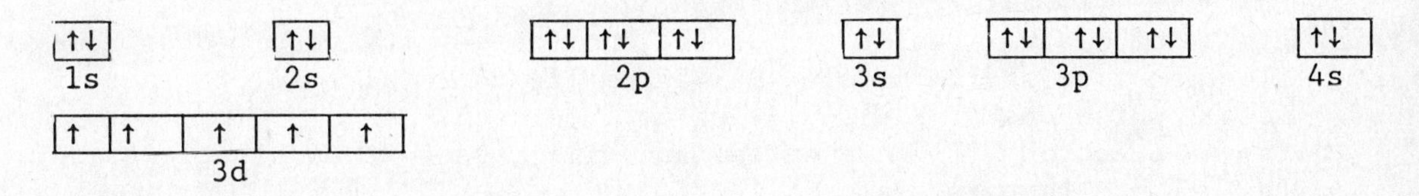

Finally with 35 electrons added, we are finished.

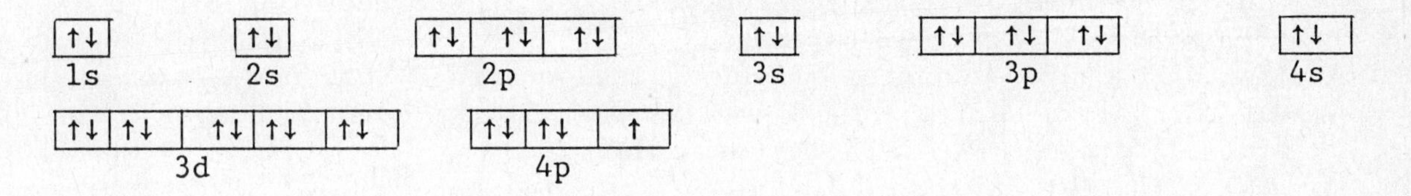

SELF TEST

Part I True or False

1. The frequency associated with orange light is larger than that related to blue light.
2. The diffraction of light is an experimental result that can be accounted for by the wave nature of light.
3. According to the Heisenberg uncertainty principle, it is impossible to say exactly how far the n = 1 electron of hydrogen atom is away from the nucleus.
4. The maximum number of electrons allowed in the n = 4 shell is 16.
5. Carbon cannot form an promoted state that has the configuration $1s^2 2s^2 2p^1 3s^1$.
6. At the equilibrium internuclear separation in chlorine molecule, there is no electron density in the internuclear region.
7. A bonding molecular orbital extends over at least two nuclei.
8. Boron atom uses all its valence electrons in forming BF_3, but carbon atom uses only two 2p electrons in forming CH_4.

Part II Multiple Choice

9. In the emission spectrum of atomic hydrogen, one of the lines in the Balmer series has a wavelength of 656 nm. What is the frequency of this line?

   (a) 2.18×10^{-5} s (b) 4.57×10^{14} s^{-1} (c) 8.23×10^{12} Hz
   (d) 4.36×10^{-15} s (e) 1.2×10^{10} Hz

10. What is the energy in joules of a photon of light of wavelength 656 nm?
(a) 8.7×10^{-31} (b) 6.6×10^{-34} (c) 3.0×10^{-19}
(d) 8.0×10^{-20} (e) 3.0×10^{-22}

11. Light of wavelength 200 nm is associated with what region of the electromagnetic spectrum?
(a) ultraviolet (b) visible (c) infra-red (d) X-ray (e) microwave

12. A sample containing potassium is heated in a bunsen flame and emits a characteristic color. The frequency of the light is 7.40×10^{14} Hz. The wavelength (in nm) of this light is:
(a) 911 (b) 247 (c) 405 (d) 222 (e) 127.

13. Light of 519 nm wavelength has the smallest energy required to remove an electron from sodium metal. If 450 nm light is allowed to hit sodium metal, which **one** of the following is incorrect?
(a) An electron will be ejected, and will possess some kinetic energy.
(b) 450 nm light is visible light.
(c) The longest wavelength that can remove an electron is 519 nm.
(d) Part of the energy of a 450 nm photon provides the energy needed for an electron to overcome attractive forces in the metal.
(e) No electron will be ejected.

14. One of the hydrogen atom emission lines occurs at 656 nm. It is a member of the Balmer series ($n_{final} = 2$). What is the quantum number n for the initial state? $E_n = -1312/n^2$ kJ mol^{-1}.
(a) 1 (b) 2 (c) 3 (d) 4 (e) 5

15. Which of the following equations correctly describes the wave nature of a particle? (v = velocity of particle).
(a) $\lambda = hc/v$ (b) $\lambda = h/mv$ (c) $\lambda = h/cv$ (d) $\lambda = m/hv$ (e) $\lambda = hm/v$

16. How many electrons occupy 3p orbitals for the ground state of chlorine (Z = 17)?
(a) 3 (b) 5 (c) 6 (d) 10 (e) 17

17. What is the ground state electron configuration for zinc (Z = 30)?
(a) $1s^22s^22p^63s^23p^64s^23d^{10}$ (b) $1s^22s^22p^63s^23p^64s^24p^64d^4$
(c) $1s^22s^22p^63s^23p^64s^24d^{10}$ (d) $1s^22s^22p^63s^23p^63d^{10}4p^2$
(e) $1s^22s^22p^63s^23p^63d^{12}$

18. Which **one** statement concerning the alkaline earth elements is **not** true?
(a) They all contain electrons in s and p orbitals, but none in d orbitals.
(b) They are all metals.
(c) None of them contain singly-occupied orbitals in the ground state.
(d) They all have two electrons in s orbitals in their outer shell.
(e) They all have the same number of valence electrons.

19. Which **one** of the following ground state orbital diagrams is **not** correct?

(a) Be [↑↓] 1s [↑↓] 2s

(b) N [↑↓] 1s [↑↓] 2s [↑ | ↑ | ↑] 2p

(c) Na [↑↓] 1s [↑↓] 2s [↑↓ | ↑↓ | ↑↓] 2p [↑] 3s

(d) O [↑↓] 1s [↑↓] 2s [↑↓ | ↑↓ |] 2p

(d) Ne [↑↓] 1s [↑↓] 2s [↑↓ | ↑↓ | ↑↓] 2p

20. A diatomic molecule has a bond energy of 400 kJ mol^{-1}. What is the longest wavelength of light which can dissociate it?
a) 299 nm b) 412 nm c) 573 nm d) 1400 nm

21. The wavelength associated with an N_2 molecule travelling through space at 100 km hr^{-1} is
a) 8.5×10^{-22} pm b) 0.14 pm c) 0.51 pm d) 1.0 pm e) 2.0 pm
22. The number of spherical nodes associated with a 4p orbital is
a) 0 b) 1 c) 2 d) 3 e) 4.

Answers to Self Test

1. F; 2. T; 3. T; 4. F; 5. F; 6. F; 7. T; 8. F; 9. b; 10. c;
11. a; 12. c; 13. e; 14. c; 15. b; 16. b; 17. a; 18. a; 19. d; 20. a;
21. c; 22. c.

CHAPTER 8

PHOSPHORUS AND SULFUR

SUMMARY REVIEW

Phosphorus, [Ne]$3s^2$ $3p^3$ in group 5 and **sulfur,** [Ne]$3s^2$ $3p^4$, in group 6, are important industrially. **Sulfur** is found as the free element in volcanic regions and in underground deposits (from which it is extracted by the **Frasch process),** as sulfide and sulfate ores and as hydrogen sulfide, H_2S, and sulfur dioxide, SO_2. **Phosphorus** is too reactive to occur as the free element, but occurs as phosphate ion, PO_4^{3-}, in phosphate rock, $Ca_3(PO_4)_2$, from which the element is obtained industrially by reduction with carbon in the presence of silica, SiO_2. It is essential to living organisms and occurs as $Ca_5(PO_4)_3OH$, hydroxyapatite, in bones and teeth and as polyphosphate ions in ADP and ATP.

Both elements occur as different **allotropes;** sulfur in orthorhombic, monoclinic, and plastic forms, and phosphorus in white, red and black forms. Allotropes have different formulas and/or different crystal structures. The S_8 molecules in orthorhombic and monoclinic sulfur are arranged differently, while plastic sulfur is a polymer containing long chains of sulfur atoms. White phosphorus contains P_4 molecules; red and black phosphorus have polymeric sheet structures which differ slightly. Both S and P react readily with oxygen; sulfur ignites in air to $SO_2(g)$ which is further oxidized to $SO_3(g)$ in the presence of a catalyst. White phosphorus reacts spontaneously with air to give P_4O_6 and P_4O_{10}.

Elements of the third and ensuing periods have valence shells that can hold more than eight electrons and consequently they have more than one value for their valence. Sulfur has valences of 2, 4 and 6 and phosphorus has valences of 3 and 5. The lowest valences correspond to elements in their ground state electron configurations; higher valences correspond to the promotion of one or more electrons.

Sulfur trioxide, SO_3, is the anhydride of **sulfuric acid,** H_2SO_4. Industrially, SO_2 is converted to SO_3, using V_2O_5 as the catalyst, and 98% H_2SO_4 is obtained by dissolving the SO_3 in aqueous H_2SO_4. Addition of more SO_3 gives pure sulfuric acid and oleum (H_2SO_4 + excess SO_3). The latter contains polysulfuric acids, $H_2S_nO_{3n+1}$. Sulfuric acid is a very important industrial chemical. In its first stage of ionization in water it is a strong acid. This property and its low volatility are used to prepare more volatile acids from their salts. It reacts with basic oxides and hydroxides to give **sulfates,** containing the sulfate ion, SO_4^{2-}, and **hydrogen sulfates,** containing HSO_4^-. As a **dehydrating agent,** H_2SO_4 removes water from hydrated salts and the elements of water from many compounds. As an **oxidizing agent,** it oxidizes carbon to CO_2, Br^- to Br_2, and I^- to I_2, and is itself reduced to SO_2. Many unreactive metals, such as copper, that do not react with H_3O^+ in aqueous acids, are oxidized by H_2SO_4.

The preparation of HCl illustrates **Le Chatelier's principle.** Removal of HCl as it is formed in the NaCl/H_2SO_4 reaction leads to the formation of more HCl, so that the reaction goes eventually to completion.

Oxidation numbers (states) can be assigned to each of the atoms of a molecule or ion and are used to identify the species that are oxidized and reduced in reactions. The most electronegative atom, F, is -1 in all of its compounds; oxygen is -2, except in peroxides and F_2O; chlorine, bromine and iodine are generally -1, and hydrogen is +1 (except in the H^- ion). All uncombined elements have oxidation numbers of zero. Any monatomic ion has an oxidation number equal to the charge on the ion. **Given that the oxidation numbers of all the atoms in a molecule or ion add up to its overall charge,** the oxidation numbers of atoms other than those listed above are readily calculated. Oxidation is associated with an increase in oxidation number and reduction with a decrease.

Sulfur displays oxidation numbers of **+6, +4, +2, zero** and **-2; phosphorus** displays **+5, +3, zero** and **-3**.

SO_2 is very soluble in water and is in equilibrium with H_3O^+, SO_3^{2-} and HSO_3^- ions. SO_2, HSO_3^- and SO_3^{2-} (S in the +4 oxidation state) are readily oxidized to SO_3, HSO_4^- and SO_4^{2-}, respectively (S in the +6 oxidation state). SO_2 can also behave as an oxidizing agent in reactions such as that with H_2S, where it is reduced to sulfur.

P_4O_{10}(s) is a better dehydrating agent than H_2SO_4; with water the triprotic acid H_3PO_4 is formed, a weak acid. **Orthophosphoric acid** (H_3PO_4) molecules readily condense and eliminate water molecules to give polymeric chains and rings containing P-O-P linkages. Two H_3PO_4 molecules give diphosphoric acid ($H_4P_2O_7$), three give triphosphoric acid ($H_5P_3O_{10}$) or the six membered ring $(HPO_3)_3$. Other forms of **metaphosphoric acid,** $(HPO_3)_n$, contain larger rings or long chains of P-O-P linked HPO_3 groups.

Phosphates are essential to animal and plant life. Reacting insoluble $Ca_3(PO_4)_2$ with H_2SO_4 gives a mixture of $CaSO_4 \cdot H_2O$ and $Ca(H_2PO_4)_2 \cdot H_2O$ (superphosphate of lime), or $Ca(H_2PO_4)_2$ (triple phosphate), which are important **fertilizers.** Reaction of H_3PO_4 with ammonia gives $NH_4H_2PO_4$, a fertilizer rich both in P and N.

P_4O_6 is the anhydride of the weak diprotic **phosphorous** acid, H_3PO_3, which has the structure $(HO)_2P(H)O$ and has two ionizable hydrogens. Phosphorous acid and its salts are readily oxidized to P(V) compounds.

Phosphine, PH_3 is flammable in air and behaves as a much weaker base than ammonia; its conjugate acid, PH_4^+, is a stronger acid than NH_4^+ and is found only in the salts PH_4Cl and PH_4I.

Hydrogen sulfide, H_2S, is a soluble poisonous gas which is a weak diprotic acid in water. When H_2S is passed into a solution of a metal salt the metal sulfide is precipitated (except for group 1 and 2 sulfides that are soluble and dissolve sulfur to give yellow solutions containing **polysulfide** ions, S_n^{2-}). H_2S is less basic than water and H_3S^+ is unknown in aqueous solution. H_2S is a strong reducing agent; it is normally oxidized to sulfur.

Hydride acidity increases from left to right across the periodic table, as the $X^{\delta-}$ - $H^{\delta+}$ polarity of the X-H bond increases with increasing electronegativity difference between X and H. It also increases with the increasing size of X in descending any group, due to the decreased basicity of X^-, despite the diminishing polarity of the X-H bond. Thus HF is more acidic than H_2O and H_2S is more acidic than water. For a hydride to behave as a **base,** it must have an unshared pair of electrons that can accept a proton. Basicity decreases across any period (e.g. $NH_3 > H_2O > HF$) due to the decrease in the size of lone pairs on X as its electronegativity increases. In descending any group, the lone pairs on X become more spread out and attract a proton less strongly; thus basicity decreases ($PH_3 < NH_3$ and $H_2S < H_2O$).

Oxoacids, $XO_m(OH)_n$, have one or more OH groups attached to an electronegative atom (nonmetal). X attracts electrons from the X-OH bond, which increases the electronegativity of oxygen and increases the polarity of the O-H bond; oxoacid strength increases in going from left to right along any period (e.g. $Si(OH)_4 < PO(OH)_3 < SO_2(OH)_2 < ClO_3(OH)$). When X is less electronegative than O, the X-OH bond becomes polar in the opposite sense and $X(OH)_n$ can behave as a **base.** Under these conditions X is normally a metal. The oxidation state of X also influences acidity. H_2SO_4 is a much strong acid than H_2SO_3 because S(VI) is more electronegative than S(IV). In an oxoacid, the number of OH groups attached to X is limited by its size and electronegativity.

REVIEW QUESTIONS

1. Assign sulfur and phosphorus to their appropriate positions in the periodic table. Are they metals or nonmetals?
2. How is sulfur prepared commercially?
3. What is an allotrope?
4. By what reaction is phosphorus obtained industrially?
5. Starting with powdered sulfur, how can the following forms be made? (a) orthorhombic (b) monoclinic (c) plastic sulfur.
6. Write the ground state electron configurations of phosphorus and sulfur.
7. What electron configuration of P accounts for its valence of 5?
8. What electron configurations of S account for its valences of 4 and 6?
9. What are the formulas for the oxides of sulfur with sulfur in the two, four and six valence states? Draw their Lewis structures.
10. What is the maximum number of electrons that can be accommodated in the $n = 3$ shell?
11. What is the product when SO_3 reacts with water? Draw its Lewis diagram.
12. Why is sulfuric acid referred to as a strong acid, and as a diprotic acid?
13. How is sulfuric acid manufactured commercially by the Contact process?
14. Draw Lewis structures for disulfuric acid and tetrasulfuric acid.
15. Write equations for each of the stages of ionization of sulfuric acid in aqueous solution.
16. Write equations for each of the stages of ionization of phosphoric acid in aqueous solution.
17. Write equations for the reactions with sulfuric acid of (a) NaCl, (b) NaBr, (c) NaI.
18. What salts can result from the reaction of sulfuric acid with calcium hydroxide?
19. How can anhydrous $CuSO_4$ be obtained from its hydrate, $CuSO_4 \cdot 5H_2O$?

20. How can the presence of sulfate ion be tested for in aqueous solution?
21. What products result from heating ethanol with concentrated H_2SO_4?
22. What products result from heating carbon with concentrated H_2SO_4?
23. What reaction occurs between magnesium and dilute sulfuric acid?
24. Why does copper react with hot concentrated H_2SO_4 but not with dilute aqueous H_2SO_4?
25. What is the oxidation state of an atom in the free or uncombined state of the element?
26. What oxidiation state has F in each of the following? F_2, F^-, F_2O, CF_4, SF_6.
27. What is the oxidation state of oxygen in each of the following? O_2, O^{2-}, F_2O_2, O_2^-, O_2^{2-}, H_2O_2.
28. What is the oxidation number of hydrogen in each of the following? H_2, H_2O, H_2O_2, LiH.
29. What is the oxidation number of chlorine in each of the following? Cl_2, ClO_4^-, ClO_3^-, HCl, HOCl.
30. What is the oxidation number of sulfur in each of the following? SO_4^{2-}, SO_2, HSO_3^-, S_8, H_2S.
31. Write the balanced equation for the reduction of Cl_2 to Cl^- by sulfite ion in acidic aqueous solution.
32. Write the equation for the oxidation of H_2S by SO_2 to give sulfur.
33. Draw the Lewis structures for each of the following: PH_3, P_4O_6, H_3PO_4, HPO_3^{2-}.
34. Draw a structure for $(HPO_3)_3$, which contains a six membered ring.
35. What salts can result from the reaction of phosphoric acid with potassium hydroxide?
36. How can phosphorous acid be made from P_4O_6?
37. Draw a structure for phosphorous acid which is consistent with its behavior as a diprotic acid.
38. Why is H_2S a stronger acid in aqueous solution than water and why is it a much weaker base?
39. Draw the Lewis structures of PH_3 and PH_4^+.
40. Why has methane, CH_4, no tendency to behave either as an acid or a base?
41. Why is phosphoric acid a weaker acid than sulfuric acid?
42. Why is hydrogen fluoride, HF, a weak acid in aqueous solution, while lithium hydride, LiH, behaves as a strong base?
43 Why has carbonic acid the structure $(HO)_2C{=}O$, rather than $C(OH)_4$?
44. Write formulas for (a) silicon tetrachloride, (b) iron(II) sulfate, (c) chromium(III) sulfate, (d) dinitrogen tetroxide.
45. Name the following:
(a) $FeCl_3$ (b) BaO (c) Ca_3N_2 (d) $AlPO_4$ (e) Fe_2O_3 (f) $HClO_4$ (g) $HClO_3$.

Answers to Selected Questions

10. 18 26. 0, rest are -1 27. 0,-2,+1,-1/2,-1,-1 28. 0,+1,+1,-1
29. 0,+7,+5,-1,+1 30. +6,+4,+4,0,-2 44. $SiCl_4$, $Fe_2(SO_4)_3$, $Cr_2(SO_4)_3$, N_2O_4

OBJECTIVES

Be able to:

1. Describe how sulfur and phosphorus are obtained industrially.

2. List the names and properties of the allotropic forms of sulfur and phosphorus.
3. Write formulas for compounds of sulfur that exhibit valences of 2, 4, 6.
4. Draw Lewis structures for higher-valence compounds of sulfur and phosphorus; give the AX_nE_m classification for such compounds.
5. List the reactions involved in the industrial preparation of sulfuric acid.
6. State Le Chatelier's principle.
7. Give four important properties of concentrated sulfuric acid and show how these properties are used in chemical reactions.
8. Define the term oxidation number and assign oxidation numbers to atoms in molecules and ions.
9. Define the terms oxidation and reduction of an element using a change in its oxidation number.
10. List typical chemical reactions involving sulfur in oxidation states +6, +4, -2.
11. List typical chemical reactions involving phosphorus in oxidation states +5, +3, 0, -3.
12. State how acid-base strengths of nonmetal hydrides vary across and down the periodic table.
13. Rewrite common oxoacid formulas as $XO_m(OH)_n$, and relate m to acid strength.
14. Name binary compounds, aqueous solutions of nonmetal hydrides, and oxoacids.

PROBLEM SOLVING STRATEGIES

8.2 Reactions and Compounds of Sulfur

Higher Valences of Sulfur and Phosphorus

In drawing Lewis structures for molecules and ions mentioned in this chapter, keep in mind that if the valence of sulfur is four or six, or if that of phosphorus is five, more than an octet of electrons is present at that atom. From the box diagrams for promoted electron configurations given in the text, sulfur with a valence of four has one nonbonding pair and four electrons in singly occupied orbitals, while sulfur with a valence of six has no nonbonding pairs and six electrons in singly occupied orbitals. These electrons in singly occupied orbitals can be used to form four or six covalent bonds, respectively. Similarly, phosphorus with a valence of five has no nonbonding pairs and five electrons in singly occupied orbitals available for covalent bonding. Fluorine and oxygen always have an octet, and commonly occur in Lewis structures as $-\ddot{\underset{..}{F}}:$ and as $\ddot{\underset{..}{O}}=$ or $-\ddot{\underset{..}{O}}-$. If the molecule is charged, oxygen can occur commonly as O^-, i.e. $^-:\ddot{\underset{..}{O}}-$.

Example 8.1

Draw the Lewis structure for SF_6, sulfur hexafluoride. Classify SF_6 according to the AX_nE_m notation.

Solution 8.1

Since fluorine always has a valence of one, the valence of sulfur must be six and thus it possesses no lone pairs. Each F then is $-\ddot{\underset{..}{F}}:$, and the complete structure is

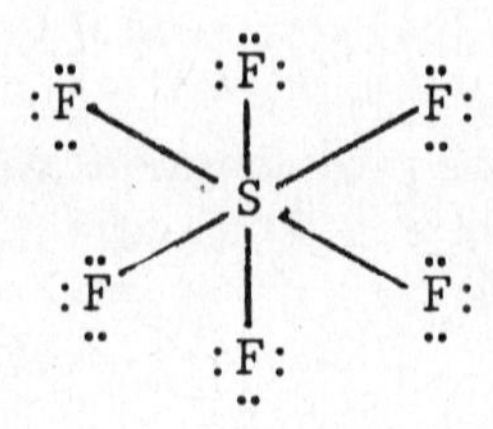

Since S of SF_6 is surrounded by 6 bonding pairs and no nonbonding pairs, SF_6 is AX_6.

Properties of Concentrated Sulfuric Acid

Four important properties of concentrated sulfuric acid are listed in the text. It has a **high boiling point,** and can act as a **strong acid,** a **dehydrating agent,** and an **oxidizing agent.** Just because sulfuric acid is an **acid** (i.e., a proton donor), it does not follow that it behaves as an acid in **all** its reactions. Also note that the high boiling point of concentrated sulfuric acid is a **physical** property. Use is made of this property in preparing a volatile acid from the salt of that acid reacting with H_2SO_4. Next, since sulfuric acid is a **proton donor,** any base present can accept the proton. In fact since H_2SO_4 can donate **two** protons, two types of salts can be formed in acid-base reactions, either hydrogen sulfate salts like $NaHSO_4$, or sulfate salts like Na_2SO_4. When concentrated sulfuric acid acts as a dehydrating agent, water is removed from a substance -- either water of crystallization or two hydrogen atoms and one oxygen atom not already present as H_2O. As an oxidizing agent, S of H_2SO_4 is reduced from O.N. = +6 to some lower O.N. (Oxidation numbers are discussed in the next paragraph.) If the O.N. for S remains +6 in all sulfur-containing products, then H_2SO_4 is not acting as an oxidizing agent in a reaction.

Example 8.2

Decide whether concentrated sulfuric acid is acting as an oxidizing agent, a dehydrating agent, or an acid in each of the following:

(a) $H_2SO_4(conc) + OH^-(aq) \rightarrow HSO_4^-(aq) + H_2O$

(b) $H_2SO_4(conc) + 2OH^-(aq) \rightarrow SO_4^{2-}(aq) + 2H_2O$

(c) $H_2SO_4(conc) + 2Br^-(aq) + 2H_3O^+(aq) \rightarrow SO_2(g) + Br_2(l) + 4H_2O$

(d) $H_2SO_4(conc) + NaHSO_4 \cdot H_2O \rightarrow NaHSO_4 + H_3O^+(soln) + HSO_4^-(soln)$

Solution 8.2

(a) The first point to note is that the O.N. of sulfur does not change. H_2SO_4 is not acting as an oxidizing agent here. Next, see that although H_2O is produced, H_2SO_4 is not a dehydrating agent here. H_2O has not been eliminated from OH^-. Thus we conclude that this is an acid-base reaction, producing a salt plus water by transfer of a proton from H_2SO_4 to OH^-. (The cation, such as Na^+, has been omitted from the reaction.)

(b) This example is like the first; in this case the salt formed is a sulfate. Two protons have been transferred from H_2SO_4.

(c) The oxidation number of one of the sulfur containing products is not +6. It is +4 in SO_2. Thus H_2SO_4 is behaving as an oxidizing agent here, since it is reduced.

(d) Here water of crystallization is removed. $NaHSO_4 \cdot H_2O$ is being dehydrated. The H_2O given off by the hydrated salt receives a proton from H_2SO_4, forming H_3O^+ and HSO_4^-. Both of these ions are surrounded by unionized H_2SO_4, which is the predominant species in concentrated sulfuric acid.

Oxidation Numbers

Oxidation numbers (O.N.) signify the relative extent to which an atom in a compound or ion is oxidized. Very useful generalizations about the chemistry of various elements can be made with use of oxidation numbers. They also provide an easy way to balance oxidation-reduction equations. **The rules for assigning oxidation numbers are given in the text in Section 8.2.** Let's see some examples.

Example 8.3

Assign oxidation numbers (O.N.'s) to each atom in the following:

(a) N^{3-} (b) P_4 (c) MnO_4^- (d) $H_2S_2O_3$

Solution 8.3

(a) Any monatomic ion has an O.N. equal to the charge on the ion. Thus N^{3-}, nitride ion, has an O.N. = -3.

(b) Since the molecule P_4 is an element, each P atom in P_4 has an O.N. = 0.

(c) Each of the 4 oxygen atoms in MnO_4^- has an O.N. = -2. Since the sum of oxidation numbers of manganese and the four oxygens must equal the charge on the MnO_4^- ion

O.N. of manganese + 4(O.N. of oxygen) = -1.

O.N. of manganese = -1-4(-2) = +7 in MnO_4^-

(d) First assign O.N. = -2 for each oxygen and O.N. = +1 for each hydrogen. Since the molecule is neutral, the sum of oxidation numbers for all the atoms must equal zero.

2(O.N. of sulfur) + 2(O.N. of hydrogen) + 3(O.N. of oxygen) = 0

2(O.N. of sulfur) = 0-2(+1)-3(-2) = +4

O.N. of sulfur = +2 in $H_2S_2O_3$, thiosulfuric acid.

In later chapters we will encounter **fractional** oxidation numbers.

8.3 Reactions and Compounds of Phosphorus

Acid-Base Reactions of Polyprotic Acids

Some acids, like HCl, can donate only one proton to a base, forming a salt plus water. Thus $HCl + NaOH \rightarrow NaCl + H_2O$. When acids that can donate

more than one proton react with a base, several **different** salts can be formed. Sulfuric acid, H_2SO_4, with two ionizable hydrogen atoms, can form hydrogen sulfate salts like $KHSO_4$ and sulfate salts, like K_2SO_4. Phosphoric acid, H_3PO_4, can donate one, two or three protons, and can form sodium salts like NaH_2PO_4, Na_2HPO_4, Na_3PO_4. However, not all acids containing three hydrogens have three ionizable hydrogens. H_3PO_3, for example, only has two ionizable hydrogens, and can form $H_2PO_3^-$ salts and HPO_3^{2-} salts. When polyprotic acids react with bases, the number of moles of base reacting with a given number of moles of the acid must be known, in order to determine the salt that will be formed.

Example 8.4

Give the product of the reaction of 1 mole of H_3PO_3 (phosphorous acid, 2 ionizable protons) with 1 mole of NaOH. Draw the Lewis structure of the product. Assign an oxidation number to phosphorus in this salt.

Solution 8.4

$$H_3PO_3 + NaOH \rightarrow NaH_2PO_3 + H_2O$$

Although phosphorous acid has two ionizable hydrogens, the acid and base are reacted in equimolar amounts, so the salt NaH_2PO_3 is formed. For the Lewis structure of $H_2PO_3^-$, first recognize that phosphorus is the central atom, and that one H is directly bonded to it. Thus one oxygen is bonded without an H atom to P, and must form a double bond to P. The other two O's are bonded to H's, and thus are singly bonded to P. The final structure therefore is

```
              :O:
               |
      ..       |     ..
   H——O———P———O:⁻
      ..       |     ..
               |
               H
```

Since P has a valence of five, it has no nonbonding pairs.

To calculate the O.N. for P in $H_2PO_3^-$, use the relationship

2(O.N. for hydrogen) + 3(O.N. for oxygen) + O.N. for phosphorus = -1

O.N. for phosphorus = -1-2(1)-3(-2) = +3

Example 8.5

For the reaction of H_3PO_3 and NaOH, forming Na_2HPO_3, calculate the volume in mL of 0.100 M NaOH needed for neutralization of 125 mL of 0.200 M H_3PO_3.

Solution 8.5

The balanced equation is $H_3PO_3 + 2NaOH \rightarrow Na_2HPO_3 + 2H_2O$

Thus **2 moles** of NaOH are required for 1 mole of H_3PO_3. From the volume of acid of the stated molarity we calculate how many moles of acid are present.

$$\text{moles } H_3PO_3 = \frac{0.200 \text{ moles } H_3PO_3}{1 \text{ L acid solution}} \times \frac{1 \text{ L}}{10^3 \text{ mL}} \times 125 \text{ mL acid solution} = 0.0250$$

$$\text{moles NaOH} = 0.0250 \text{ moles } H_3PO_3 \times \frac{2 \text{ moles NaOH}}{1 \text{ mol } H_3PO_3} = 0.0500$$

Thus 0.0500 moles NaOH are needed.

$$\text{Volume NaOH} = \frac{1 \text{ L basic solution}}{0.100 \text{ mole NaOH}} \times \frac{10^3 \text{ mL}}{1 \text{ L}} \times 0.0500 \text{ moles NaOH}$$
$$= 500 \text{ mL basic solution}$$

8.4 Structure and Acid-Base Strengths

Oxoacids are of the general form $XO_m(OH)_n$. In Chapter 8 you have seen X = S and P. Other examples include X = Cl, Br, I. For a given atom X, the larger m is, the stronger the acid is. These m O's are **doubly-bonded** to X. The n O's are singly-bonded to X and singly-bonded to H.

Example 8.6

Draw Lewis structures for the acids HClO and $HClO_2$. Which is stronger?

Solution 8.6

HClO has 14 valence electrons. It is of the form $ClO_0(OH)_1$. We draw the oxygen as singly-bonded to Cl and to H, and arrange the electrons in pairs so that there is an octet around oxygen. This uses all the electrons. Note the formal charges are all zero.

:C̈l- Ö - H
‥ ‥

$HClO_2$ has 20 valence electrons, and is $ClO_1(OH)_1$. We draw one oxygen as doubly-bonded to Cl, and one as singly-bonded to Cl and to H. We fill in the remaining electrons, ensuring that each oxygen has an octet. Any remaining electrons are arranged around chlorine.

Ö = C̈l —— Ö —H
‥ ‥ ‥

Note the formal charges are all zero.

$HClO_2$ is the stronger acid. The more doubly-bonded oxygens there are attached to chlorine, the more electron density can be pulled to Cl from H via the singly-bonded O.

SELF TEST

1. The process that produces sulfur on an industrial scale is called (a) the Contact process (b) the Le Chatelier process (c) the Frasch process (d) The Haber process (e) the Lewis process.
2. Which one(s) of the following is(are) the main allotropic forms of phosphorus?
   (a) red phosphorus (b) black phosphorus (c) blue-green phosphorus
   (d) brown phosphorus (e) white phosphorus
3. In which of the following molecules does sulfur exhibit a valence of 6?
   (a) SO_2 (b) SO_3 (c) H_2S (d) H_2SO_4 (e) $NaHSO_4$
4. Which is the correct Lewis structure for PCl_5?

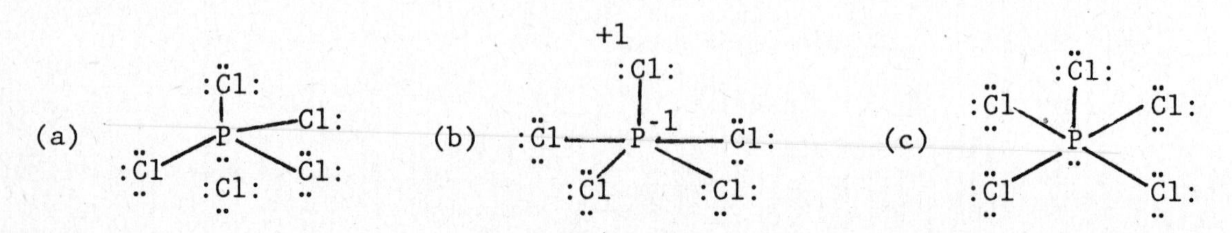

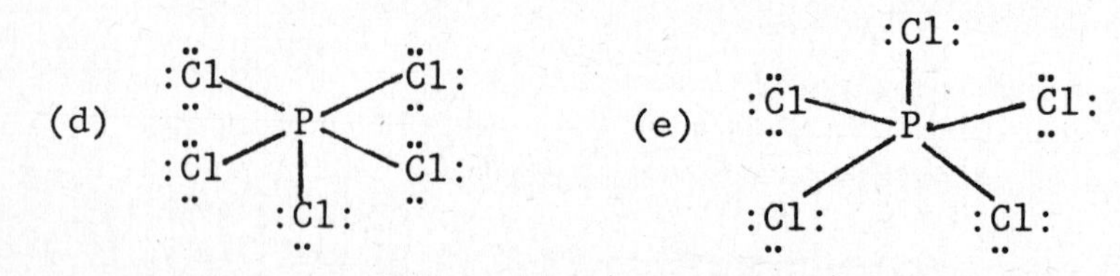

5. Which is the world's most important industrial chemical?
   (a) sodium chloride (b) sulfur (c) hydrogen sulfide
   (d) sulfuric acid (e) phosphoric acid
6. Choose all the reactions that are part of the industrial production of sulfuric acid.
   (a) $S + H_2 + 2O_2 \rightarrow H_2SO_4$
   (b) $2SO_2 + O_2 \xrightarrow{catalyst} 2SO_3$
   (c) $H_2S + 2O_2 \rightarrow H_2SO_4$
   (d) $2S + 3O_2 \rightarrow 2SO_3$
   (e) $H_2SO_4(conc) + H_2O \rightarrow H_2SO_4(dilute)$
7. When concentrated sulfuric acid reacts with CH_3CH_2OH, ethyl alcohol, to give diethyl ether, the acid is acting as
   (a) an oxidizing agent (b) a reducing agent (c) a hydrating agent
   (d) a dehydrating agent (e) a catalyst
8. The oxidation number for S in H_2S is
   (a) -2 (b) -1 (c) 0 (d) +1 (e) +2
9. The oxidation number for H in LiH is
   (a) -2 (b) -1 (c) 0 (d) +1 (e) +2
10. For the following unbalanced equation, choose the one correct statement.
   $MnO + PbO_2 + HNO_3 \rightarrow HMnO_4 + Pb(NO_3)_2 + H_2O$
   (a) PbO_2 is oxidized and MnO is reduced (b) MnO is oxidized and PbO_2 is reduced (c) MnO is oxidized and HNO_3 is reduced (d) HNO_3 is oxidized and MnO is reduced (e) PbO_2 is oxidized and HNO_3 is reduced

11. The change in oxidation number for arsenic in the equation
$H_3AsO_4 + H_2S \rightarrow H_3AsO_3 + S + H_2O$
is equivalent to:
(a) 0 electrons lost (b) 1 electron lost (c) 2 electrons lost
(d) 1 electron gained (e) 2 electrons gained

12. For the following half-reaction $MnO_4^- + 2H_2O + 3e^- \rightarrow MnO_2 + 4OH^-$, decide which one statement about the oxidation number change for manganese is correct.
(a) it decreases from 8 to 4 (b) it decreases from 7 to 2 (c) it decreases from 9 to 2 (d) it decreases from 7 to 4 (e) it decreases from 9 to 4

13. Choose the strongest acid among the following:
(a) HBr (b) HBrO (c) $HBrO_2$ (d) $HBrO_3$ (e) $HBrO_4$

14. The correct ordering for the acid strengths of the hydrides of row 2 of the periodic table is
(a) $CH_4 > NH_3 > H_2O > HF$ (b) $HF > CH_4 > NH_3 > H_2O$ (c) $HF > NH_3 > H_2O > CH_4$ (d) $CH_4 < NH_3 < H_2O < HF$ (e) $NH_3 < H_2O < CH_4 < HF$

15. Aqueous solutions of PH_4^+ cannot be obtained because
(a) it reacts with water to give PH_3 and H_3O^+
(b) it reacts with OH^- to give PH_4OH.
(c) it reacts with water to give PH_5^{2+} and OH^-.
(d) it reacts with water to give P_4O_6.
(e) it reacts with water to give P_4O_{10}.

16. Which of the following have 2 (and only 2) ionizable hydrogens?
(a) $SO_2(OH)_2$ (b) H_3PO_4 (c) H_3PO_3 (d) H_2SO_3 (e) PH_3

17. The correct name for K_2HPO_4 is
(a) dipotassium phosphate (b) dipotassium hydrogen phosphate
(c) potassium hydrogen phosphate (d) potassium phosphate
(e) dipotassium hydrogen phosphite

18. The correct name for $HClO_3$ is
(a) chlorous acid (b) hypochlorous acid (c) hydrochlorous acid
(d) chloric acid (e) perchloric acid

19. The common oxidation numbers for sulfur are
(a) -3 (b) -2 (c) +1 (d) +4 (e) +6

20. The common valences for phosphorus are
(a) 1 (b) 2 (c) 3 (d) 4 (e) 5

Answers to Self Test

1. c; 2. a,b,e; 3. b,d,e; 4. d; 5. d; 6. b; 7. d; 8. a; 9. b; 10. b; 11. e; 12. d; 13. e; 14. d; 15. a; 16. a,c,d; 17. b; 18. d; 19. b,d,e; 20. c,e.

CHAPTER 9

MOLECULAR GEOMETRY

SUMMARY REVIEW

The equilibrium positions of the atoms in a molecule or ion define its **molecular shape**. The distances between atoms forming bonds are called **bond lengths**; the angles formed by the bonds to an atom are called **bond angles**. Experimentally, molecular geometry is determined using techniques such as X-ray and neutron diffraction for solids, and electron diffraction for gases.

The **Lewis structure** of a molecular species is the starting place for predicting molecular geometry and the central atom of the species must be known. In determining Lewis structures, the following rules are followed:

1. Connect all of the atoms to the central atom by single bonds.
2. Count up all of the valence electrons of the atoms forming the molecule, allowing for any charges if the species is an ion, and convert these to **electron pairs.**
3. Subtract the number of electron pairs used to form single bonds to the central atom, and use the remainder to complete the octets around atoms (other than H), starting with the most electronegative atoms. If any electron pairs remain, add these to the central atom. Calculate the formal charges on atoms.
4. If any atom still has an incomplete octet, use nonbonding (lone) electron pairs on atoms with complete octets to form double or triple bonds to achieve octet completion, and recalculate the formal charges on the atoms.
5. If rule 4 creates additional formal charges, use the structure derived in 3.
6. If the central atom is from period 3 or a later period the octet may be exceeded. Delocalize additional lone pairs of electrons into the bonding region to form additional multiple bonds to remove as many formal charges as possible.

Formal charges may be calculated as described in Chapter 4, but since oxygen commonly occurs it is useful to remember that for oxygen forming one electron pair bond the formal charge on O is -1, as in $-\ddot{\underset{\cdot\cdot}{O}}:^-$; for oxygen forming two bonds, the formal charge is 0, as in $=\ddot{O}:$; and for oxygen forming three bonds, the formal charge is +1, as in $\equiv O:^+$.

The **VSEPR** (Valence Shell Electron Pair Repulsion) **model** of molecular geometry assumes that in most molecular species the electron pairs can be described as occupying either bonding orbitals or nonbonding orbitals, and that there is a charge cloud for each electron pair occupying a separate region of space, such that the electron pair charge clouds in the valence shell of any atom stay as far apart from each other as possible (as a consequence of the Pauli exclusion principle). For a molecule AX_nE_m with **n** ligands forming single bonds to A, and with **m** unshared (nonbonding or lone) pairs in the valence shell of A, the geometry predicted by the VSEPR model is summarized in the table on the following page.

total electron pairs on A n + m	ideal geometry	Molecule AX_nE_m	Molecular geometry	$\widehat{XAX}$
2	linear	AX_2	linear	180°
		AXE	linear	-
3	trigonal planar	AX_3	trigonal planar	120°
		AX_2E	angular	<120°
		AXE_2	linear	-
4	tetrahedral	AX_4	tetrahedral	109.5°
		AX_3E	trigonal pyramidal	<109.5°
		AX_2E_2	angular	<109.5°
		AXE_3	linear	-
5	trigonal bipyramidal	AX_5	trigonal bipyramidal	90°, 120° and 180°
		AX_4E	disphenoid	<90°, <120° <180°
		AX_3E_2	T shape	<90°
		AX_2E_3	linear	180°
6	octahedral	AX_6	octahedral	90^0 and 180^0
		AX_5E	square pyramid	<90°
		AX_4E_2	square	90°

For AX_n species, with all the X ligands the same, the ideal geometries are observed; with non-identical ligands, bond angles smaller than the ideal angles are observed. Deviations from the ideal angles are found when one or more unshared pairs, E, are present, because lone pairs in the valence shell of A occupy more space than do bonding pairs. Similarly, as the electronegativity of X increases relative to that of A, A-X bonding pairs take up progressively less space relative to lone pairs and the bond angles become progressively smaller.

For ligands forming double and triple bonds, the approximate geometry of the species is that for an AX_nE_m species, where all ligands are treated as forming single bonds; however, because two electron pairs constitute a double bond, and three electron pairs constitute a triple bond, their presence distorts the ideal shapes according to the rule that the single electron pair of a single bond occupies less space in a valence shell of the central atom A than do the two electron pairs of a double bond, which in turn occupy less space in the valence shell of A than do the three electron pairs of a triple bond. The shapes of molecules containing double and triple bonds can be predicted on the basis that the double bond and triple bond charge clouds are arranged at a maximum distance from the other single bond and lone pair charge clouds.

The observed structures of many molecules and ions do not conform to that predicted by a single Lewis structure. For example, the Lewis structure for the carbonate ion, CO_3^{2-}, predicts two C-O single bonds and a C=O double bond,

and the O-C-O bond angle to be less than the O=C-O bond angle. Yet, experimentally CO_3^{2-} is found to have all CO bonds the same (intermediate in length between a single bond and a double bond), with a regular triangular planar shape and all bond angles exactly 120°. It is not possible to represent the structure of CO_3^{2-} using a single Lewis structure. Three Lewis structures may be drawn for CO_3^{2-}, none of which represents the **actual** structure. However, a good representation of the

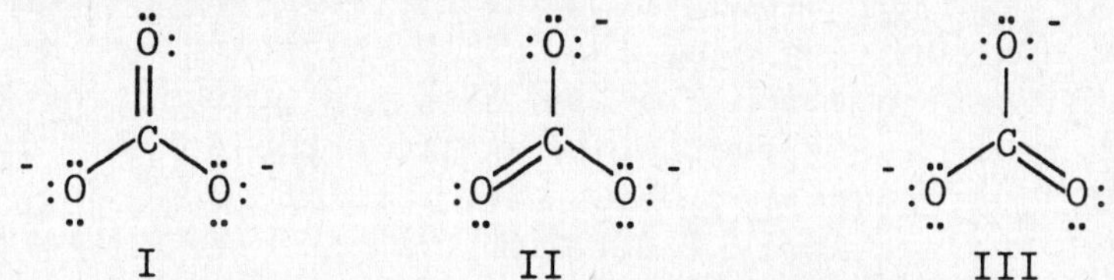

observed structure is obtained by taking the average of the above Lewis structures. This amounts to a structure in which each CO bond is a single bond, with an additional pair of electrons delocalized over all three O atoms and the C atom, giving 1 1/3 bonding pairs for each CO bond, or a bond order of 1 1/3. Formally, the CO bond order of each bond is obtained from the Lewis structures by selecting any of the CO bonds (say the top bond in structures I, II, and III above), adding up the bond orders, and dividing by the total number of structures. For example, the CO bond order $= \frac{2+1+1}{3} = 1\frac{1}{3}$.

Similarly, the formal charge on each O atom is obtained by averaging the formal charges in each of the Lewis structures. Thus, the formal charge on O $= \frac{0-1-1}{3} = -\frac{2}{3}$. The resulting structure is referred to as a **resonance hybrid**.

Resonance is not a phenomenon; rather, it is the name for a method of obtaining a better representation of a molecular structure than is given by any single Lewis structure. For any bond in a particular position, the bond order is obtained by adding together the bond orders for **this** bond in all of the equivalent Lewis structures and then dividing by the number of structures. Similarly, the formal charge on any atom is obtained by summing the formal charges on the particular atom in each Lewis structure and dividing by the number of structures. These procedures give fractional bond orders consistent with the observed bond lengths, and fractional formal charges.

For molecules with five and six electron pairs in the valence shell of the central atom, the basic geometries are AX_5, **trigonal bipyramidal**, and AX_6, octahedral. In AX_5 molecules, the three **equatorial** bonds and the two **axial** bonds are nonequivalent. Because the axial positions are more crowded than the equatorial positions, the axial bonds in molecules such as PCl_5 are longer than the equatorial bonds. In AX_6 molecules all of the bonds are equivalent and have identical lengths. In AX_4E, AX_3E_2, and AX_2E_3 type molecules, the lone pairs E occupy the less crowded equatorial positions of the trigonal bipyramid and such species are described as **disphenoid, T-shaped,** and **linear**, respectively. An AX_5E species has the shape of a **square pyramid,** and in an AX_4E_2 species, the two lone pairs maximize their distance apart by adopting a linear arrangement, to give a **square planar** geometry. In AX_4E, AX_3E_2, and AX_5E species, the bond angles are slightly smaller than the ideal angles due to the relatively larger charge cloud of a lone pair compared to that of a bond pair. AX_2E_3 and AX_4E_2 molecules have regular geometries but relatively long bond lengths, due to large interactions between the smaller bond pairs and the larger lone pairs, and a similar effect causes the axial bond in a AX_5E molecule to be slightly shorter than the four equivalent equatorial bonds.

In terms of **molecular orbital theory,** overlap of two singly occupied p orbitals on O with hydrogen 1s orbitals gives a structure for water where the O-H bond orbital charge clouds make an angle of 90°, which is much smaller than the observed HOH bond angle of 104.5°. And starting with the valence state electron configuration $2s^1 2p_x{}^1 2p_y{}^1 2p_z{}^1$ for carbon, formation of localized bonding orbitals by overlap of each of these orbitals with hydrogen 1s orbitals gives a structure for CH_4 with three HCH bond angles of 90°, and a fourth HCH bond angle that is indeterminate. According to the VSEPR model, H_2O and CH_4 have AX_2E_2 and AX_4 structures, respectively, based on a **tetrahedral** arrangement of four electron pairs in the valence shell of the central atom, which are inconsistent with the results of the molecular orbital theory given above.

To obtain four tetrahedrally directed orbitals on the central atoms, the one s and three p orbitals are combined mathematically, in an operation called **hybridization**, to obtain four equivalent atomic orbitals, called sp^3 **hybrid orbitals**, each with 1/4 s and 3/4 p character. In CH_4, overlap of each of the four sp^3 hybrid orbitals with a hydrogen 1s orbital gives tetrahedral CH_4. By a similar procedure a set of suitable hybrid orbitals can be obtained that corresponds to the observed geometric arrangement of the bonds in each of the AX_2, AX_3, AX_5, and AX_6 geometries:

Valence State	Hybrid Orbitals	Arrangement of Orbitals
$sp_x{}^1$	sp	linear
$sp_x{}^1 p_y{}^1$	sp^2	triangular
$sp_x{}^1 p_y{}^1 p_z{}^1$	sp^3	tetrahedral
$sp_x{}^1 p_y{}^1 p_z{}^1 d^1$	sp^3d	trigonal bipyramidal
$sp_x{}^1 p_y{}^1 p_z{}^1 d^1 d^1$	sp^3d^2	octahedral

Hybridization is not a physical phenomenon but a mathematical operation that converts a set of atomic orbitals into an equivalent set of hybrid orbitals with an identical total electron density distribution.

In CH_4, each of four sp^3 hybrid orbitals is used to form four C-H bonds by overlap with a hydrogen 1s orbital; in NH_3, three of the sp^3 hybrid orbitals are used to form three N-H bonds, leaving an sp^3 hybrid orbital to accommodate the lone pair on nitrogen; in H_2O, two of the sp^3 hybrid orbitals are used to form two O-H bonds, leaving two sp^3 hybrid orbitals on oxygen to accommodate the two lone pairs. However, this is only an approximate description because it predicts that the bond angles should all be exactly the tetrahedral angle of 109.5°. In reality the lone pairs take up more room in the valence shell of the central atom than the bond pairs, which means that they have more s character than the bond pairs. Thus, a hybrid orbital model is only approximate for molecules in which the central atom has both bond pairs and lone pairs.

Multiple bonds can be described by two equivalent approximate models - the **bent bond model** and the **σ-π model,** and these are most conveniently discussed using ethene, $H_2C{=}CH_2$, and ethyne, $HC{\equiv}CH$, as examples. In the **bent bond model,** four sp^3 orbitals on each C atom are used as the basic set of hybridized orbitals. In **ethene**, two of these on each C atom are overlapped with H 1s orbitals to form the C-H bonds and the remaining pair on each C atom are overlapped to form two bent bonds. In **ethyne**, a single sp^3 orbital of each C is overlapped with a H 1s orbital to form the C-H bonds and the remaining three sp^3 hybrid orbitals on each C atom are overlapped to form three bent CC bonds. **In the σ-π model,** a σ-bonded framework is first constructed by end-on overlap of appropriate hybrid orbitals, leaving p orbitals perpendicular to this framework that can be overlapped sideways to form the requisite number of π bonds. In **ethene**, three sp^2 orbitals on each C are used to form four C-H σ bonds and a C-C σ bond, leaving a p orbital on each C atom that is perpendicular to the plane of the molecule, and these two p orbitals are overlapped sideways to form a π bond. In **ethyne**, two sp orbitals on each C atom are used to form two C-H σ bonds and a C-C σ bond, leaving two p orbitals on each C atom that are overlapped to form two π bonds.

Benzene, C_6H_6, may be described by two resonance (Kekulé) structures containing alternate C-C and C=C bonds in the six membered carbon ring, for a CC bond order of 1 1/2, consistent with the observed bond angle of 140 pm, intermediate in length between a C-C bond (154 pm) and a C=C bond (134 pm). Alternatively, a set of sp^2 hybrid orbitals on each C atom is used to form a C-H σ bond and two C-C σ bonds, leaving a singly occupied 2p orbital on each C atom that is perpendicular to the plane of the ring. These six orbitals overlap sideways to form π-type molecular orbitals extending over all six C atoms. Three such orbitals are occupied by three pairs of electrons, to give a ring of electron density above and below the plane of carbon atoms.

The partial charges on the atoms of a **polar** diatomic molecule, such as HCl, $^{\delta+}$H-Cl$^{\delta-}$, constitute a **dipole**, the magnitude of which is given by

$$\mu = Qr$$

where μ is the **dipole moment,** +Q and -Q are the partial charges δ+ and δ-, and **r** is their distance apart (the H-Cl bond length). μ has the units C m. The bonds in a polyatomic molecule may also be polar, but a molecule is polar and has a dipole moment ($\mu \neq 0$) only if **its center of negative charge does not coincide with its center of positive charge.** Among AX_nE_m molecules, all AX_n type molecules, with all X atoms identical, are **nonpolar** ($\mu = 0$), and all

AX_nE_m type molecules are **polar** ($\mu \neq 0$), except for those of the AX_2E_3 and AX_4E_2 type.

REVIEW QUESTIONS

1. Draw a diagram of the water molecule, given that each O-H bond has a length of 97 pm and the HOH bond angle = 104.5°.
2. What are the arrangements of 2, 3, and 4 pairs of electrons in the valence shell of a central atom A in a molecule?
3. Write Lewis structures for each of the following, and express the number of electron pairs in the valence shell of the central atom in terms of the AX_nE_m nomenclature: $BeCl_2$, BCl_3, CCl_4, NCl_3, OCl_2, and HCl.
4. What are the geometries of each of the molecules in question 3?
5. What are the approximate bond angles in the molecules in question 3?
6. Describe each of the following in terms of the AX_nE_m nomenclature and predict their geometries: NH_4^+, H_2O, NH_3, OH^-, NH_2^-, and H_3O^+.
7. What is meant by the Lewis structure of a molecule or ion?
8. What is the relationship between the group number of a main group element and its number of valence electrons?
9. How many pairs of valence electrons are there in each of the following species? CO, NO_2^+, SO_2, SO_3^{2-}, SO_4^{2-}, PO_4^{3-}, ClO_4^-.
10. How many pairs of valence electrons are there in each of the following species? CO_2, H_2CO, HCN, SO_3, HSO_3^- and HPO_3^{2-}.
11. Using NO_3^- and CO_3^{2-} as examples, list the rules that should be applied in writing their Lewis structures.
12. Which of the species in questions 9 and 10 have atoms where an octet of electrons is exceeded?
13. On the basis of the given rules, deduce the Lewis structures of CO_3^{2-}, NO_3^-, and SO_3^{2-}, showing the formal charges on the appropriate atoms.
14. Deduce the Lewis structures of PO_4^{3-}, SO_4^{2-}, and ClO_4^-.
15. On the basis of their Lewis structures, what are the expected geometries of CO_3^{2-}, NO_3^-, SO_3^{2-}, PO_4^{3-}, SO_4^{2-}, and ClO_4^-?
16. Explain why the HNH bond angle in NH_3 is 107° and the HOH bond angle in H_2O is 104.5°, rather than the ideal tetrahedral angle of 109.5°.
17. Arrange the molecules H_2O, Cl_2O and F_2O in order of the expected decrease in bond angle.
18. Deduce the expected approximate geometries of H_2CO, CO_2, and CO_3^{2-} and estimate the approximate values of their bond angles.
19. What is the observed geometry of the CO_3^{2-} ion? Are the lengths of the CO bonds equal? Draw the resonance structures of CO_3^{2-}. What is the bond order of each of the CO bonds? Explain how this bond order is consistent with the observed bond lengths.
20. Repeat question 19 for (a) the NO_3^- ion, and (b) the PO_4^{3-} ion.
21. Draw the two resonance structures for the methanoate (formate) ion HCO_2^-. What is the CO bond order and the formal charge on each atom? Would the CO bond length be expected to be longer than, or shorter than, the length of the CO bond in the carbonate ion, CO_3^{2-}?
22. Explain why in a molecule such as PCl_5 the axial bonds are found to be longer than the equatorial bonds.
23. Draw Lewis structures for each of the following, express their structures in terms of the AX_nE_m nomenclature, and deduce their approximate geometries: PCl_4^+, PCl_6^-, AsF_5, SeF_4, ICl_3, SF_6, IF_5 and ICl_2^-.

24. In terms of atomic orbitals on nitrogen, what are the expected geometries of NH_3 and NH_4^+?
25. In terms of the VSEPR model, deduce the approximate geometries of NH_3 and NH_4^+. What is the appropriate set of hybrid orbitals on nitrogen that should be used to give these geometries?
26. What appropriate combination of 2s and 2p orbitals gives (a) two mutally linear hybrid orbitals, (b) three orbitals in the same plane with trigonal geometry, and (c) four tetrahedral orbitals?
27. Can hybrid orbitals be used to predict the geometry of molecules?
28. What appropriate set of hybrid orbitals on the central atom of each of the following is consistent with the observed geometry? $BeCl_2$, BF_3, SiH_4, PF_5, SF_6, ClF_5, ClF_3, ICl_2^-.
29. What are the basic sets of hybrid orbitals on carbon that are used to describe the structure of ethene, C_2H_4, in terms of (a) the bent bond model, and (b) the σ-π model?
30. Give an appropriate description of the bonding in $H_3C\text{-}C{\equiv}N{:}$, in terms of (a) the bent bond model, and (b) the σ-π model.
31. Which of the following molecules will have a dipole moment? Cl_2, CO, HF, CO_2, SCO, HCN, BF_3, CCl_4?
32. Arrange the hydrogen halides in order of the expected increase in dipole moment.
33. Which of the following molecules are expected to have a dipole moment? H_2O, NH_3, CH_4, PCl_5, SF_6, ClF_3, SF_4, BrF_5, XeF_2, SeF_4.
34. Would the molecule H_2SiCl_2 be expected to have a dipole moment?

Answers to Selected Review Questions

2. Linear, trigonal planar, tetrahedral. 3. AX_2, AX_3, AX_4, AX_3E, AX_2E_2, AXE_3. 4. Linear, trigonal planar, tetrahedral, trigonal pyramidal, angular, linear. 5. 180°, 120°, 109.5°, < 109.5°, < 109.5°, 180°. 6. AX_4 tetrahedral; AX_2E_2, angular; AX_3E, trigonal pyramidal; AXE_3, linear; AX_2E_2, angular; AX_3E, trigonal pyramidal. 8. The number of valence electrons is the same as the group number. 9. 5,8,9,13,16,16,16. 10. 8,6,5,12,13,13. 12. SO_2, SO_3^{2-}, SO_4^{2-}, PO_4^{3-}, ClO_4^-, SO_3, HSO_3^-, HPO_3^{2-}. 15. AX_3, trigonal planar; AX_3, trigonal planar; AX_3E, trigonal pyramidal; AX_4, tetrahedral; AX_4, tetrahedral; AX_4, tetrahedral. 23. AX_4, tetrahedral; AX_6, octahedral; AX_5, trigonal bipyramidal; AX_4E, disphenoidal; AX_3E_2, T-shaped; AX_6, octahedral; AX_5E, square pyramidal; AX_2E_3, linear. 25. trigonal pyramidal; tetrahedral; sp^3; sp^3. 26. (a) sp, (b) sp^2, (c) sp^3. 27. No. 28. sp, sp^2, sp^3, sp^3d, sp^3d^2, sp^3d^2, sp^3d, sp^3d. 31. CO, HF, SCO, HCN. 33. H_2O, NH_3, ClF_3, SF_4, BrF_5, SeF_4. 34. Yes.

OBJECTIVES

Be able to:

1. State the main factors that affect the size and shape of electron pair charge clouds.
2. Give examples of molecules with 4 electron pairs in the valence shell of the central atom that have no nonbonding pairs (AX_4), 1 nonbonding pair (AX_3E), 2 nonbonding pairs (AX_2E_2), and 3 nonbonding pairs (AXE_3); draw their Lewis structures; draw and describe their shapes.

3. Give examples of molecules with 3 electron pairs in the valence shell of the central atom that have no nonbonding pairs (AX_3), and 1 nonbonding pair (AX_2E); draw their Lewis structures; draw and describe their shapes.
4. Give examples of molecules containing double and triple bonds; classify them as AX_nE_m; draw their Lewis structures; draw and describe their shapes.
5. Give examples of molecules with 5 electron pairs in the valence shell of the central atom that have no nonbonding pairs (AX_5), 1 nonbonding pair (AX_4E), 2 nonbonding pairs (AX_3E_2), and 3 nonbonding pairs (AX_2E_3); draw their Lewis structures; draw and describe their shapes.
6. Give examples of molecules with 6 electron pairs in the valence shell of the central atom that have no nonbonding pairs (AX_6), 1 nonbonding pair (AX_5E), and 2 nonbonding pairs (AX_4E_2); draw their Lewis structures; draw and describe their shapes.
7. Define the term bond order.
8. Draw resonance structures and calculate bond orders and formal charges on atoms for molecules where a single Lewis structure is insufficient.
9. Use hybrid orbitals to describe the bonds in simple AX_nE_m molecules of known geometry, where (n + m) equals 2,3, or 4.
10. List the hybrid orbitals that can be constructed from s, p, and d atomic orbitals to describe bonding for AX_5 and AX_6 types of molecules.
11. Use hybrid orbitals to describe multiple bonds.
12. Vectorially add bond moments of a molecule to determine if a molecular dipole moment exists.

PROBLEM SOLVING STRATEGIES

9.1 Lewis Structures

We saw in Chapter 4 some Lewis structures for relatively simple molecules, and in Chapter 8 several Lewis structures for higher-valence compounds of sulfur and phosphorus that violated the octet rule. Chapter 9 presents a convenient set of rules for drawing Lewis structures. Learn it. As well, at this point you may wish to review the formal charge concept in Chapter 4 and the drawing of Lewis structures for sulfur and phosphorus compounds with double bonds (Chapter 8).

Example 9.1
Draw the Lewis structure for ClO_3^- ion.

Solution 9.1

Chlorine is the central atom, and contributes 7 valence electrons; each oxygen has 6 valence electrons, and the negative charge of the ion provides one additional electron, for a total of 26 valence electrons. Draw **single** bonds from oxygen to chlorine, and have an octet of electrons around each oxygen. Any remaining electrons are arranged around chlorine. So far we have

```
     ..   ..   ..
    :O----Cl----O:
     ..   |    ..
          |
         :O:
          ..
```

If we calculate formal charges for this structure, we get +7 -2 -1/2(6) = +2 for chlorine and +6 -6 -1/2(2) = -1 for each oxygen. This formal charge situation can be improved by forming 2 **double bonds** from O to Cl, giving

```
     ..   ..   ..
    :O =  Cl = O
          |    ..
          |  -
         :O:
          ..
```

In this structure there is just one formal charge, -1, on one oxygen atom. Incidentally, after we treat the concept of resonance later in the chapter, you should recognize that this Lewis structure is only one of the three resonance forms for ClO_3^-.

9.2 The VSEPR Model

VSEPR theory, as you recall, makes use of the **arrangement** of electron pairs about the central atom. We have already seen in Chapter 4 how to predict shapes and bond angles for AX_2, AX_3 and AX_4 molecules (where all the X atoms are the same), using the VSEPR model. When molecules contain **nonbonding** electron pairs around the central atom, we need to modify our predictions, since nonbonding pairs take up more space than bonding pairs in the valence shell of the central atom. The measured bond angles in AX_3E and AX_2E_2 molecules are less than the predicted 109.5° for AX_4 molecules. For AX_2E, the XAX angle is found to be less than the 120° prediction for AX_3 molecules.

Five electron pairs are situated in a **triangular bipyramidal** arrangement, and six electron pairs in an **octahedral** arrangement. Just as you have seen in the above paragraph, **it is the arrangement of bonding pairs** (i.e., the arrangement of **bonds**) that gives the molecule's shape or geometry. So be sure to distinguish between the arrangement of all of the electron pairs and the arrangement of the **bonding** pairs around a central atom. Figures 9.14 - 9.17 in the textbook show the shapes of molecules with five and six electron pairs.

Here is a set of rules to follow in determining molecular geometries for some of these molecules.

1. Draw the Lewis structure for the molecule (if A is from period 3 or beyond, it may violate the octet rule). Remember to reduce, if possible, the formal charge. If there are double or triple bonds, assume that they can be treated as single bonds. Thus 2 bonding pairs forming a double bond are treated as if they were 1 bonding pair.
2. Count the number of electron pairs around the central atom. If it is 6, go to step 3; if it is 5, go to step 4. If it is less than 5, you can use the methods of Chapter 4.

3. For 6 electron pairs about the central atom, determine if there are 0, 1 or 2 nonbonding pairs (AX_6, AX_5E, AX_4E_2). AX_6 is octahedral; AX_5E is square pyramidal; AX_4E_2 is square planar.
4. For 5 electron pairs about the central atom, determine if there are 0, 1, 2 or 3 nonbonding pairs (AX_5, AX_4E, AX_3E_2, AX_2E_3). AX_5 is trigonal bipyramidal; AX_4E is disphenoidal (a distorted or squashed tetrahedron); AX_3E_2 is T-shaped; AX_2E_3 is linear. Notice that nonbonding pairs preferentially occupy the equatorial positions in the trigonal bipyramid.

Example 9.2

Classify the following molecules as AX_6, AX_5E, etc., and describe their shapes.
(a) BrF_3 (b) BrF_4^- (c) SCl_4 (d) ClO_4^- .

Solution 9.2

(a) Both bromine and fluorine are halogen atoms. Thus this is a 28 valence electron problem. If we use single bonds from F to Br, there will be an octet around each fluorine, but 10 electrons around Br

:F̤̈—B̤̈r—F̤̈:
F̤̈:

The formal charges are all zero. Since there are 3 bonding and 2 nonbonding pairs about Br, we identify the molecule as AX_3E_2, and thus its shape is like a T. See Figure 9.17 in the text.

(b) The same atoms are involved here as in the above example. This is a polyhalide ion. The total number of valence electrons is (7 x 5) + 1 for the negative charge, or 36 in all. By forming single F-Br bonds, there will be an octet around each fluorine, but 12 electrons around bromine.

:F̤̈ F̤̈:
B̤̈r⁻
:F̤̈ F̤̈:

The formal charge on Br is -1, but zero on all F's, so it cannot be further reduced. This is an AX_4E_2 molecule, which is square planar.

(c) SCl_4 has 34 valence electrons. By forming single Cl-S bonds, chlorine obeys the octet rule, but there are 10 electrons around sulfur.

:C̤̈l:
|
:S — C̤̈l:
— C̤̈l:
|
:C̤̈l:

The formal charges are zero. This is an AX_4E molecule, which has a disphenoid, see-saw geometry. See Figure 9.17 in the text.

(d) ClO_4^- has a total of 32 valence electrons. If we form 4 single bonds from O to Cl, with octets around oxygen, we have

```
             ..
            :O:
             |
  -  ..      | 3+      ..  -
    :O ——— Cl ——— O:
     ..      |         ..
             |
            :O: -
             ..
```

This structure has a +3 formal charge on chlorine. We reduce this formal charge to zero by forming multiple bonds. With 3 double bonds, we are left with

```
             ..  -
            :O:
             |
  ..         |         ..
   O ====== Cl ====== O
  ..         ||        ..
            :O:
```

Remember we count a double bond as one electron pair for geometry considerations. This is an AX_4 molecule, which is tetrahedral. By the way, this is one of four resonance structures for the perchlorate ion.

9.3 Resonance Structures

Some molecules and ions have more than one atom X of the same kind joined to a central atom A. When a **single** Lewis structure is drawn for such a species, it may happen that one or more of the A-X bonds is represented by a double bond, while other A-X bonds may be depicted as single bonds. If we draw a **different** Lewis structure, **without moving any atoms,** but only rearranging electrons pairs, one or more of the former A-X single bonds may now appear as a double bond, and a double bond may now be a single bond. Yet the experimentally **measured A-X bond lengths usually are all the same.** In other words, a single Lewis structure is sometimes insufficient to describe the bonding in a molecule. We say the real molecule is **none** of the extreme forms represented by the various Lewis structures, but lies somewhere in between. Since Lewis structures consider bonds to be only single, double or triple, we cannot describe a bond containing more than 1 bonding pair but less than 2 bonding pairs by a single Lewis structure. We describe the molecular bonding by writing several Lewis structures, and state that the actual structure of the molecule is a resonance hybrid of these forms. (The molecule does not oscillate or resonate between these extreme structures.) Remember when drawing resonance structures that **the atoms must remain fixed.** Only the electron pairs can be moved, with F, O, C, and N obeying the octet rule. Formal charges should be kept as small as possible in the resonance forms.

Example 9.3

Draw resonance structures for ClO_2^-. Determine the bond order and the formal charge on oxygen.

Solution 9.3

In Example 4.10 of the Study Guide we saw one Lewis structure for ClO_2^- to be

$$\underset{\cdot\cdot}{\ddot{\mathrm{O}}} = \underset{\cdot\cdot}{\ddot{\mathrm{Cl}}} - \underset{\cdot\cdot}{\ddot{\mathrm{O}}}:^-$$

However, we can just as well write the left O-Cl bond as a single bond, and the right one as a double bond.

$$^-:\underset{\cdot\cdot}{\ddot{\mathrm{O}}} - \underset{\cdot\cdot}{\ddot{\mathrm{Cl}}} = \underset{\cdot\cdot}{\ddot{\mathrm{O}}}$$

There are 2 resonance structures for ClO_2^-. We draw all the individual Lewis structures and relate them by double-headed arrows to denote resonance.

$$\underset{\cdot\cdot}{\ddot{\mathrm{O}}} = \underset{\cdot\cdot}{\ddot{\mathrm{Cl}}} - \underset{\cdot\cdot}{\ddot{\mathrm{O}}}:^- \quad \leftrightarrow \quad ^-:\underset{\cdot\cdot}{\ddot{\mathrm{O}}} - \underset{\cdot\cdot}{\ddot{\mathrm{Cl}}} = \underset{\cdot\cdot}{\ddot{\mathrm{O}}}$$

The bond order of the O-Cl bond in this ion is calculated by summing the bond orders of a given O-Cl bond (say the left one) in all the resonance structures, and dividing by the total number of contributing structures. Since a double bond has a bond order of 2 and a single bond has a bond order of 1,

$$\text{bond order of O-Cl bond} = \frac{2 + 1}{2} = 3/2.$$

The formal charge on an atom in a molecule that has resonance structures is calculated by focusing on that atom (say the left O of ClO_2^-) and taking the average of the formal charges for that atom in all the structures. Here, the formal charge on O is (0 + (-1))/2 = -1/2.

9.4 Hybrid Orbitals

In solving the Schrödinger wave equation for the energy levels of the hydrogen atom by quantum mechanics, we also obtain the mathematical forms of the wave functions, ψ, the hydrogen atomic orbitals. In contrast to the hydrogen atom, the **Schrödinger equation cannot be solved exactly for many-electron atoms and molecules.** Quantum mechanics attempts to solve these complicated problems by using the results for the hydrogen atom, and making appropriate adjustments to the hydrogen atomic orbitals. **So the model of describing many-electron atoms by modified hydrogen atomic orbitals is an approximation.** However, it is very useful. For example, we have described the electron configurations of many-electron atoms using atomic orbital designations like 2p, 3d, and so on. Filling electrons into these orbitals gave an excellent accounting for the periodicity in the chemical and physical properties of the elements.

When we attempt to use atomic orbitals, with the simplest version of the **criterion of maximum overlap** to describe **molecular geometries,** we often run into problems. This is due to the fact that s orbitals, unlike p and d, have no directional characteristics. So we mathematically combine atomic orbitals to form **hybrid orbitals,** which have the appropriate directional character.

Thus to describe the geometry and bonding in methane, we use a 2s and three 2p atomic orbitals on carbon (using a promoted state called a valence state, as described in Chapter 7), to yield four sp^3 hybrid orbitals, pointing to the corners of a tetrahedron. These hybrid orbitals are allowed to overlap with four 1s atomic orbitals on hydrogen, to form tetrahedral CH_4. **We use sp^3 hybrid orbitals to describe CH_4 because we know CH_4 is tetrahedral.**

In describing bonding situations using hybrid orbitals, you start with a knowledge of the geometry (obtained by drawing the Lewis structure and classifying the molecule as AX_nE_m). If $m + n = 2$, use sp hybrid orbitals; if 3, use sp^2; if 4, use sp^3; if 5, use sp^3d, if 6, use sp^3d^2. Form bonds using the hybrid orbitals on the central atom and atomic orbitals on the remaining atoms. In ethane, CH_3CH_3, the C-C bond involves the overlap of an sp^3 hybrid orbital on one carbon with an sp^3 hybrid orbital of the other carbon.

Example 9.4

Give a hybrid orbital description of the bonds in PF_3.

Solution 9.4

P has 5 valence electrons, and each F has 7. The Lewis diagram is

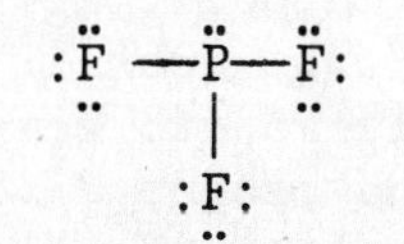

The molecule is AX_3E, and has a triangular pyramid geometry; the four electron pairs are approximately tetrahedrally-arranged around P. We use a 3s and three 3p atomic orbitals on phosphorus to form four sp^3 orbitals, and place 2 electrons (with opposite spins) into each of the four sp^3 orbitals, since there are a total of four electron pairs around phosphorus. The three P-F bonds are formed by overlap of three sp^3 hybrid orbitals on phosphorus with a 2p orbital on each of three fluorines. The nonbonding pair on phosphorus is placed in the fourth sp^3 orbital. The resultant geometry is trigonal pyramidal. Whether a molecule is AX_4, AX_3E, AX_2E_2 or AXE_3, we use sp^3 hybrid orbitals on A to discuss the bonding since there are four electron pairs in the valence shell of A in each case.

Example 9.5

Give a hybrid orbital description of the bonds in SF_6.

Solution 9.5

S has 6 valence electrons, and each F has 7. The Lewis diagram is

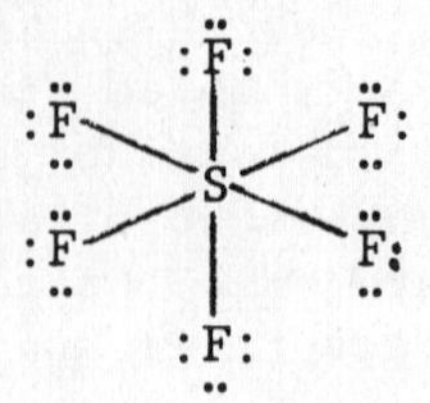

The molecule is AX_6 and has an octahedral geometry. We use one 3s, three 3p and two 3d atomic orbitals on S to form six sp^3d^2 hybrid orbitals. Since there are six electron pairs around S in the Lewis diagram, we place 2 electrons (with opposite spins) into each of the six sp^3d^2 hybrid orbitals. The six S-F bonds are formed by overlap of six sp^3d^2 hybrid orbitals on sulfur with a 2p orbital on each of the six fluorines. We use sp^3d^2 hybrid orbitals because we know SF_6 is octahedral.

9.5 Molecular Polarity and Dipole Moments

Diatomic molecules composed of 2 different atoms always have a dipole moment. A **polyatomic molecule** has a dipole moment if the center of negative charge does not coincide with the center of positive charge. We can predict whether a polyatomic molecule will have a dipole moment by **vectorially adding** its **bond dipole moments.** The individual bond dipole moments are represented by arrows of the appropriate length pointing to the more electronegative atom of the pair that are joined together. Geometry plays an important role, for the bond moments will vectorially add to **zero** for AX_2, AX_3, AX_4, AX_5 and AX_6 molecules, when all X atoms are the same. These are **nonpolar** molecules. Likewise, AX_4E_2 and AX_2E_3, with all the X the same, have zero dipole moments. On the other hand, the vectorial addition of bond moments will be nonzero for AX_2E, AX_3E, AX_2E_2, AX_4E, AX_3E_2 and AX_4E types. These have a molecular dipole moment and are called **polar** molecules. As well, for AX_n molecules where the X atoms **are not all the same,** there will generally be a net molecular dipole moment.

To determine whether a molecule is polar or nonpolar, use the following.

1. Draw the Lewis structure for the molecule.
2. Classify the molecule as AX_4, AX_3E, etc.
3. Realize that AX_n molecules, with all X the same, are nonpolar. The others, if nonbonding electrons are present, may be polar. In some cases you may have to actually vectorially add the bond moments.

Example 9.6

Which of the following molecules are polar: (a) SiH_4 (b) PBr_3 (c) $AlCl_3$ (d) SF_6 (e) SO_3?

Solution 9.6

(a) For SiH_4 there are 8 valence electrons. The Lewis structure is

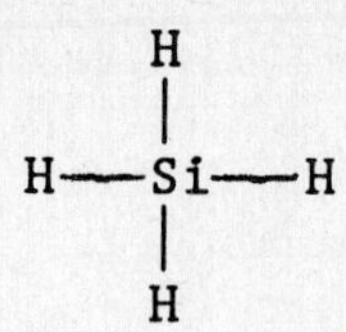

This is an AX_4 molecule, and is nonpolar.

(b) In PBr_3, P has 5 valence electrons, and each bromine has 7, for a total of 26. By forming 3 **single** bonds from Br to P, we have a Lewis structure with zero formal charge.

```
 ..    ..    ..
:Br —— P —— Br:
 ..    |    ..
      :Br:
       ..
```

PBr_3 is an AX_3E molecule (trigonal pyramidal) and the 3 P-Br bond dipole moments, which point to the bromines, vectorially add to give a nonzero result. PBr_3 is polar.

(c) Aluminum has 3 valence electrons, each chlorine has 7, for a total of 24 in $AlCl_3$. Form 3 Al-Cl single bonds, and obtain a Lewis structure with zero formal charge.

```
 ..          ..
:Cl —— Al —— Cl:
 ..    |     ..
      :Cl:
       ..
```

$AlCl_3$ is an AX_3 molecule, and is nonpolar.

(d) In SF_6 there are 48 valence electrons. Form 6 S-F single bonds. There are no formal charges.

```
        ..
       :F:
  ..    |    ..
 :F\    |   /F:
  ..  \ S /  ..
  ..  /   \  ..
 :F/    |   \F:
  ..    |    ..
       :F:
        ..
```

SF_6 is AX_6 and is nonpolar.

(e) SO_3 has 24 valence electrons. If we write 3 single S-O bonds, we get a formal charge of -1 on each oxygen and +3 on sulfur. We decrease these to zero by forming 3 **double** bonds.

```
      :O:
       ||
 ..    ||    ..
 O ==  S  == O
 ..          ..
```

By counting each double bond as a single bond we see SO_3 in an AX_3 molecule, and is nonpolar.

SELF TEST

1. Give the arrangement of 3 electron pairs around a central atom.
   (a) octahedral (b) planar triangular (c) trigonal pyramidal
   (d) trigonal bipyramidal (e) linear
2. Give the shape of gallium triiodide (gallium is in group 3).
   (a) linear (b) bent (c) trigonal pyramidal
   (d) planar triangular (e) tetrahedral
3. Which of the following is (are) **not** AX_2E_2 molecule(s)?
   (a) Cl_2O (b) BeH_2 (c) O_3 (d) CO_2 (e) SCl_2
4. Pick the correct classification for sulfur dioxide from among the following.
   (a) AX_2 (b) AX_2E (c) AX_2E (d) AX_2E (e) AX_2E_4
5. Which of the following has (have) at least one bond angle of 90°?
   (a) PH_4^+ (b) PCl_5 (c) SO_3 (d) SO_4^{2-} (e) SeF_6
6. What is the formal charge on P in PCl_4^+?
   (a) -2 (b) -1 (c) 0 (d) 1 (e) 2
7. Which of the following have a triple bond?
   (a) N_2 (b) CO_2 (c) HCN (d) C_2H_4 (e) P_4
8. Which is the correct Lewis structure for formaldehyde, H_2CO?

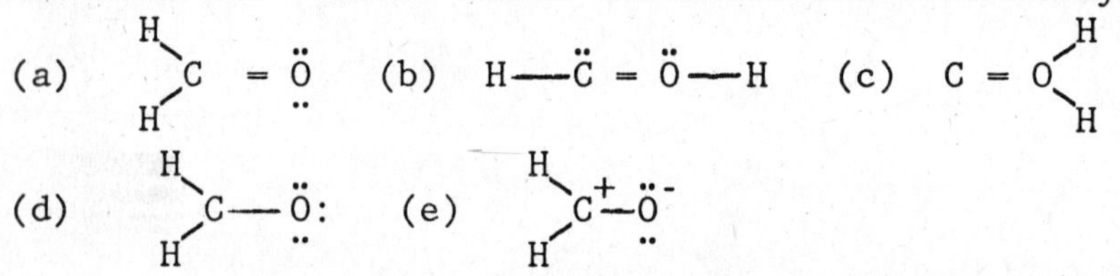

```
     H                                              H
      \                 ..    ..                   /
(a)    C  =  Ö     (b)  H——C = Ö——H    (c)  C = O
      /      ..                                    \
     H                                              H

     H                      H
      \      ..              \  + ..-
(d)    C——Ö:        (e)       C——Ö
      /      ..              /    ..
     H                      H
```

9. The fact that a correct Lewis structure for SO_4^{2-} is

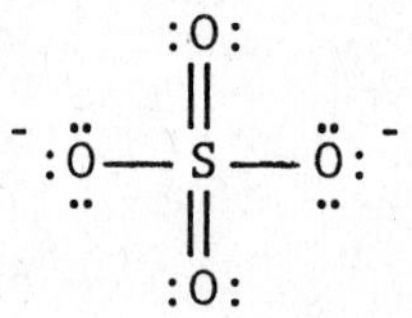

```
           :O:
            ||
   - ..     |     .. -
    :Ö——— S ———Ö:
     ..     |     ..
            ||
           :O:
```

   and yet all the bonds are identical is accounted for by
   (a) the concept of resonance (b) the Lewis octet rule (c) VSEPR theory
   (d) sulfur always has four bonds (e) sulfur is larger than oxygen.
10. How many equivalent resonance structures does ClO_4^- have?
    (a) 0 (b) 2 (c) 3 (d) 4 (e) more than 4
11. Select all the nonpolar molecules from the following.
    (a) SCO (carbon is the central atom) (b) Cl_2 (c) SiF_4 (d) SCl_4 (e) H_3PO_3
12. What is the formal charge on an oxygen atom in nitrate ion, NO_3^-?
    (a) 0 (b) +1/3 (c) -1/3 (d) +2/3 (e) -2/3
13. What is the bond order of the S-O bond in sulfate ion, SO_4^{2-}?
    (a) 1/2 (b) 3/4 (c) 1 (d) 1 1/2 (e) 2
14. Select from the following all the molecules that have a dipole moment.
    (a) $BeCl_2$ (b) H_2S (c) NH_3 (d) Br_2 (e) CO_2
15. In the cage structure of P_4O_{10}, each phosphorus atom is joined to how many oxygen atoms?
    (a) 2 (b) 3 (c) 4 (d) 6 (e) 8

16. What is the correct Lewis structure for BF_3?

(a) :F̤̈—B̈—F̤̈: with :F̤: below B (b) :F̤̈:$^+$= B̄—F̤̈: with :F̤: below B (c) :F̤̈—B—F̤̈: with :F̤: below B

(d) $^-$:F̤ =B$^+$—F̤̈: with F̤: below B (e) :F̈ = B = F̈: with :F̤ below B

17. Consider the possible geometries for SO_3 and SO_3^{2-}. Which is true?
(a) SO_3 only is planar (b) SO_3^{2-} only is planar
(c) Both SO_3 and SO_3^{2-} are planar (d) Neither of SO_3 or SO_3^{2-} is planar.

18. The hybridization at carbon in CH_3^+ is
(a) sp (b) sp^2 (c) sp^3 (d) sp^3d (e) sp^3d^2

19. The hybridization at sulfur in SF_4 is
(a) sp (b) sp^2 (c) sp^3 (d) sp^3d (e) sp^3d^2

20. The hybridization at chlorine in ClF_5 is
(a) sp (b) sp^2 (c) sp^3 (d) sp^3d (e) sp^3d^2

Answers to Self Test

1, b; 2. d; 3. b,c,d; 4. b; 5. b,e; 6. d; 7. a,c; 8. a; 9. a; 10. d; 11. b,c; 12. e; 13. d; 14. b,c; 15. c; 16. c; 17. a; 18. b; 19. d; 20. e.

CHAPTER 10

SOME COMMON METALS: ALUMINUM, IRON, COPPER AND LEAD

SUMMARY REVIEW

Aluminum, lead and the transition metals **iron** and **copper** are important and familiar. Copper occurs as the free metal and as sulfide and oxide ores. Iron is rarely found as such on the earth's surface but has many ores; the most important are hematite, Fe_2O_3, magnetite, Fe_3O_4, and siderite, $FeCO_3$. The principal ore of lead is galena, PbS. Aluminum is the most abundant metal in the earth's crust, mainly in the form of aluminosilicates, but the prime source of the metal is bauxite, $Al_2O_3 \cdot xH_2O$.

The most important ores of these elements are oxides and sulfides, which contain the metals as cations, M^{n+}. These are reduced to the metals with reducing agents, such as coke (carbon), carbon monoxide, hydrogen, or electrolytically with electrons. Oxides are normally **smelted** directly and sulfides **roasted** first in air to convert them to oxides. Ores usually contain silica, or silicates, as impurities. These are removed as a **slag** during smelting by adding a **flux** such as calcium carbonate or oxide, which reacts to form a layer of calcium silicate, $CaSiO_3$, on top of the molten metal. In the laboratory, a more reactive metal may be used as the reducing agent.

All metals, except mercury, are **crystalline** solids as room temperature and consist of spherical atoms packed together in a regular pattern. When atoms are **close-packed** in one plane, each is surrounded by six others. These planes of atoms may be stacked (with the atoms of one layer nestling in the hollows formed by adjacent layers) in two possible ways -- either with the layers in the sequence ABABAB ... -- the **hexagonal close-packed structure,** in which atoms of alternate layers are directly above each other, or in the sequence ABCABC ... -- the **cubic close-packed structure,** in which it is not until a fourth layer that atoms are directly above those of the first layer. In both structures the number of atoms nearest to any given atom is 12 (6 atoms in the same plane, 3 atoms above and 3 atoms below). Both close-packed structures are common among metals. Also common is the **monatomic body-centered cubic structure,** in which the atoms are not close-packed but surrounded by 8 others at the corners of a cube.

Metallic bonding is neither ionic nor covalent; there are insufficient electrons to covalently bond all of the atoms to all their neighbors, and a metal is represented using a large number of resonance structures. Thus the valence electrons are delocalized throughout the structure. Alternatively, metallic bonding may be pictured as an electron "cloud" in which are embedded a regular array of positively charged metal ions -- the whole being held together by eletrostatic attraction. Some metals have bonds that are intermediate between metallic and covalent.

In a metal, bonding electrons are relatively free to move under the influence of an electrical potential, or when the metal is heated; thus metals are good **conductors** of electricity and heat. In contrast, covalent substances, where the electrons are localized in covalent bonds, and ionic solids, are **insulators.** Only in a molten salt or in solution can ions move by ionic conduction. The mechanical properties of metals (malleability and ductility) are due to the possibility of distorting metallic structures without breaking bonds between atoms. The **thermionic** and **photoelectric** effects arise from the relative ease with which valence electrons are ionized from a metallic surface.

Metals are oxidized by acids to M^{n+} ions. The oxidizing agent in dilute acid is $H_3O^+(aq)$, which is a sufficiently strong oxidizing agent to oxidize Fe(s) and Al(s),

$$Fe(s) + 2H_3O^+(aq) \longrightarrow Fe^{2+}(aq) + H_2(g) + 2H_2O(l)$$
$$2Al(s) + 6H_3O^+(aq) \longrightarrow 2Al^{3+}(aq) + 3H_2(g) + 6H_2O(l)$$

Cu(s) and Pb(s) are not oxidized by the $H_3O^+(aq)$ in dilute acid but react with concentrated H_2SO_4 and dilute and concentrated $HNO_3(aq)$, which contain the stronger oxidizing agents H_2SO_4, HNO_3, and NO_3^-.

The reactivity of metals allows them to be classified in terms of decreasing strength as reducing agents:

Na, K, Ca	Mg, Zn, Al, Fe, Pb	Cu, Ag, Au
very reactive	less reactive	unreactive

Aluminum has the valence state $[Ne]3s^13p_x^13p_y^1$. Because it lies on the borderline between the metals and nonmetals of the periodic table, it can lose the 3 valence electrons to give Al^{3+}, or share them to form 3 covalent bonds. Al is a reactive metal but does not corrode in the atmosphere because of the rapid formation of a thin, hard, insoluble layer of $Al_2O_3(s)$, which protects the metal from further reaction. The bonds in $Al_2O_3(s)$ are best described as polar covalent. $Al^{3+}(aq)$ is strongly hydrated with 6 H_2O molecules and $Al(H_2O)_6^{3+}$ occurs both in aqueous solution and in hydrated salts. $AlCl_3(s)$ is ionic but covalent Al_2Cl_6 molecules are found in the gas phase.

$AlCl_3$ and molecules such as BCl_3 behave as electron pair acceptors, or **Lewis acids,** and accept electron pairs from **Lewis bases** to form **adducts.** Examples of Lewis acid-base reactions include the formation of Al_2Cl_6, formation of adducts such as $BCl_3 \cdot NH_3$ and ions such as BCl_4^- and $AlCl_4^-$, the reaction between an acidic oxide and a basic oxide, such as

$$CO_2 + O^{2-} \longrightarrow CO_3^{2-} \quad \text{and} \quad SO_3 + O^{2-} \longrightarrow SO_4^{2-}$$

and the reaction of an acidic oxide (acid anhydride) with water to give an oxoacid.

$$SO_3 + H_2O \longrightarrow H_2SO_4\ ; \quad P_4O_{10} + 6H_2O \longrightarrow 4H_3PO_4; \quad N_2O_5 + H_2O \longrightarrow 2HNO_3$$

In general the **main group metallic oxides are basic,** because they are ionic compounds containing the O^{2-} ion, and the **main group nonmetal oxides are**

acidic. Oxides on the borderline between the metals and nonmetals, such as Al_2O_3 and BeO, have both acidic and basic properties and **are amphoteric oxides.** Other oxides, such as O_2 and F_2O, have no acid-base properties.

$Al(OH)_3(s)$ is precipitated from an $Al^{3+}(aq)$ solution on addition of a base and is an amphoteric hydroxide,

$$Al(OH)_3(s) + 3H_3O^+ \longrightarrow Al^{3+} + 6H_2O; \; Al(OH)_3(s) + OH^- \longrightarrow Al(OH)_4^-$$

The preparation and reactions of aluminum compounds are summarized in Fig. 10.15.

Lead has the electron configuration $[Xe]4f^{14}5d^{10}6s^26p^2$ in its ground state and readily loses the two p electrons to form **ionic** Pb(II) compounds. It also forms less common **covalent** Pb(IV) compounds, utilizing the valence state $[Xe]4f^{14}5d^{10}6s^16p_x^16p_y^16p_z^1$. The preparation and reactions of Pb(II) compounds are summarized in Figure 10.16. $Pb(OH)_2(s)$ is an **amphoteric** hydroxide. Red lead, $Pb_3O_4(s)$, contains both Pb(II) and Pb(IV) and may be formulated as $Pb(II)_2Pb(IV)O_4$. $PbCl_4$ is an unstable yellow covalent liquid analogous to CCl_4 and $SiCl_4$.

Iron has the electron configuration $[Ar]3d^64s^2$ and readily loses the two 4s electrons to give Fe^{2+}, and an additional 3d electron, to form Fe^{3+}. In addition to the basic Fe(II) and Fe(III) oxides, FeO(s) and $Fe_2O_3(s)$, the oxide $Fe_3O_4(s)$ occurs, which is formulated as $Fe(II)Fe(III)_2O_4$. Iron reacts with aqueous acids to give Fe(II) salts and hydrogen. Addition of NaOH(aq) to a solution of a Fe(II) salt gives white $Fe(OH)_2(s)$,

$$Fe^{2+}(aq) + 2OH^-(aq) \longrightarrow Fe(OH)_2(s)$$

which is readily oxidized in air to brown $Fe(OH)_3(s)$, and the latter decomposes on heating to $Fe_2O_3(s)$. The more important reactions of iron and its compounds are summarized in Figure 10.17. The **rusting of iron** occurs when the metal is exposed to moist air,

$$4Fe(s) + 8H^+(aq) + 2O_2(g) \longrightarrow 4Fe^{2+}(aq) + 4H_2O(l)$$

$$4Fe^{2+}(aq) + O_2(g) + 4H_2O(l) \longrightarrow 2Fe_2O_3(s) + 8H^+(aq)$$

giving an overall reaction,

$$4Fe(s) + 3O_2(g) \longrightarrow 2Fe_2O_3(s).$$

Fe(II) compounds are readily oxidized to Fe(III) compounds and behave as reducing agents, and Fe(III) compounds can behave as oxidizing agents, because they are readily reduced to Fe(II) compounds. Of the halide ions, only I^- can reduce Fe^{3+} to Fe^{2+}.

Copper has the electron configuration $[Ar]3d^{10}4s^1$ and forms ionic Cu(I) and Cu(II) cmpounds, of which the latter are the more stable. Cu(II) salts contain the blue $Cu(H_2O)_2{}^{4+}$ ion. Pale blue $Cu(OH)_2(s)$ is precipitated when a base is added to a solution containing $Cu^{2+}(aq)$, which dissolves in $NH_3(aq)$ to give a solution containing the deep blue $Cu(NH_3)_4{}^{2+}(aq)$ ion. Cu(I) compounds are readily oxidized to Cu(II) compounds and are formed under reducing conditions. Cu(I) compounds are generally insoluble; $Cu^{+}(aq)$ is unstable in aqueous solution and readily **disproportionates** to $Cu^{2+}(aq)$ and Cu(s). The preparation and reactions of copper compounds are summarized in Figure 10.18.

The **important oxidation states** of the metals are:

Al	Fe	Pb	Cu
+3	+3	+4	+2
	+2	+2	+1

In their higher oxidation states, metals have a stronger tendency to form covalent bonds than in their lower states. Fe(II) compounds are readily oxidized to Fe(III) compounds; Pb(IV) compounds, such as PbO_2, are good oxidizing agents. In aqueous solution, Cu(I) compounds rapidly disproportionate to Cu and Cu(II). Thus, except for Cu(I) and Pb(IV), the aqueous solution chemistry is that of the respective hydrated cations. Highly charged cations such as $Al(H_2O)_6^{3+}$ and $Fe(H_2O)_6^{3+}$, behave as weak acids. Most salts of Al(III), Fe(III), Fe(II) and Cu(II) are soluble in water. By contrast, most salts of Pb(II) are insoluble or only slightly soluble; exceptions are $PbCl_2$ which is quite soluble in hot water but not in cold, and $PbNO_3$ which, like most nitrates, is soluble.

REVIEW QUESTIONS

1. Write the ground state electron configurations of Al, Pb, Fe and Cu.
2. Why is copper called a coinage metal and why is it suitable for this use?
3. Define and distinguish between the terms mineral and ore.
4. Name five reducing agents that are used to obtain metals from their ores.
5. Write equations to illustrate the production of metallic copper from cuprous oxide (cuprite) and from malachite.
6. In the production of iron, why is limestone used as a flux and what is the nature of the slag that it forms with silica?
7. In the production of steel, what is the basic oxygen process?
8. What reaction occurs between iron oxide, Fe_2O_3, and aluminum metal at high temperature?
9. In the above reaction, does Al behave as an oxidizing agent or as a reducing agent?
10. What number of atoms in the same layer surrounds a given atom in a close-packed layer of atoms?
11. How does hexagonal close-packing differ from cubic close-packing?
12. In terms of the number of atoms directly in contact with any particular atom, how does a monatomic body-centered cubic structure differ from a close-packed structure?
13. Classify each of the crystal structures of Al, Fe, Pb and Cu as hexagonal close-packed, cubic close-packed, or monatomic body-centered cubic.
14. Why is it possible for two alkali metal atoms to form a diatomic molecule in the gas phase? How would the metal-metal bond in such a molecule be described?
15. What alternative models are used to describe the bonding in a crystal of sodium metal?

16. Why is a metal not expected to have an ionic structure composed of positive and negative ions?
17. Why are metals good conductors of heat and electricity?
18. How does the model of metallic bonding explain the difference in melting points between sodium, calcium, and iron?
19. In terms of metallic bonding, why are metals malleable and ductile?
20. Write a general equation for the reaction between a metal and an aqueous acid to give a tripositive cation and hydrogen.
21. In the above reaction, which species is oxidized and which is reduced?
22. Arrange Cu, Fe, Al and Pb in order of their chemical reactivities to oxidation by H_3O^+ ions (very reactive, reactive, or unreactive).
23. Assuming that all of the water of crystallization in the salts $AlCl_3.6H_2O$ and $FeCl_2 \cdot 4H_2O$ is bonded to the metal, write formulas for the hydrated ion in each salt.
24. When a metal reacts with nitrogen, is it behaving as an oxidizing agent or a reducing agent?
25. Write equations to describe the reactions of Cu and Pb, respectively, with concentrated sulfuric acid to give a sulfate salt, sulfur dioxide, and water.
26. What is the change in the oxidation number of Cu in its reaction with dilute HNO_3 to give $Cu(NO_3)_2$ and nitrogen monoxide, NO? What is the change in the oxidation number of nitrogen?
27. What are the changes in the oxidation numbers of Cu and N in the reaction of copper with concentrated nitric acid, to give $Cu(NO_3)_2$ and NO_2?
28. Copper forms two oxides of formula CuO and Cu_2O, respectively. Which is described as cuprous oxide and which as cupric oxide?
29. What are the oxidation states of the metal atoms in each of the following compounds?
Al_2O_3, $Al(OH)_4^-$, $Fe_2(SO_4)_3$, $FeSO_4$, Cu_2O, $CuCl_2 \cdot 2H_2O$, $Pb(NO_3)_2$, $PbCl_4$, $Pb(C_2H_5)_4$
30. Formulate Pb_3O_4 as a compound containing both Pb(II) and Pb(IV).
31. Why is alumina insoluble in water but dissolves in both aqueous acid and aqueous base?
32. Write equations for the reaction of Al_2O_3 with (a) aqueous hydrobromic acid, and (b) aqueous potassium hydroxide.
33. Draw Lewis structures for $AlCl_3$ and Al_2Cl_6; using the VSEPR model predict their shapes.
34. In the formation of Al_2Cl_6, does $AlCl_3$ behave as a Lewis acid or a Lewis base?
35. How may aluminum chloride be purified?
36. What is meant by the term "hydroscopic salt"?
37. What is a Lewis acid?
38. Which of H_3O^+ and H^+ can be described as a Lewis acid?
39. In the reaction between BF_3 and NH_3, which molecule is the Lewis acid and which the Lewis base?
40. What term is used to describe the product of the above reaction?
41. Draw the Lewis structures of BF_3NH_3.
42. Define the terms basic and acidic oxide.
43. Categorize each of the following oxides as acidic, basic or amphoteric:
K_2O CaO P_4O_{10} SO_3 BeO_2 SiO_2 Al_2O_3 Cl_2O_7
44. Write an equation for the reaction between CaO(s) and $SO_3(g)$ expressed as an oxide ion transfer reaction.

45. To what acids do the following anhydrides correspond? N_2O_5, SO_3, SO_2, P_4O_{10}, P_4O_6, Cl_2O_7.
46. Why are aqueous solutions of $AlCl_3$ and $FeCl_3$ acidic?
47. What are the formulas of potassium alum and ammonium alum?
48. Of what ions are the above two salts composed?
49. How is a sample of potassium alum prepared?
50. Why is $Al(OH)_3$ described as an amphoteric oxide?
51. Write an equation for the reaction of red lead, Pb_3O_4, with nitric acid.
52. Is the type of bonding in $PbCl_2$ different from that of $PbCl_4$?
53. What reaction occurs between $PbCl_4$ and water?
54. Why does $Pb(OH)_2$ dissolve in excess alkali?
55. What are the expected molecular geometries of $PbCl_4$ and the $Pb(H_2O)_6^{2+}$ ion?
56. Write the common oxidation states of iron. Which is the more stable?
57. Write an equation for the oxidation of green $Fe(OH)_2$ to brown $Fe(OH)_3$ by oxygen in aqueous acid solution.
58. Both aluminum and iron are "reactive" metals, yet iron rusts in moist air and continues to do so, while aluminum can be used for cooking utensils. Explain.
59. What is a disproportionation reaction?

OBJECTIVES

Be able to:

1. State the forms in which the elements Al, Fe, Cu and Pb are found in nature and their major uses.
2. State the various methods by which metals are extracted from their ores.
3. Draw close packed (both hexagonal and cubic) and body centered cubic structures for metals.
4. Describe a simple model for metallic bonding and use that model to account for the characteristic physical properties of metals.
5. Give the products of the reactions of metals with acids.
6. Balance oxidation-reduction equations using oxidation numbers.
7. Classify the common metals according to their reactivity with acids.
8. List the common oxidation states for oxides and chlorides of Fe, Al, Cu and Pb.
9. Define the terms Lewis acid and Lewis base and give examples of Lewis acid-base reactions.
10. Classify the oxides of Periods 2 and 3 as basic, acidic, or amphoteric and give the species formed when these oxides dissolve in water.
11. Write equations illustrating the formation of, and reactions of, compounds of Al, Fe, Cu and Pb.

PROBLEM SOLVING STRATEGIES

10.5 Reactions and Compounds of Aluminum, Iron, Copper and Lead

Balancing Redox Reactions

When **metals** like the ones discussed in Chapter 10 of the text react with oxygen, halogens, or acids, they are **oxidized.** There are also many instances where **compounds** undergo oxidation-reduction. The **oxidation** process for a substance is described by a **loss of electrons,** or as an **increase in oxidation number. One of the other substances** taking part in the reaction must **gain electrons,** and be **reduced.** This substance is the **oxidizing agent;** it undergoes a **decrease in oxidation number** of one or more of its elements.

You can use oxidation number assignments to help balance redox reactions. However, you need to master the reactions presented in Chapter 10, or else you won't know what are the **products** of a reaction. A systematic set of rules for balancing redox reactions is presented in the text. At the outset, you should be able to assign oxidation numbers to the substances involved in the reaction. See Chapter 8 of the study guide for review. Let's try some examples of balancing redox reactions.

Example 10.1

When nitric acid reacts with hydrogen sulfide, the substances sulfur and nitrogen monoxide (NO) are produced. Write a balanced equation for the reaction.

Solution 10.1

Since the products (except possibly water, H^+ and OH^-) are given, we can write the beginning of 2 half-reactions, using the principle that atom types are conserved in any process.

$$NO_3^- \rightarrow NO$$

$$H_2S \rightarrow S$$

To determine which half-reaction is an oxidation, and which is a reduction, we assign oxidation numbers to N in one half-reaction and to S in the other. N in NO_3^- has an oxidation number of +5, and it decreases to +2 in NO. So N is being reduced; this is a 3 electron transfer. As well, S has an oxidation number of -2 in H_2S and is oxidized to oxidation number 0 in the element. This corresponds to a 2 electron transfer.

$$NO_3^- + 3e^- \rightarrow NO$$

$$H_2S \rightarrow S + 2e^-$$

Note that we don't attempt to balance charge yet. In fact, we added electrons to the left side of the first, incomplete, half-reaction even though NO_3^- also appeared on the left side. Three electrons were added because NO_3^- is being **reduced.** We balance charge in the next step. To accomplish this, since we have an acid solution, add H^+ to the appropriate side of each half-reaction.

$$NO_3^- + 3e^- + 4H^+ \rightarrow NO$$

$$H_2S \rightarrow S + 2e^- + 2H^+$$

Now we are ready to balance atoms. In the first half-reaction this is accomplished by adding 2 H_2O to the right side, since it is deficient by two in the number of oxygen atoms.

$$NO_3^- + 3e^- + 4H^+ \rightarrow NO + 2H_2O$$

For the second half-reaction, the atoms are already balanced. We are now ready to multiply the two balanced half-reactions by appropriate numbers so that the total number of electrons taken on by one half-reaction are released by the other.

$$2 \times (NO_3^- + 3e^- + 4H^+ \rightarrow NO + 2H_2O)$$

$$3 \times (H_2S \rightarrow S + 2e^- + 2H^+)$$

Finally, we **add the 2 half-reactions** and simplify where necessary.

$$2NO_3^- + 2H^+ + 3H_2S \rightarrow 2NO + 4H_2O + 3S$$

Next let's try an example with a basic solution.

Example 10.2

When chlorine gas is bubbled through an aqueous solution of hydroxide ions, chloride ion and hypochlorite ion (OCl^-) are produced. Write a balanced equation for the reaction.

Solution 10.2

You are told the reactants and products. Note that some Cl_2 is oxidized and some is reduced. The oxidation number of Cl_2 decreases from 0 to -1 for each Cl^- produced, and increases from 0 to +1 for each OCl^- formed. So we write the beginning of 2 half-reactions.

$$2e^- + Cl_2 \rightarrow 2Cl^-$$

$$Cl_2 \rightarrow 2OCl^- + 2e^-$$

Note that we have atoms balanced for species undergoing oxidation and reduction at the very beginning. In Example 10.1, N and S were already balanced at the start, so this concern was not present. Next we balance charge. The first half-reaction is already charge-balanced. For the second one, we add $4OH^-$ to the left side, since there are 4 negative charges on the right side, and the solution is basic. This gives

$$4OH^- + Cl_2 \rightarrow 2OCl^- + 2e^-$$

Our next consideration is an atom balance for H and O. For the second half-reaction we add 2 H_2O to the right side, since it is deficient in two oxygens.

$$4OH^- + Cl_2 \rightarrow 2OCl^- + 2e^- + 2H_2O$$

Finally we can directly add the 2 half-reactions, since each involves 2 electrons. We find, after dividing each term by 2,

$$2OH^- + Cl_2 \rightarrow Cl^- + OCl^- + H_2O$$

You may have learned other methods for balancing redox equations, using oxidation numbers but not formally writing half-reactions. Use any correct method that you desire for balancing these equations. One benefit of using half-reactions is that you become more acquainted with a way of thinking that will be used later on in the treatment of electrochemistry.

Lewis Acids and Bases

The Bronsted-Lowry definition of acids and bases emphasized the role of the proton; an acid is a **proton donor** and a base is a **proton acceptor.** However, there is an even more general definition of acids and bases, due to Lewis. A **Lewis acid** is an **electron pair acceptor,** and a **Lewis base** is an **electron pair donor.** Acid-base reactions can be considered even when there are no protons involved. In the Lewis description of the reaction of a strong acid with a strong base, we have

$$\underset{\text{Lewis acid}}{H^{+}} \quad + \quad \underset{\text{Lewis base}}{{}^{-}:\ddot{\underset{..}{O}}\text{-}H} \quad \rightarrow \quad H\text{-}\ddot{\underset{..}{O}}\text{-}H$$

Example 10.3

Give products, if there are any, for the following reactions. Balance the equations. Which of the reactions are Lewis acid-base reactions?

(a) $Zn(s) + HBr(aq) \rightarrow$
(b) $Cu(s) + HCl(dil.aq) \rightarrow$
(c) $Al(OH)_3(s) + OH^-(aq) \rightarrow$

Solution 10.3

(a) When Zn is placed in acid, H_2 is evolved, and a salt of the acid is formed.

$$Zn(s) + 2HBr(aq) \rightarrow Zn^{2+}(aq) + 2Br^-(aq) + H_2$$

The salt $ZnBr_2$ forms ions in solution. This is an oxidation-reduction reaction, since Zn loses electrons (and H gains them). It is not an acid-base reaction.

(b) No reaction takes place when Cu metal is placed in dilute HCl solution.

(c) Solid aluminum hydroxide, $Al(OH)_3$ dissolves in excess base to form a soluble aluminate ion $Al(OH)_4^-$.

$$Al(OH)_3 + OH^- \rightarrow Al(OH)_4^-$$

Al, in group 3, has only 6 valence electrons surrounding it in $Al(OH)_3$.

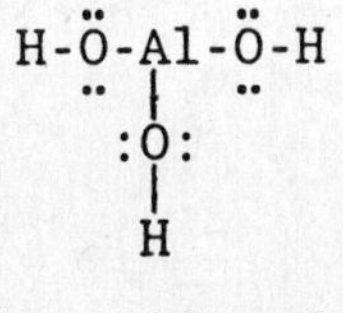

$$\begin{array}{c} H\text{-}\ddot{\underset{..}{O}}\text{-}Al\text{-}\ddot{\underset{..}{O}}\text{-}H \\ | \\ :O: \\ | \\ H \end{array}$$

Thus it can act as a Lewis acid and accept an electron pair from ${}^{-}:\ddot{\underset{..}{O}}H$, forming

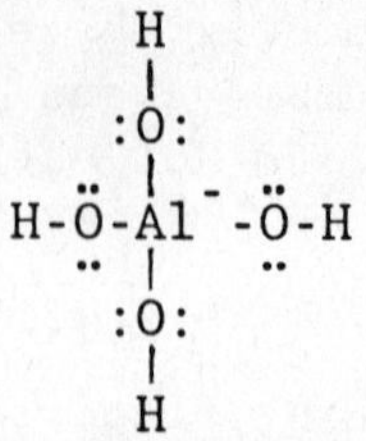

```
        H
        |
       :O:
        |  -  ..
  H-Ö-Al  -O-H
    ..  |    ..
       :O:
        |
        H
```

In the restricted Bronsted-Lowry definition, we could not refer to this reaction as an acid-base reaction, but it is one according to the Lewis definition.

Acidic and Basic Oxides

Soluble **oxides of metals** give **basic** solutions in water. Soluble **oxides of nonmetals** yield aqueous solutions that are **acidic.** These 2 types of oxides are acting as **Lewis bases** and **Lewis acids**, respectively.

Example 10.4

Give balanced equations for the dissolving in water of Na_2O and SO_2. Describe these in terms of Lewis acid-base reactions.

Solution 10.4

Since Na_2O is an ionic oxide, in water it dissociates into Na^+ ions and O^{2-} ions. It yields a basic solution since O^{2-} reacts with water:

$$O^{2-} + H_2O \rightarrow 2OH^-$$

In a Lewis description, the oxide ion is acting as a Lewis base. It donates an electron pair to H of H_2O, forming 2 OH^- ions.

```
 ..  2-      ..           ..  -
:O:    + H-O-H → 2 H-O:
 ..      ..         ..
```

In the case of SO_2, we have an oxide of a nonmetal which gives an acid solution.

$$SO_2 + 2H_2O \rightarrow HSO_3^- + H_3O^+$$

This acid oxide is an electron pair acceptor. Remember the Lewis structure for SO_2.

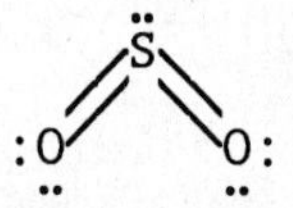

```
      ..
      S
    //  \\
  :O      O:
   ..     ..
```

Also recall that S can accommodate more than 8 electrons in its valence shell. The O atom of 1 H_2O donates an electron pair to S, forming

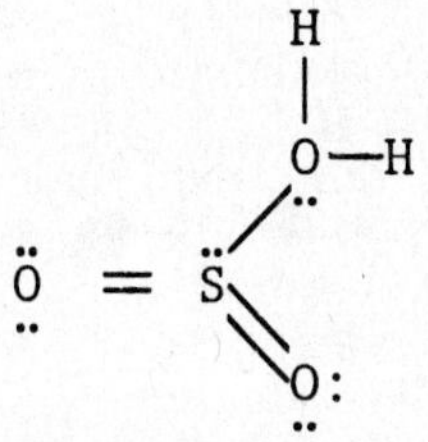

```
            H
            |
            O—H
           /..
 ..     ..
 O  =  S
 ..      \\
           O:
           ..
```

At the same time the oxygen atom of another H_2O donates an electron pair to an H atom, giving H_3O^+ and leaving

```
              ..  H
            - O/
  ..       ../  ..
  O  =  S
  ..       \\
            O:
            ..
```

A better Lewis structure puts the negative charge on oxygen.

```
   -  ..   ..   ..
    : O -- S -- O -- H
      ..   ||   ..
          :O:
```

This is one of 2 resonance forms.

SELF TEST

Part I True or False

1. In the blast furnace, iron ore is oxidized to pig iron with carbon monoxide.
2. Brass is an alloy of iron and zinc.
3. Aluminum has a higher melting point than sodium.
4. The thermionic effect is explained by stating that the speeds of ions in a molten salt increase with increasing temperature.
5. Silver is more reactive than iron in aqueous acid.
6. The 6 water molecules in $Al(H_2O)_6{}^{3+}$ have an octahedral arrangement around the aluminum ion.
7. Magnesium oxide is insoluble in water.
8. The common oxidation states for lead are +2 and +3.

Part II Multiple Choice

9. In a body centered cubic structure for metals, how many nearest neighbors does a given atom have?
   (a) 4 (b) 6 (c) 8 (d) 10 (e) 12
10. Give the main nitrogen-containing gaseous product when copper reacts with **concentrated** nitric acid.
    (a) N_2 (b) NO (c) NO_2 (d) N_2O (e) N_2O_2
11. $NO_3{}^-$ can be converted to NH_3 in basic solution. The change in oxidation number for N is from
    (a) 5 to 3 (b) 7 to 3 (c) 5 to -3 (d) 7 to -3 (e) 6 to 3
12. Balance the following equation using the smallest possible integers.
    $$Fe^{2+} + Cr_2O_7{}^{2-} + H^+ \rightarrow Fe^{3+} + Cr^{3+} + H_2O$$
    The coefficient of Fe^{2+} in this balanced equation is
    (a) 1 (b) 2 (c) 3 (d) 6 (e) 8
13. Which of the following are products of the reaction of aluminum metal and gaseous hydrogen chloride?
    (a) H^+ (b) $AlCl_2$ (c) Cl_2 (d) H_2 (e) $AlCl_3$

14. When the equation

$$Br_2 \rightarrow Br^- + BrO_3^-$$

is balanced for the reaction which occurs in basic solution, the ratio of Br^- to BrO_3^- is

(a) 5 to 1 (b) 3 to 1 (c) 1 to 1 (d) 1 to 3 (e) 1 to 5

15. Which of the following does not act as a Lewis acid when dissolved in water?

(a) CaO (b) SO_2 (c) CO_2 (d) P_4O_{10} (e) SO_3

16. Which of the following, when added to water, give basic solutions?

(a) CO_2 (b) KOH (c) Li_2O (d) Cl_2 (e) SiO_2

17. Give the formula and color of the product formed when $Pb^{2+}(aq)$ reacts with $S^{2-}(aq)$.

(a) PbS, black (b) Pb_2S, black (c) PbS, white (d) Pb_2S, white
(e) Pb, silvery-grey

Answers to Self Test

1. F; 2. F; 3. T; 4. F; 5. F; 6. T; 7. T; 8. F; 9. c; 10. c;
11. c; 12. d; 13. d,e; 14. a; 15. a; 16. b,c; 17. a.

CHAPTER 11

THE SOLID STATE

SUMMARY REVIEW

Under ordinary conditions more substances are **solids** than are liquids or gases, but solids show a greater variety of properties. For example, as a consequence of their quite different structures, diamond and graphite, the allotropes of carbon, are physically and chemically quite different substances. Solids are usefully classified according to the type of structure into **molecular solids** and **network solids. Molecular solids** consist of covalent molecules (or nonmetal atoms) held together in some regular pattern by weak intermolecular forces. **Network solids** have continuous networks of atoms or ions in which no individual molecules can be distinguished; strong interatomic or interionic bonding extends throughout the structure and they are subdivided on the basis of the type of bonding into **covalent, ionic,** or **metallic** network solids. No sharp division is possible between covalent and ionic network solids; the bonds may be polar, depending on the electronegativity differences between the atoms involved. Similarly, many solids with metallic properties have bonding intermediate between covalent and metallic. The exact nature of the bonding greatly influences properties such as melting point, boiling point, and solubility in various solvents. Network solids are also classified according to the number of dimensions into which the network extends: 1D, 2D or 3D.

Another class of solids is the **amorphous solids,** which have a more disordered random arrangement of atoms or molecules than in a crystalline solid. They lack sharp melting points. Examples include plastic sulfur, glasses and amorphous polymers.

Crystal structures are classified according to their **space lattices,** which describe the possible ways that regularly repeating points can be arranged in space. In two-dimensions such arrangements are called **2D lattices or nets.** A net is constructed by placing a chosen point of the repeating pattern (called the **motif**) at each lattice point. Only a small part of a lattice needs to be described to specify it completely. Four lattice points are chosen and connected to give a four-sided figure with pairs of parallel sides -- the **unit cell.** Unit cells may be chosen in many ways; the most convenient to choose is the smallest that gives the full symmetry of the lattice. All 2D repeating patterns fit one of just **five types** and there are **five 2D unit cells. Primitive cells** contain no interior points; cells with interior points are called **centered cells.** A full description of any 2D pattern includes the type of unit cell, its dimensions (lengths of its edges), the position of the motif and the angle between the nonparallel edges.

In three dimensions there are fourteen possible 3D arrangements of points (**space lattices** or Bravais lattices). The most symmetrical of these are (a) **the primitive cubic lattice,** (b) **the body-centered cubic lattice,** and (c) **the face-centered cubic lattice.** Descriptions of the unit cells for these are contained in the following table.

	a primitive cubic	b body-centered cubic	c face-centered cubic
Unit Cell			
Description	points at corners of cube	points at corners of a cube, plus a point at the center of the cube	points at corners of a cube, plus a point at the center of each face
Lattice points per unit cell	(8 x 1/8) = 1	(8 x 1/8) + 1 = 2	(8 x 1/8) + (6 x 1/2) = 4
Example	cesium chloride	iron	copper

Note that in the copper structure, each Cu atom is twelve coordinated and the face-centered cubic structure is one of the two close-packed structures (the one with the ABCABC ... arrangement of close-packed layers). The other close-packed structure (with the ABABAB ... arrangement) has a unit cell that is less symmetrical than the cube.

Structures of metals are determined by the tendency of identical metal atoms to close-pack, those of covalent networks by the preferred geometries of the covalent bonds. In contrast, the structures of **ionic crystals** depend on the tendency of each ion to attract to itself as many ions of opposite charge as possible in a network structure. Among ionic compounds of empirical formula MX, there are three important **1:1 structures** that are named after common examples, and based on cubic unit cells.

	Coordination number of each ion	unit cell	formula units per unit cell
NaCl	6	face-centered cubic	4NaCl
CsCl	8	primitive cubic	CsCl
ZnS	4	face-centered cubic	4ZnS

The zinc sulfide (sphalerite) structure is related to the diamond structure with alternate Zn^{2+} and S^{2-} ions replacing the C atoms.

Ionic crystals may also be described as based on a close packing of larger anions, with smaller cations occupying holes in the close-packed arrangement, in which the cations touch the anions but the anions do not necessarily touch each other. **Sodium chloride** and **sphalerite** both have structures based on the face-centered cubic lattice. In such a lattice the anions form two kinds of "holes" -- those at the center of an octahedral arrangement of six anions are called **octahedral holes,** and those at the center of four anions are called **tetrahedral holes.** Within a unit cell there are **eight tetrahedral holes,** one octahedral hole at the center and an additional twelve octahedral holes halfway along each edge and shared between a total of four unit cells, for a total of **4 octahedral holes** per unit cell. **The NaCl structure** may be described as a cubic close-packed arrangement of Cl^- ions with all of the octahedral holes occupied by Na^+ ions. **The ZnS structure** is a cubic close-packed arrangement of Cl^- ions with one-half of the tetrahedral holes occupied by Zn^{2+} ions.

Among **ionic 1:2 and 2:1 structures,** the **fluorite** (CaF_2) and **antifluorite** (M_2X) structures have face-centered cubic lattices of their larger ions. In CaF_2, the larger Ca^{2+} ions are at the lattice points and the smaller F^- ions occupy all of the eight tetrahedral holes -- there are four Ca^{2+} ions and eight F^- ions per unit cell, with the F^- ions each tetrahedrally coordinated by four Ca^{2+} ions and each Ca^{2+} ion coordinated by eight F^- ions arranged at the corners of a cube. In the **antifluorite** structure the cations and anions are interchanged.

The coordination number of an ion depends on the relative sizes of cation to anion, i.e., on the radius ratio r_+/r_-.

X-ray diffraction of crystals gives accurate interionic distances from which the radii of individual ions may be estimated. Ionic radii are useful for predicting structures of ionic salts, but not infallibly so, because of their approximate nature and other factors, such as the degree to which the bonding in ionic crystals is polar or partially covalent.

REVIEW QUESTIONS

1. How are the remarkably different properties of diamond and graphite related to their structures?
2. Why is diamond a poor conductor of electricity while graphite is a relatively good conductor?
3. Why does graphite possess useful lubricating properties?
4. Give a possible reason why silicon carbide, SiC, occurs only with a 3D network structure similar to diamond and no graphite analogue exists.
5. What is the principal difference between molecular solids and network solids?
6. What are the three main types of network solids?
7. Give an example of each of the three main types of network solids and describe the bonding in each.
8. How does the structure of NaCl in the gas phase differ from that in the solid?
9. Give one reason why the melting points of metals cover a very large range (from 3400°C for tungsten to -39°C for mercury).
10. Give two examples of two-dimensional network solids.
11. Why is plastic sulfur an amorphous solid rather than a crystalline solid?

12. Explain why crystalline solids have sharp melting points while amorphous solids "melt" over a relatively large temperature range.
13. What is a two-dimensional net?
14. What is a motif?
15. What is a unit cell?
16. What 2D unit cells correspond to each of the five possible types of net?
17. What is the difference between a primitive unit cell and a centered unit cell?
18. How many lattice points are associated with (a) a primitive cubic unit cell, (b) a body-centered cubic unit cell, and (c) a face-centered cubic unit cell?
19. In each of the above cases, where are the lattice points located?
20. What is the coordination number of each Fe atom in an iron crystal?
21. Describe the iron structure in terms of the lattice and motif.
22. In the iron structure, how many Fe atoms are associated with each unit cell?
23. Repeat questions 20, 21 and 22 for crystalline copper.
24. Repeat questions 20, 21 and 22 for crystalline titanium which is hexagonal close-packed.
25. In the diamond structure, what is the lattice and how many carbon atoms are associated with the unit cell?
26. What is the space lattice for NaCl and what is the coordination number for each ion?
27. How many NaCl formula units are there per unit cell?
28. What is the space lattice for CsCl and what is the coordination number for each ion?
29. How many CsCl formula units are there per unit cell?
30. What is the space lattice for sphalerite (ZnS) and what is the coordination number of each ion?
31. How is the ZnS structure related to that of diamond?
32. Why is the description of ZnS as an ionic crystal only an approximate description?
33. What is the space lattice of the fluorite (CaF_2) structure and what is the coordination number of each ion?
34. How does the structure of Li_2O differ from that of CaF_2?
35. Why is it difficult to obtain precise values for the radii of individual ions in crystals?
36. For a given atom, why is its anion larger than the neutral atom?
37. For a given atom, why is its cation smaller than the neutral atom?
38. What quantity is useful for predicting the expected coordination number of one ion by another?
39. Why are there many crystals where the structure is not that expected in terms of the radius ratio of cation and anion?

OBJECTIVES

Be able to:

1. List the physical properties, and describe the structures, of diamond and graphite.

2. Classify solids into molecular and network types and give examples of each.
3. Describe and give examples of the three categories of network solids.
4. Contrast melting points, boiling points and solubilities for molecular solids versus network solids.
5. Describe the nature of bonding and the type of network (1D, 2D or 3D) for network solids.
6. List properties that distinguish amorphous solids from crystalline solids.
7. Describe the 5 two-dimensional lattices in terms of their unit cells.
8. Describe the 3 cubic lattices, and state the number of lattice points per unit cell in each of these types.
9. Relate density data with unit cell dimensions for **metals** with differing cubic lattices.
10. Describe the diamond structure and various layer structures in terms of cubic lattices.
11. Describe the sodium chloride, cesium chloride, and sphalerite structures.
12. Relate density data with unit cell dimensions for **ionic crystals** with differing cubic lattices.
13. Describe ionic crystal structures in terms of close-packed arrangements of one type of ion, with the other type of ion occupying tetrahedral or octahedral holes.
14. Use radius ratio (r_+/r_-) arguments to describe coordination of one type of ion about another in ionic crystals.

PROBLEM SOLVING STRATEGIES

11.4 Metallic Crystals

Three common structures for metals are cubic close-packed, body-centered cubic, and hexagonal close-packed. We will consider here only the first 2 of these.

In a **cubic close-packed structure,** the space lattice is **face-centered cubic** and the lattice points are occupied by atoms. In a unit cell there is the equivalent of 1/8 of a metal atom at each corner, plus 1/2 of a metal atom at each face, for a total of 8 x 1/8 + 6 x 1/2 = 4 atoms per unit cell. It is the atom in the face of the unit cell which touches an atom at the corner of that face. The corner atoms of the unit cell do not touch each other. The figure shows a unit cell edge of length **a** and a face diagonal of length **b** = 4r, where r is the atomic radius of the metal atom.

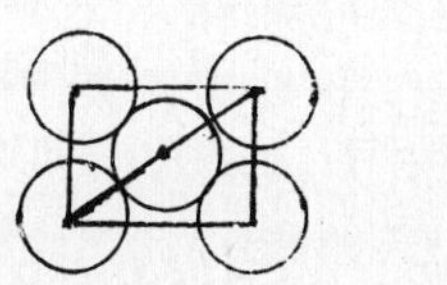

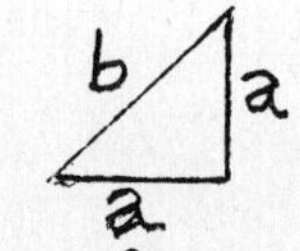

From the Pythagorean theorem we get $\mathbf{a}^2 + \mathbf{a}^2 = (4r)^2 = 16r^2$, or $a = 2\sqrt{2}r$

In a **body-centered cubic structure** the space lattice is **body-centered cubic** and there is an **atom** at each lattice point. The unit cell consists of 2 atoms, that is, 1/8 of an atom at each of 8 corners plus 1 atom in the body center. In this case the atom at the body center is touching the atoms at the corners of the unit cell. The corner atoms do not touch each other. The geometrical relationship between the body diagonal **c**, the face diagonal **b** and the unit cell edge **a** is derived in Example 11.3 in the text.

$$\mathbf{c}^2 = \mathbf{a}^2 + \mathbf{b}^2$$
$$= \mathbf{a}^2 + (\mathbf{a}^2 + \mathbf{a}^2) = 3\mathbf{a}^2$$

Since the body diagonal equals 4 atom radii, $(4r)^2 = 3\mathbf{a}^2$, and $a = (4/\sqrt{3})r$.

The main points to consider in quantitative calculations with **metallic crystals** are the following:

1. Know the relationships between atom radius and unit cell edge.
2. Know how many atoms are contained in the unit cell, and from the mass of one atom, calculate the mass and volume of the unit cell.
3. Be able to relate density, mass and volume.

Various combinations of data are provided in different problems.

Example 11.1

Silver atoms have a radius of 144 pm. Metallic silver exhibits a cubic close-packed structure. What is the density of silver, in g cm^{-3}, given a molar mass of 107.9 g?

Solution 11.1

We base the solution on the calculation of the density of a **unit cell** of silver, which is the same as the density of the overall crystal.

$$\text{density of unit cell} = \frac{\text{mass of unit cell}}{\text{volume of unit cell}}$$

First we calculate the mass of the unit cell. For a cubic close-packed structure there are 4 silver atoms in a unit cell.

$$\text{Mass of unit cell} = \frac{107.9\ \text{g}}{1\ \text{mole}} \times \frac{1\ \text{mole}}{6.022 \times 10^{23}\ \text{atoms}} \times 4\ \text{atoms}$$
$$= 7.167 \times 10^{-22}\ \text{g}$$

To get the volume of the unit cell, we use the relationship

$$\text{volume of unit cell} = (\text{unit cell edge})^3 = a^3$$

Let r equal the radius of a silver atom. The relationship we have seen between r and a is

$$a^2 + a^2 = (4r)^2 = 16r^2;\quad 2a^2 = 16r^2;\quad a = 2\sqrt{2}\ r$$

$$\text{Thus volume of unit cell} = a^3 = (2\sqrt{2}\ r)^3 = [2\sqrt{2}(144\ \text{pm} \times \frac{10^{-10}\ \text{cm}}{1\ \text{pm}})]$$
$$= 6.757 \times 10^{-23}\ \text{cm}^3$$

Finally, density of unit cell $= \dfrac{7.167 \times 10^{-22}\ \text{g}}{6.757 \times 10^{-23}\ \text{cm}^3} = 10.6\ \text{g cm}^{-3}$

Example 11.2

A metal X has a density of 5.96 g cm^{-3} and an atom radius of 132 ppm. It is known that the crystal structure of the metal is either cubic close-packed (ccp) or body-centered cubic (bcc). Identify the metal, using only a table of atomic masses.

Solution 11.2

Given 1.32 pm as the radius r of metal X, the unit cell edge, a, is given by either

$$(4r)^2 = 3a^2 \text{ for a bcc}$$

or

$$(4r)^2 = 2a^2 \text{ for a ccp}$$

Thus by substitution we have

$$a = 305 \text{ pm if the structure is bcc}$$

or

$$a = 373 \text{ pm if the structure is ccp}$$

and then the volume V is either

$$V = (305\ \text{pm})^3 \times \left(\frac{10^{-10}\ \text{cm}}{1\ \text{pm}}\right)^3 = 2.84 \times 10^{-23}\ \text{cm}^3 \text{ if bcc}$$

or

$$V = (373\ \text{pm})^3 \times \left(\frac{10^{-10}\ \text{cm}}{1\ \text{pm}}\right)^3 = 5.19 \times 10^{-23}\ \text{cm}^3 \text{ if ccp}$$

Mass of unit cell = density of unit cell x volume of unit cell. A unit cell contains 2 atoms if bcc or 4 atoms if ccp.

If bcc, 2 atoms x $\dfrac{\text{molar mass in g}}{6.022 \times 10^{23}\ \text{atoms}} = 5.96\ \text{g/cm}^3 \times 2.84 \times 10^{-23}\ \text{cm}^3$

$$\text{molar mass} = \frac{5.96\ \text{g/cm}^3 \times 2.84 \times 10^{-23}\ \text{cm}^3 \times 6.022 \times 10^{23}\ \text{atoms}}{2\ \text{atoms}} = 50.96\ \text{g}$$

If ccp, 4 atoms x $\dfrac{\text{molar mass in g}}{6.022 \times 10^{23}\ \text{atoms}} = 5.96\ \text{g/cm}^3 \times 5.19 \times 10^{-23}\ \text{cm}^3$

$$\text{molar mass} = \frac{5.96\ \text{g/cm}^3 \times 5.19 \times 10^{-23}\ \text{cm}^3 \times 6.022 \times 10^{23}\ \text{atoms}}{4\ \text{atoms}} = 46.57\ \text{g}$$

From a table of atomic masses we find that scandium has a molar mass of 44.96, titanium has a molar mass of 47.90, and vanadium has a molar mass of 50.94. The metal is most likely vanadium, with the bcc structure.

11.7 Ionic Crystals

Let us look at 3 **ionic** crystal structures where there are equal numbers of positive ions and negative ions. These are called 1:1 structures. In all 3 of these structures, whatever the space lattice is, the **lattice points are occupied by one type of ion.** The number and positions of the other type of ion differ from structure to structure. Note carefully that a metallic crystal and an ionic crystal may have the same space lattice, but the number of particles per unit cell is different.

In the **sodium chloride structure,** the space lattice is face-centered cubic, and there is a Na^+ at each lattice point. As well, Cl^- ions are located half-way along each unit cell edge; there is one Cl^- ion in the middle of the unit cell. See Figure 11.25 in the text. The mass of the unit cell is 4 times a Na^+ mass plus 4 times a Cl^- mass, since there are 8 x 1/8 Na^+ ions at the corners of unit cell, 6 x 1/2 Na^+ at the faces of the unit cell, 12 x 1/4 Cl^- at the middle of the unit cell edges, and 1 Cl^- at the center of the unit cell. The unit cell edge equals 2 times the distance from the center of a Na^+ to the center of a Cl^-.

Example 11.3

MgO has a unit cell edge of 420 pm and crystallizes in the sodium chloride structure. What is the density of MgO, in g cm^{-3}?

Solution 11.3

$$\text{density of unit cell} = \frac{\text{mass of unit cell}}{\text{volume of unit cell}}$$

$$\text{mass of unit cell} = \frac{4 \text{ atoms x } (24.31 + 16.00) \text{ g}}{1 \text{ mole}} \text{ x } \frac{1 \text{ mole}}{6.022 \text{ x } 10^{23} \text{ atoms}}$$

$$= 2.68 \text{ x } 10^{-22} \text{ g}$$

$$\text{volume of unit cell} = (420 \text{ pm})^3 \text{ x } \left(\frac{10^{-10} \text{ cm}}{1 \text{ pm}}\right)^3 = 7.41 \text{ x } 10^{-23} \text{ cm}^3$$

$$\text{density of unit cell} = \frac{2.68 \text{ x } 10^{-22} \text{ g}}{7.41 \text{ x } 10^{-23} \text{ cm}^3} = 3.62 \text{ g cm}^{-3}$$

The density of MgO is 3.62 g cm^{-3}. Note that this is a 1:1 structure, even though the ion charges are +2 and -2.

For the **cesium chloride** structure, the space lattice is primitive cubic and the lattice points are occupied by Cl^- ions. There is one Cs^+ in the center of the unit cell, whose edge length is equal to the Cl^--Cl^- distance. The unit cell thus contains 1 Cl^- (that is, 8 x 1/8 Cl^-) and 1 Cs^+. Once you can calculate the mass of the unit cell and its volume, problems involving density should be straightforward. Study Example 11.8 in the text as an illustration of a typical cesium chloride structure problem.

For the **sphalerite (ZnS) structure,** the space lattice is face-centered cubic, and lattice points are occupied by S^{2-} ions. Zn^{2+} ions are situated on each of the 4 body diagonals of the unit cell, a distance of 1/4 of the way along that diagonal. Refer to Figure 11.27. We have already seen that there are 4 ions of a given type at the lattice points of the unit cell of a face-centered cubic ionic crystal. These are the 8 x 1/8 + 6 x 1/2 S^{2-} ions. In addition there are 4 Zn^{2+} ions within the cell.

Example 11.4

You are told that CuBr has either a sodium chloride structure or a sphalerite structure. What is the shortest distance, s, between the centers of a Cu^+ and a Br^- ion, in terms of the unit cell edge, a, for these two structures?

Solution 11.4

In the NaCl structure the unit cell edge equals twice the cation center to anion center distance.

$$a = 2s$$

$$s = 0.5a \text{ if CuBr has the NaCl structure.}$$

For the sphalerite structure, the closest distance between Cu^+ and Br^- is 1/4 the body diagonal. We have already seen that, for a face-centered cubic structure (like the sphalerite structure), the relationship between the body diagonal c and the unit cell edge a is

$$c^2 = 3a^2$$

So here $s = c/4$, or $4s = c$, and

$$(4s)^2 = 3a^2$$

$$16s^2 = 3a^2$$

$$s = \sqrt{3/16}\ a$$

$$s = 0.433a \text{ if CuBr has the ZnS structure.}$$

SELF TEST

Part I True or False

1. All metals have crystal structures that are either hexagonal close-packed, cubic close-packed, or body-centered cubic.
2. If ions are considered hard spheres the ions at the corners of the unit cell in a cesium chloride type structure must touch.
3. Anions are usually larger than cations.
4. The 3-dimensional lattice for diamond is face-centered cubic.
5. Molecular solids generally have lower melting points than network solids.
6. The radius ratio is the ratio of the radius of the cation to that of the anion in an ionic crystal.
7. One can equally well describe the NaCl structure by having Na^+ ions at the corners of the unit cell, or by having Cl^- ions at the corners of the unit cell.

8. Diamond has a much lower density than graphite.
9. The noble gases form network solids.
10. In all network solids, only lattice points are occupied by atoms or ions or molecules.
11. Amorphous solids generally have sharp melting points.
12. Unit cells for two dimensional lattices need not have 90° angles, but the sides must have equal lengths.
13. There can be different numbers of particles in the unit cell for 2 substances that have the same space lattice.
14. One description of the sphalerite structure uses the concept of tetrahedral holes.

Part II. Multiple Choice

15. How many nearest neighbors does a metal atom have in a cubic close-packed arrangement?
(a) 4 (b) 6 (c) 8 (d) 9 (e) 12
16. What is the relationship between the face diagonal b and the unit cell edge a for a metal exhibiting a cubic close-packed structure?
(a) $b = 2a$ (b) $b = \sqrt{2}\,a$ (c) $b = \sqrt{2}\,a^2$ (d) $b = (2a)^2$ (e) $b = 2a^2$
17. In the sodium chloride structure, each Cl^- ion is closest to
(a) 6 Cl^- (b) 6 Na^+ (c) 4 Cl^- (d) 4 Na^+ (e) 8 Na^+
18. In the cesium chloride structure, the number of CsCl units per unit cell is
(a) 1 (b) 2 (c) 4 (d) 6 (e) 8
19. If you are given data for atomic molar masses and the type of crystal structure an ionic soilid possesses, what further information is required to calculate the unit cell edge length for that solid?
(a) the value of the Avogadro constant (b) the density of the solid
(c) the radius ratio (d) the charges on the ions
(e) the Pythagorean theorem.
20. Which of the following solids have a face-centered cubic lattice?
(a) graphite (b) CsCl (c) NaCl (d) Fe (e) Si
21. Gold exhibits cubic close packing. The shortest distance between centers of gold atoms is 288 pm. What is the unit cell edge length, in pm?
(a) 144 (b) 288 (c) 333 (d) 407 (e) 576
22. Potassium fluoride has an empirical formula mass of 58.10 g and possesses the sodium chloride structure. Its density is 2.48 g cm^{-3}. What is the closest distance between centers of opposite ions, in pm?
(a) 156.0 (b) 78.0 (c) 538 (d) 269 (e) 135

Answers to Self Test

1. F; 2. F; 3. T; 4. T; 5. T; 6. T; 7. T; 8. F; 9. F; 10. F; 11. F; 12. F; 13. T; 14. T; 15. e; 16. b; 17. b; 18. a; 19. a,b; 20. c,e; 21. d; 22. d.

CHAPTER 12

WATER, LIQUIDS, SOLUTIONS, AND INTERMOLECULAR FORCES

SUMMARY REVIEW

Water occurs naturally on earth as liquid, solid (ice and snow), and gas (water vapor). **Seawater** contains large amounts of dissolved substances, from which sodium chloride and magnesium are extracted commercially. Water is purified for domestic use by adding lime, CaO(s), to make it slightly basic, and then adding $Al_2(SO_4)_3(s)$, or alum, to precipitate $Al(OH)_3(s)$, which carries with it much of the suspended solids and bacteria. After filtration and treatment with activated charcoal, the water is sprayed into air, to speed up the oxidation of organic impurities. The final stage of purification involves treatment with ozone or chlorine.

Transformation of a substance from one state to another (solid to liquid, liquid to gas, etc.,) is called a **phase change.** With increasing temperature atoms and molecules increase their average kinetic energy. On heating a solid, the greater atomic and molecular vibrations eventually allow their ordered arrangement to be replaced by the more random arrangement of the liquid, and the solid **melts.** At the **melting point,** melting and freezing occur at the same rate and the system is in dynamic equilibrium. Liquids normally have greater molar volumes than solids (water is an exception). The **molar enthalpy of fusion,** ΔH_{fus}, is the energy required to melt 1 mole of substance at 1 atm pressure; melting is an **endothermic** process. The magnitude of ΔH_{fus} is related to the strength of the intermolecular forces in the solid. The molecules in a liquid have a distribution of kinetic energies similar to those in a gas. Those with sufficient energy can **evaporate** and give rise to a **vapor pressure.** A liquid **boils** when its vapor pressure becomes equal to the external pressure. The **molar enthalpy of vaporization,** ΔH_v, is the energy required to transform 1 mole of liquid to vapor at 1 atm pressure at a specified temperature; vaporization is an **endothermic** process. To form a vapor the molecules must be widely separated and ΔH_v is a good measure of intermolecular forces in the liquid.

When the **vapor pressure** of a liquid is equal to the atmospheric pressure, the liquid **boils.** The **normal boiling point** is defined as the temperature at which the vapor pressure equals 1 atm. When a gas is collected over water, the total pressure is that of the gas plus the vapor pressure of water at the particular temperature. Solids normally exert very small vapor pressures; **sublimation** is the process by which a solid is directly converted to vapor.

Intermolecular forces are due to electrostatic interactions between (in order of decreasing strength) ions, dipoles and induced dipoles. **Ion-ion** interactions predominate in ionic salts and in molten salts; **ion-dipole** interactions are responsible for ion solvation and account for the solubilities of salts. Next in strength are **dipole-dipole** forces between polar molecules. When an ion or polar molecule is adjacent to a **nonpolar molecule,** a dipole is induced in the latter. Such **induced dipoles** arise as a result of the **polarizability** of the electron cloud of the nonpolar molecule. The **ion-induced dipole** and **dipole-induced dipole** interactions so created make important contributions to the intermolecular forces that affect solute solubilities and the properties of ions in solution. Finally, the weakest intermolecular forces are **induced dipole-induced dipole** interactions, or **London forces.** As a result of the constant motion of electrons in molecules, instantaneous fluctuating dipoles are created which interact with each other to give weak intermolecular attractions. The magnitude of such London forces depend on the **polarizibility** of atoms and molecules, which is a measure of the ease of distortion of their electron clouds. For a given core charge, atom polarizability increases with atomic size; large molecules have greater polarizabilities than smaller ones. London forces make an important contribution to the overall attractions between all atoms and molecules.

The existence of intermolecular forces explains many observations, including the non-ideal behavior of gases, especially at high pressures and when their molecules are large, and the condensation of gases to liquids as the temperature is lowered.

When molecules are brought together, they at first attract each other as a result of the intermolecular (van der Waals) forces, but in close proximity their filled orbitals repel each other, as a consequence of the Pauli exclusion principle. At some intermediate "equilibrium" distance the forces balance and the atoms just touch each other. The closest distance of approach of nonbonded atoms is their **"nonbonded", or van der Waals, radii.** They are approximately constant for the same atom in going from molecule to molecule. The hard sphere electron-pair model predicts:

$$r_{vdw} \simeq 2r_{cov} \quad \text{and} \quad r_{vdw} \simeq r_{ion}$$

where r_{vdw} is the van der Waals radius, r_{cov} is the covalent radius, and r_{ion} is the radius of the corresponding anion.

In contrast to most solids, where the density decreases on melting, the density of **water** increases from 0°C to 4°C before assuming normal behavior at higher temperatures. Also, water has an anomalously high melting point and boiling point (0°C and 100°C) in comparison, for example, with those of H_2S and H_2Te. These and other anomalous properties suggest that in addition to the expected van der Waals forces, some additional intermolecular forces act between H_2O molecules; they are referred to as **hydrogen bonds.** In such bonds, the H atom of one molecule containing a very polar X-H bond can interact with a lone pair of an adjacent molecule. Hydrogen bonds are highly polar; unlike other dipole-dipole interactions they have the directionality normally associated with covalent bonds. An especially strong hydrogen bond is found in the linear HF_2^- ion, in which the H atom is symmetrically placed between the two F atoms. Hydrogen bonds more commonly have approximately 5% the strength of normal covalent bonds. Hydrogen bonding in ice accounts for its anomalous properties and open cage-like structure.

The two important factors that influence the **solubility** of a solute in a solvent are: 1) The tendency for all systems to move towards greater disorder, and 2) the relative strengths of the intermolecular forces in the solvent and in the solute, and between the solvent and the solute. The disorder of a system is measured by a property called entropy; **entropy** increases with increasing disorder or randomness. Substances such as water and ethanol that mix completely with each other in all proportions are said to be **miscible**; substances that do not dissolve in each other, such as hexane and water, are **immiscible.** In general, "like dissolves like". Polar substances tend to be soluble in polar solvents but are insoluble in nonpolar solvents; nonpolar substances tend to dissolve in nonpolar solvents but are insoluble in polar solvents, and hydrogen bonded substances are soluble in other hydrogen bonded substances. Similarly, covalent network solids are insoluble in water because their interactions with solvent molecules are far too weak to break the covalent bonds; metals are insoluble in water because the strong metallic bonds prevent their atoms from separating to mix with the solvent, and many ionic solids are insoluble in water, unless strong **hydration** of their ions is sufficient to pull the ions apart.

There is a natural tendency for one substance to dissolve in another because a solution is more disordered (has greater entropy) than the solute or solvent, but this tendency to disorder is opposed by attractive forces between the atoms or molecules of the solute or solvent that may prevent them from mixing. The direction in which a change occurs in an isolated system is determined by the tendency for the **entropy to increase** and for its **energy to decrease;** thus, two substances will not mix if this causes the energy of the system to increase too much, which is the case if too much energy is required to overcome the attractive intermolecular forces between the solvent or solute molecules. The **molar enthalpy of solution,** ΔH_{soln}, is the sum of the three enthalpy terms; $\Delta H_{soln} = \Delta H_1 + \Delta H_2 + \Delta H_3$, where ΔH_1 is the enthalpy change involved in separating the solvent molecules, ΔH_2 is the enthalpy change in separating the solute molecules, and ΔH_3 is the enthalpy change when the solute and solvent molecules mix with each other. For a solid, ΔH_1 is called the **lattice energy.** If ΔH_{soln} is large and positive (as it would be, for example, for hexane mixing with water) such processes tend not to occur, but when ΔH_{soln} has a small positive value or is negative, there is a good chance that the process will occur, provided the change in entropy is sufficiently large. A number of examples are illustrated in Table 12.7.

Colligative properties of solutions depend only on the concentrations of solutes (molecules and ions in solution) and not on their nature. The **vapor pressure** of the solvent in a solution at a given temperature is given by

$$P_{solvent} = P^{o}_{solvent} \cdot X_{solvent}$$

where $P^{o}_{solvent}$ is the vapor pressure of the pure solvent and X is its **mole fraction.** This is because a solution has a higher entropy than the pure solvent relative to the entropy of the vapor, so that solvent molecules have a smaller tendency to leave a solution than to leave the pure solvent. As a consequence, the **boiling point** of a solution is **higher** than that of a pure solvent because the temperature of a solution must be raised to a temperature higher than the normal boiling point in order that its vapor pressure is equal to 1 atm. For a solute concentration **m molal,** the **boiling point elevation,** ΔT, is given by $\Delta T = K_b m$, where K_b is the **boiling point elevation constant** of the solvent. Similarly the freezing point of a solution is **lower** than that of the pure solvent because in the solution the pure solid solvent is in equilibrium with the solution, which has a lower vapor pressure than the pure solvent, and to maintain the equilibrium the vapor pressure of the solid solvent must be lowered, which is achieved by lowering the temperature. For a solute of concentration **m molal,** the **freezing point depression** is given by $\Delta T = K_f m$, where K_f is the **freezing point depression constant** of the solvent. Strictly, **m** is the **concentration of solute species,** and the above equations apply to solutions of **nonelectrolytes.** When a solute ionizes in solution, or reacts, the equations become $\Delta T = iK_b m$ and $\Delta T = iK_f m$, where **i** is the number of solute species (ions or molecules) formed from one solute molecule. Elevation of boiling point and depression of freezing point experiments are useful in determining molar masses, especially for solutions of nonelectrolytes.

In **osmosis, osmotic pressure** is due to the ability of more solvent molecules to pass through a semipermeable membrane from a solvent to its solution than vice-versa. This is because solvent molecules have a greater entropy in the solution than they do in the pure solvent. The pressure that is just sufficient to stop osmosis is called **osmotic pressure.** This osmotic pressure, π, is given by $\pi V = nRT$, where V is the volume of the solution in liters, n is the number of moles of solute, T is the temperature in kelvins, and R is the gas constant (0.0821 atm L mol^{-1} K^{-1}).

In **aqueous solution** the three most important reaction types are **acid-base** (Bronsted-Lowry and Lewis), **precipitation,** and **oxidation-reduction.** In **Bronsted acid-base reactions,** water can behave both as an acid (proton donor) and as a base (proton acceptor). Water exerts a **leveling effect** on the strengths of acids and bases since $H_3O^+(aq)$ and OH^- (aq) are the strongest acids and bases that can exist in aqueous solution. The only common strong acids are HCl, HBr, HI, HNO_3, H_2SO_4, and $HClO_4$. The only common strong bases are O^{2-}, H^-, NH_2^-, and the group 1 and 2 metal hydroxides, $M(OH)_n$. Conjugate acids of weak bases, such as $NH_4^+(aq)$, are **weak acids,** and conjugate bases of weak acids, such as F^-, CN^-, or $CH_3CO_2^-$, are **weak bases.** The range of available acid-base strength in water lies between that of the acid H_3O^+ and the base OH^-. Acid-base reactions are always the **neutralization** reaction $H_3O^+(aq) + OH^-(aq) \rightleftharpoons 2H_2O(l)$, and such reactions are essentially quantitative because only very small concentrations of H_3O^+ and OH^- can coexist in aqueous solution. In **Lewis acid-base reactions,** the water molecule behaves as a **Lewis base.** Reactions of water with nonmetal oxides (acid anhydrides) are reactions where H_2O behaves as a Lewis base and gives a solution of an oxoacid, and related reactions are those between water and a nonmetal halide, to give a solution of an oxoacid and a hydrogen halide. Water also behaves as a Lewis base in the formation of strongly hydrated ions such as $Al(H_2O)_6^{3+}(aq)$, $Be(H_2O)_4^{2+}(aq)$, and $Mg(H_2O)_6^{2+}(aq)$, which behave as weak Bronsted acids in aqueous solution. In **oxidation-reduction** reactions, water can behave as a weak oxidizing agent and a very weak reducing agent. Such reactions involve the half reactions:

$$2H_2O(l) + 2e^- \longrightarrow 2OH^-(aq) + H_2(g) \qquad \textbf{reduction}$$

or

$$2H_2O(l) \longrightarrow 4H^+(aq) + O_2(g) + 4e^- \qquad \textbf{oxidation}$$

The reactions of water are summarized in Table 12.11. By first identifying the nature of the other reactant (an element, oxide, halide, salt, oxidizing agent or reducing agent) a good start is made in predicting the products of reactions involving water.

REVIEW QUESTIONS

1. What proportion of the earth's water occurs as salt water?
2. Why does natural water generally contain large amounts of dissolved substances?
3. Describe the four stages in the purification of water.
4. What cations and anions are the major constituents of the salts found in sea water?
5. What substances are obtained commercially from sea water?
6. Why does a liquid, or a solid, exert a vapor pressure when placed in a closed container?
7. What is the difference between evaporation and boiling?
8. Why does a liquid cool (lose heat) as it evaporates?
9. Define the normal boiling point of a liquid.
10. Why is the normal boiling point of water greater than that of tetrachloromethane, which in turn is greater than that of n-pentane?
11. What is meant by the term change of phase?
12. Is the molar enthalpy of vaporization of a liquid an exothermic or an endothermic quantity?
13. Define the molar enthalpy for the sublimation of a solid.
14. What correction has to be applied to find the partial pressure of a gas that is collected over water at 25°C and an atmospheric pressure of 1 atm?
15. What forces exist between the ions in an ionic solid?
16. Write the expression for the force between two ions of charge +Q and -Q, separated by the distance r.
17. Why is the enthalpy of vaporization of an ionic solid a very large quantity?
18. What is the nature of sodium chloride in the gas phase?
19. What is a polar molecule, and what is a dipole?
20. Why is there an overall force of attraction between an ion and polar molecules?
21. Draw diagrams to illustrate the solvation by water molecules of (a) a cation, for example Al^{3+}, and (b) an anion, for example SO_4^{2-}.
22. Why do the molecules of a polar solvent attract each other overall?
23. What is a nonpolar molecule?
24. How does an ion, or a polar molecule, induce a dipole in an adjacent nonpolar molecule?
25. Why do nonpolar molecules attract each other?
26. What is meant by an induced dipole?
27. What is meant by the polarizability of an atom or molecule?
28. How does the polarizability of an atom depend on its size?
29. What are the units of polarizability?
30. Arrange the molecules HF, HCl, HBr and HI in order of (a) increasing polarizability; (b) increasing dipole moment.
31. Why is the boiling point of n-pentane greater than that of its branched isomer 2,2-dimethylpropane?
32. Why do covalent fluorides generally have lower boiling points than the analogous covalent chlorides?
33. Arrange CH_4, CF_4 and CCl_4 in order of increasing normal boiling point.
34. Which of the gases H_2 and SO_2 is expected to show the greater deviations from the ideal gas law?

35. Distinguish between the term van der Waals force and the term London force.
36. Why is it not possible for filled shells of electrons to overlap with each other?
37. What is the van der Waals radius of an atom?
38. What approximate relationships connect the covalent radius, the ionic radius, and the van der Waals radius of an atom?
39. How can values for the covalent radius, the ionic radius, and the van der Waals radius of an atom be obtained?
40. How do the covalent radii, the ionic radii, and the van der Waals radii of atoms change (a) across any period of the periodic table? (b) down any group?
41. What is the nature of the additional strong forces that act between molecules such as H_2O, NH_3 and HF, in their liquids and solids?
42. Why would you conclude that strong hydrogen bonds would not exist between HCl molecules in liquid HCl?
43. How do the energies of typical hydrogen bonds compare to normal bond energies?
44. How is the fact that water has its maximum density at 3.98°C accounted for?
45. Which of the following pairs of molecules might you expect to interact through hydrogen bonds? H_2O and NH_3; H_2O and CH_4; C_2H_5OH and HF; H_3O^+ and H_2O
46. Describe the structure of the HF_2^- ion.
47. What is the arrangement of H atoms around each O atom in the structure of ice?
48. Describe in structural terms the changes that occur when ice melts, when the temperature is raised to the boiling point, and when liquid water forms vapor.
49. What takes place at the atomic and molecular levels when a salt, such as $NaNO_3$, dissolves in water?
50. What takes place at the molecular level when HCl(g) dissolves in water?
51. Why are solids such as (a) diamond, (b) SiO_2, (c) Al_2O_3 quite insoluble in water?
52. Would ethanol be expected to be miscible with water? Explain.
53. Explain the solubility of pentane in water.
54. Explain the solubility of iodine (a) in water, (b) in a solution of KI(aq), and (c) in CCl_4.
55. Which of the ions Na^+ and Al^{3+} would be expected to be more strongly hydrated in aqueous solution?
56. Why is a solution of $Al_2(SO_4)_3$(aq) slightly acidic?
57. What occurs when blue crystals of $CuSO_4 \cdot 5H_2O$ are heated?
58. What happens when $MgCl_2 \cdot 6H_2O$ crystals are heated?
59. Why is the proton in water written as H_3O^+(aq), rather than as H^+(aq)?
60. How would you formulate the proton in liquid ammonia?
61. Draw structures for the H_3O^+ and $H_9O_4^+$ ions.
62. In the reaction of sodium with water, what species is oxidized and which is reduced?
63. Is the reaction of ammonia with water an oxidation-reduction reaction, a Bronsted acid-base reaction, or a Lewis acid-base reaction?
64. What reaction occurs between lithium hydride and water?

65. Write equations for the reactions of the following oxides with water, and classify each oxide as acidic or basic; SO_2, SO_3, P_4O_{10}, Na_2O and BaO.
66. What is meant by the mole fraction of ethanol in an aqueous solution of ethanol?
67. Why does a solute lower the vapor pressure of a solvent?
68. Define molality, m.
69. Why does a solute depress the freezing point of a solvent?

OBJECTIVES

Be able to:

1. Describe how NaCl and Mg are recovered from sea water, and how water is purified.
2. State the names and the signs of ΔH associated with various phase changes.
3. Draw plots of distributions of kinetic energies of liquids as a function of temperature.
4. Define the terms vapor pressure, boiling point, and normal boiling point.
5. Describe the nature of intermolecular forces between polar molecules, as well as those between nonpolar molecules.
6. Define the term polarizability.
7. Relate melting points and boiling points to strengths of intermolecular forces.
8. Distinguish between the terms covalent radius and van der Waals radius.
9. Define the term hydrogen bond and use hydrogen bonding to account for some unusual properties for water and other hydrides.
10. Describe the solubilities of liquids in other liquids and solids in liquids in terms of disorder and the relative strengths of intermolecular forces.
11. Rationalize the fact that enthalpies of solution for some reactions are negative, and for other reactions are positive.
12. Describe the colligative properties of solutions containing a dissolved nonelectrolyte.
13. Perform simple calculations involving colligative properties.
14. State how the treatment of colligative properties of a solution is changed when the dissolved material is an electrolyte.
15. Give examples of the following types of reactions of water: acid-base (both Bronsted-Lowry and Lewis) and oxidation-reduction.

PROBLEM SOLVING STRATEGIES

12.3 Intermolecular Forces

We know that forces **between** molecules exist in nature, since all matter is not gaseous. These **intermolecular** forces are strong enough at low temperatures to overcome the random chaotic motion of individual gas molecules. Forces **between** molecules are electrostatic in nature, just like the forces that hold the individual atoms together **in** a molecule. Beginning students in chemistry often have difficulty in distinguishing between **inter-**

and **intramolecular forces,** so try to keep these terms clear. Realize that, in general, intramolecular forces (the forces between bonded atoms within a molecule) are stronger than the intermolecular forces. Due to the different strengths of intermolecular forces in various systems, different substances have different melting points and boiling points. The greater the intermolecular forces, the greater are the melting points and boiling points. Polar molecules in general have higher melting points and boiling points than nonpolar molecules. See Chapter 9 in this study guide to review how to determine if a molecule is polar or not. Nonpolar molecules have intermolecular forces that are of the induced dipole-induced dipole type. This type of force exists between **all** molecules, since all molecules have electrons with distortable electron clouds. However, polar molecules also have dipole-dipole intermolecular forces. So, for two different molecules, if one is polar and the other is nonpolar, we can assume, all other factors being equal, that a sample containing the polar molecule will have the higher melting point and boiling point.

Example 12.1

Which has the higher melting point, $SnCl_4$ or $SnCl_2$? Why?

Solution 12.1

The key to determining the polar or nonpolar character of a molecule is a knowledge of the Lewis structure. Tin is in group 4 and has 4 valence electrons. Each chlorine has 7 valence electrons. For $SnCl_4$ we have

```
           ..
          :Cl:
      ..   |   ..
     :Cl--Sn--Cl:
      ..   |   ..
          :Cl:
           ..
```

This is an example of an AX_4 molecule with a tetrahedral geometry. The vectorial addition of the 4 **bond dipole moments,** each pointing to chlorine from tin, results in a zero molecular dipole moment. $SnCl_4$ is nonpolar. For $SnCl_2$ the Lewis structure is

```
      ..  ..
      Sn--Cl:
      |   ..
     :Cl:
      ..
```

This is an AX_2E molecule, and is bent. The 2 **bond dipole moments** yield a nonzero molecular dipole moment. $SnCl_2$ is polar. We predict $SnCl_4$ to have a lower melting point than $SnCl_2$. The actual data are: $SnCl_4$, -33°C; $SnCl_2$, 246°C. The data agrees with our predictions.

When comparing two molecules that are **both** nonpolar, we need to determine which of the two has a greater polarizability, or electron cloud deformability, since the larger the polarizability, the greater the intermolecular forces are, and hence the higher the melting point and the boiling point. Polarizability increases with increasing molecular size -- actually with increasing number of electrons.

Example 12.2

Which molecule has the higher boiling point, H_2 or Cl_2? Why?

Solution 12.2

Both H_2 and Cl_2 are nonpolar molecules. There are more electrons in Cl_2 than in H_2. Cl_2's electron cloud is more spread out than H_2's. Cl_2 is larger than H_2 (the covalent radii are given in Figure 4.9 of the textbook). Thus Cl_2 has the greater polarizability and greater induced dipole-induced dipole intermolecular forces. We predict Cl_2 to boil higher than H_2. Note that both of these boiling points are below room temperature, since you know both H_2 and Cl_2 are gases at room temperature. The boiling point data are: H_2, -252.5°C; Cl_2, -34.6°C.

12.7 Colligative Properties

When a solid solute with a negligible vapor pressure is added to a liquid solvent, forming a solution, several properties of that **solution** differ from those of the pure **solvent.** These include a vapor pressure lowering, a boiling point elevation, a freezing point depression, and the ability to exhibit an osmotic pressure. All of these properties depend on **how many** solute particles are present, and not on the detailed nature of **what** the solute is. Since the vapor pressure of the solution is **lowered,** the **solution** will require a higher temperature for boiling to take place, compared to the pure **solvent.** See Figure 12.23 in the textbook. Thus the boiling point of the solution is greater than that for the pure solvent by an amount ΔT, which only depends on how much nonvolatile solute is present in the solution. The relation is $\Delta T = K_b m$, where K_b is a constant characteristic of the **solvent,** and the molality m, a concentration unit that is independent of temperature, equals moles of solute in 1.000 kg solvent. Likewise, the **freezing point** of the **solution** is lower than that for the pure solvent by an amount $\Delta T = K_f m$. Again, K_f is a constant characteristic of the solvent. Since $K_f \neq K_b$, a given solution has a boiling point elevation different from its freezing point depression. You should note that when **electrolytes** are added to the solvent water, larger boiling point elevations and freezing point depressions are observed than for nonelectrolyte solutions of the same concentration. See Example 12.12 in the textbook.

Example 12.3

What is the boiling point of a solution formed if 10.00 g solid naphthalene, $C_{10}H_8$, is dissolved in 250.0 g liquid benzene? Pure benzene has a normal boiling point of 80.08°C. $K_b = 2.53°C\ kg\ mol^{-1}$ for benzene.

Solution 12.3

Since $\Delta T = K_b m$, $T_{boiling,solution} = 80.08°C + \Delta T$.

The molality determination requires us to calculate the number of moles of solute in 1.000 kg solvent.

Thus

$$m = \frac{1}{250\ g} \times \frac{10\ g\ solute}{128\ g\ mol^{-1}} \times \frac{250\ g}{0.250\ kg\ solvent} = 0.312\ mol\ solute\ (kg\ solvent)^{-1}$$

Note the use of the unit conversion factor $\frac{250\ g}{0.250\ kg} = 1.$

Then $\Delta T = K_b m = 2.53°C\ kg\ mol^{-1} \times 0.312\ mol\ kg^{-1} = 0.79°C.$

and $T_{boiling,solution} = 80.08 + 0.79 = 80.87°C.$

The same type of calculation is used for the freezing point depression. Note carefully that since K_f and m are both positive, ΔT is always positive, and refers to the freezing point of the **solvent** minus that of the **solution.** In other words, for boiling point elevation, $\Delta T = T_{boiling,\ solution} - T_{boiling,\ solvent}$, while for freezing point depression, $\Delta T = T_{freezing,\ solvent} - T_{freezing,\ solution}$.

The last colligative property, osmotic pressure, is a very sensitive tool for measuring molar masses of large molecules. Like the other colligative properties, osmotic pressure only depends on the **number** of solute particles present. From the expression $\pi V = nRT$ relating the osmotic pressure and the number of moles of solute, n, dissolved in the volume V of **solution,** one can get the molar mass of the solute since

$$n = \text{moles solute} = \frac{g\ solute}{g\ mol^{-1}}$$

Examples 12.10 amd 12.11 in the text illustrate molar mass calculations from osmotic pressure measurements.

Example 12 4

If 6.00 g of urea, $(NH_2)_2CO$, is dissolved in water to make 1000 mL solution, what osmotic pressure, in atm, will be developed at 27°C?

Solution 12.4

Since $\pi V = nRT$, to get π in atm we must use $R = 0.0821\ L\ atm\ mol^{-1}\ K^{-1}$ and V in L. Don't forget 27°C = 300 K. You can calculate the molar mass of urea to be 60.60 g mol^{-1} from its formula. Then

$$\pi = \frac{\left(\frac{6.00\ g}{60.06\ g\ mol^{-1}}\right) \times 0.0821\ L\ atm\ mol^{-1}\ K^{-1} \times 300\ K}{1000\ mL \times \frac{1\ L}{10^3\ mL}}$$

$$\pi = 2.46\ atm$$

SELF TEST

Part I True or False

1. H_2S has a higher boiling point than H_2O.
2. Helium can exist only in the gaseous state, no matter what the temperature is.
3. Hydrogen bonding is very important for alkanes.
4. The boiling point elevation constant, K_b, for water is smaller than the freezing point depression constant, K_f, for water.
5. The average kinetic energy for water molecules at 20°C is larger than at 10°C.
6. Ion-ion interactions are independent of distance.
7. Water acts as a Lewis base in its reaction with sulfur dioxide.
8. Even though the normal boiling point of water is 100°C, water can boil over a rather wide range of temperatures.
9. Addition of 1 mole of sodium sulfate will depress the freezing point of 1.000 kg of water twice as much as the addition of 1 mole of sodium chloride.

Part II Multiple Choice

10. Which one of the following characteristics would you expect in a good solvent for NaCl?
(a) polar (b) high boiling point (c) low molecular weight
(d) high ionization potential (e) low polarizability
11. Which **one** of the following statements about London forces is false?
(a) are relatively weak (b) increase with increasing molecular size
(c) are electrostic in nature (d) are operative only in the solid phase
(e) involve induced dipole interactions
12. Which one of the following would you expect to have the lowest melting point?
(a) NaCl (b) Cu (c) H_2 (d) SiO_2 (e) Cl_2
13. A solid composed of nonpolar molecules is held together primarily by
(a) London forces (b) ionic bonds (d) metallic bonds
(d) covalent bonds (e) none of these
14. The phase change from solid to vapor is called
(a) sublimation (b) vaporization (c) crystallization
(d) fusion (e) none of these
15. The interactions between C_6H_6 (benzene) and $CHCl_3$ (chloroform) molecules are called
(a) dipole-dipole (b) ion-dipole (c) dipole-induced dipole
(d) ion-induced dipole (e) van der Waals
16. The reaction of hydride ion, H^-, with water is an example of which one or more of the following types?
(a) precipitation (b) acid-base (c) oxidation-reduction
(d) addition (e) elimination
17. The product of the reaction of P_4O_{10} with water is
(a) H_3PO_3 (b) H_2PO_2 (c) H_2PO_3 (d) H_3PO_4 (e) PH_3

18. The enthalpy charge for an ionic solid like NaCl dissolving in water has contributions from which of the following?
(a) energy needed to pull ions apart
(b) energy needed to form the ion Na^+ from Na(s)
(c) energy given off when the ions are hydrated
(d) energy needed to pull water molecules apart
(e) energy needed to form H^+ and OH^- from H_2O

19. Propylene glycol, $CH_2OH\text{-}CHOH\text{-}CH_2OH$, is used as an antifreeze for automobile radiators (that is, it depresses the freezing point of the water in the radiator). What is the freezing point of a 10% (by weight) solution of propylene glycol in water? The molar mass of propylene glycol is 92.0 g mol^{-1}. K_f for water is 1.86°C kg mol^{-1}
(a) +2.0°C b) +0.2°C c) -0.2°C d) -2.0°C e) -2.2°C

20. One tenth of a mole of sodium phosphate, Na_3PO_4, is dissolved in 500 g of water (K_f = 1.86°C kg mol^{-1}). The freezing point of the resulting solution will be close to
(a) -0.19°C (b) -0.37°C (c) -0.74°C (d) -1.12°C (e) -1.49°C

Answers to Self Test

1. F; 2. F; 3. F; 4. T; 5. T; 6. F; 7. T; 8. T; 9. F; 10. a; 11. d; 12. c; 13. a; 14. a; 15. c,e; 16. b,c; 17. d; 18. a,c,d; 19. e; 20. e.

CHAPTER 13

CHEMICAL EQUILIBRIUM

SUMMARY REVIEW

For any chemical reaction, the **forward reaction** (reactants → products) is accompanied by a **reverse reaction** (products → reactants). All reactions in closed systems eventually reach a state of **dynamic equilibrium** in which there is no change in any of the concentrations of reactants or products. At equilibrium, the forward reaction and the reverse reaction still proceed, but **at the same rate.**

When an equilibrium greatly favors the products, the reaction is said to go to completion, and this is indicated by connecting the reactants and products in the equation for the reaction by a **single arrow.** More generally, when significant amounts of the products and the reactants are present at equilibrium, and reactants and products are joined by a **double arrow.**

For any reaction in a **homogeneous system** in a state of equilibrium at a given temperature, e.g.,

$$aA + bB + cC + \ldots. \rightleftharpoons pP + qQ + rR + \ldots.$$

it is found experimentally that the expression

$$\left(\frac{[P]^p[Q]^q[R]^r\ldots.}{[A]^a[B]^b[C]^c\ldots.}\right)_{eq}$$

where [P], [Q], [R], ..., and [A], [B], [C], ... are the equilibrium **molar** concentrations of products and reactants, respectively, and p, q, r, ... and a, b, c, ... are the number of moles of each in the balanced equation **as written,** is a **constant,** called the **equilibrium constant,** K_{eq}, for the reaction at the given temperature.

A large value for K_{eq} signifies that the position of equilibrium favors products over reactants; in other words, that the reaction goes essentially to completion. A small value for K_{eq} indicates that the reaction **as written** proceeds in the forward direction to a limited extent. The magnitude of K_{eq} gives, however, no information about the rate (speed) at which the equilibrium is achieved, or how long it takes to achieve equilibrium.

For reactions in the gas phase, the equilibrium constant expression is often written in terms of **partial pressures** of reactants and products, rather than molar concentrations.

$$K_p = \left(\frac{p_P^{\,p}\cdot p_Q^{\,q}\cdot p_R^{\,r}\ \ldots.}{p_A^{\,a}\cdot p_B^{\,b}\cdot p_C^{\,c}\ \ldots.}\right)_{eq}$$

Because the molar concentration of any reactant or product, [X], is related to its partial pressure, p_X, by the expression

$$p_X V = n_X RT; \; p_X = \frac{n_X}{V} RT = [X]RT \quad \text{or} \quad [X] = \frac{p_X}{RT}$$

The K_{eq} for a gaseous reaction is related to its K_p by the expression

$$K_p = K_c (RT)^{\Delta n}$$

where K_c is the value of K_{eq} with the concentrations in molar units and Δn = (p + q + r + ...) - (a + b + c + ...), the difference between the number of moles of gaseous products and the number of moles of gaseous reactants in the balanced equation.

If we multiply the coefficients in a balanced equation by a factor **n**, then the equilibrium constant of the original equilibrium is raised to the power n to obtain the value of the new equilibrium constant. When the equilibrium constant is needed for a reaction that is the sum of two or more reactions, with individual equilibrium constants K_1, K_2, etc., it can be obtained from the expression

$$K_{overall} = K_1 K_2 \; ....$$

When a reaction (reactants → products) is reversed, (products → reactants),

$$K_{reverse} = \frac{1}{K_{forward}}$$

For any reaction not at equilibrium, we can write

$$\frac{[P]^p [Q]^q [R]^r \; ....}{[A]^a [B]^b [C]^c \; ....} = Q$$

where the concentrations are the non-equilibrium concentrations and **Q** is called the **reaction quotient.**

When $Q < K_{eq}$, the reaction proceeds with an increase in the concentrations of the products until equilibrium is established; if $Q > K_{eq}$, the reaction proceeds with an increase in the concentrations of the reactants until equilibrium is established, and when $Q = K_{eq}$, the system is at equilibrium.

In calculating equilibrium concentrations, or equilibrium constants, the **first step** is to write the balanced equation for the reaction; the **second step** is to write the equilibrium constant expression from the balanced equation; the **third step** is to write expressions for the concentrations present at equilibrium, and the **fourth step** is to substitute these concentrations in the expression for K_{eq} and solve for any unknown.

All of the expressions discussed so far refer to equilibria in **homogeneous systems**, that is, systems where all of the reactants and products are in the same phase (usually gas or liquid). For **heterogeneous systems**, where the equilibria involve reactants and products in different phases, for example, solids and gases, or solids and liquids and gases, the **pure** solids and **pure** liquids taking part in **heterogeneous equilibria** are **not** included in the equilibrium constant expressions.

Le Chatelier's principle states that if any of the conditions affecting an equilibrium are changed, **the equilibrium shifts to a new position of equilibrium that minimizes the change,** and is used to predict the effect of changes of concentration, temperature, or pressure on the position of equilibrium. For example, **an increase in the concentration** of any reactant or product favors whichever of the forward or reverse reaction that will **reduce** that concentration; **an increase in temperature** (addition of heat to the system) favors whichever of the forward or reverse reaction is an **endothermic** reaction, and an **increase in pressure** favors whichever of the forward or reverse reaction (if any) **decreases** the pressure by decreasing the number of gaseous molecules in the system.

REVIEW QUESTIONS

1. Two reactions are written as A + B $\rightarrow$ C + D, and A + B $\rightleftharpoons$ C + D, respectively. What is the significance of the single and double arrows?
2. Write the equilibrium constant expressions, K_c and K_p, for the reaction $N_2(g) + 3H_2(g) \rightleftharpoons 2NH_3(g)$.
3. What is meant by the term homogeneous equilibrium?
4. In general, if the value of an equilibrium constant is very large, what does it signify about the position of equilibrium? What if it is small?
5. In terms of RT, what expression relates K_p and K_c for the equilibrium $2NO(g) + Cl_2(g) \rightleftharpoons 2NOCl(g)$?
6. If the equilibrium constant for the reaction $2HCl(g) \rightleftharpoons H_2(g) + Cl_2(g)$ is K_1, and the equilibrium constant for $2ClBr(g) + H_2(g) \rightleftharpoons 2HCl(g) + Br_2(g)$ is K_2, what in terms of K_1 and K_2 is the equilibrium constant for the reaction $Br_2(g) + Cl_2(g) \rightleftharpoons 2BrCl(g)$, at the same temperature?
7. If the equilibrium constant for the reaction $I_2(g) + Cl_2(g) \rightleftharpoons 2ICl(g)$ has the value 2.1×10^5 at 25°C, what is the value of the equilibrium constant for the reaction $2ICl(g) \rightleftharpoons I_2(g) + Cl_2(g)$, at 25°C?
8. Write the balanced equation for the reaction of NO(g) with $O_2(g)$ to give two moles of $NO_2(g)$, and the equilibrium constant expression for this reaction. Suppose that initially a reaction vessel contains 0.10 mol L^{-1} NO(g), 0.10 mol L^{-1} $O_2(g)$ and 0.10 mol L^{-1} $NO_2(g)$ at 1000 K. Is the system at equilibrium, given that K_c = 1.20 L mol^{-1} for the reaction at 1000 K?
9. In question 8, when equilibrium is achieved at 1000 K, is the equilibrium concentration of $NO_2(g)$ greater than or smaller than 0.10 mol L^{-1}?
10. In question 8, if the equilibrium concentration of $O_2(g)$ is (0.1 + x) mol L^{-1}, what are the equilibrium concentrations of NO(g) and $NO_2(g)$, in terms of their initial concentrations and x?
11. The equilibrium constant for the reaction $2HI(g) \rightleftharpoons H_2(g) + I_2(g)$ has the value 1.84×10^{-2} at 698 K. What are the equilibrium concentrations

of HI, H_2, and I_2, if 0.2000 mol HI(g) are placed in a 1.00 L vessel and equilibrium established at 698 K?

12. What is the percentage dissociation of the HI(g) at equilibrium in question 11?
13. Write the equilibrium constant expressions for the reactions
    (a) $CaCO_3(s) \rightleftharpoons CaO(s) + CO_2(g)$, and
    (b) $CuSO_4 \cdot 5H_2O(s) \rightleftharpoons CuSO_4(s) + 5H_2O(g)$.
14. State Le Chatelier's principle.
15. For the equilibrium
    $3H_2(g) + N_2(g) \rightleftharpoons 2NH_3(g) \quad \Delta H° = -92.4$ kJ
    predict the effect on the equilibrium concentration of $NH_3(g)$ of,
    (a) increasing the concentration of $H_2(g)$;
    (b) increasing the pressure;
    (c) increasing the volume of the containing vessel;
    (d) increasing the temperature; and
    (e) adding an inert gas, such as He, to the equilibrium mixture.

Answers to Selected Review Questions

5. $K_p = K_c/RT$. 6. $(K_1K_2)^{-1}$. 7. 4.8×10^{-6}. 8. $2NO(g) + O_2(g) \rightleftharpoons 2NO_2(g)$; $K_c = \frac{[NO_2]^2}{[NO]^2[O_2]}$; $Q = 10$ mol^{-1} L $> K_c = 1.2$ mol^{-1}L; not at equilibrium. 9. $[NO_2] < 0.10$ mol L^{-1}. 10. $[NO] = (0.1 + 2x)$ mol L^{-1}; $[NO_2] = (0.1-2x)$ mol L^{-1}. 11. $[HI] = 0.110$; $[H_2] = [I_2] = 0.045$ mol L^{-1}. 12. 22.5% 13. (a) $K_c = [CO_2]$; (b) $K_c = [H_2O]^5$. 15. (a) increase; (b) increase; (c) decrease; (d) decrease; (e) no effect.

OBJECTIVES

Be able to:

1. Write the form of the equilibrium constant K(or K_c), given a balanced chemical equation.
2. Write the form of the equilibrium constant K_p, given a balanced chemical equation.
3. Convert K_p to K_c and vice versa for a gaseous equilibrium, given a balanced chemical equation.
4. Relate the effect of multiplying a balanced chemical equation by a constant on the numerical value of the equilibrium constant.
5. Calculate the numerical value of the equilibrium constant for an overall reaction from a knowledge of the equilibrium constant for the individual reactions that yield the overall reaction.
6. Relate the magnitude of the equilibrium constant to the approximate position of equilibrium.
7. Calculate the reaction quotient Q from product and reactant concentrations, and determine if the system is at equilibrium.
8. Use the equilibrium constant and initial concentrations of reagents to calculate equilibrium concentrations.
9. Define the terms homogeneous equilibrium and heterogeneous equilibrium.
10. Write the equilibrium constant expression for a system involving heterogeneous equilibrium.

11. Restate Le Chatelier's principle in terms of concentration changes, and use the principle to predict how concentration changes affect the position of equilibrium.
12. State how pressure changes on a gas phase equilibrium affect the position of equilibrium.
13. Appreciate that the equilibrium constant for a chemical reaction depends on the temperature.
14. State how changing the temperature affects the position of chemical equilibrium, knowing whether the reaction is exothermic or endothermic.

PROBLEM SOLVING STRATEGIES

13.1 The Equilibrium Constant

There is a wide range of examples of calculations associated with chemical equilibrium, but there are only a few basic principles involved. A system at equilibrium can be described by an equilibrium constant K, which is a constant at a given temperature for a **given balanced equation.** Even with a balanced chemical equation and a value for K given, there are in general an infinite number of possibilities for reactant and product equilibrium concentrations. For example, in Table 13.3 in the textbook you will see 5 different sets of **equilibrium** concentrations that all yield the same K. Carefully contrast **initial** concentrations and **equilibrium** concentrations. It is only the latter which describe an equilibrium situation, and only they can be used to calculate the equilibrium constant. Also realize that the stoichiometry of a chemical reaction dictates how much product is formed from a given change in a reactant concentration. It is convenient to attack problems involving **equilibrium** concentrations or numbers of moles from **initial** concentrations or numbers of moles by constructing an array with entries for the **initial conditions,** for the **change,** and for the **equilibrium situation**.

Example 13.1

At 500°C, H_2, N_2, and NH_3 gases are mixed and allowed to come to equilibrium. If the equilibrium concentrations, in moles L^{-1}, are 1.35, 1.15 and 0.412, respectively, calculate the equilibrium constant, with H_2 and N_2 as reactants and NH_3 as product.

Solution 13.1

We do **not** know what the balanced equation is so we **cannot** give a unique answer to this question. If the balanced equation is

$$3H_2(g) + N_2(g) \rightleftharpoons 2NH_3(g)$$

then

$$K_c = \frac{[NH_3]^2}{[H_2]^3[N_2]} = \frac{(0.412)^2 \text{ mol}^2 \text{ L}^{-2}}{(1.35)^3 \text{ mol}^3 \text{ L}^{-3} \ (1.15) \text{ mol L}^{-1}} = 6.00 \times 10^{-2} \text{ L}^2 \text{ mol}^{-2}$$

However, if we write

$$3/2\ H_2(g) + 1/2N_2(g) \rightleftharpoons NH_3(g)$$

then

$$K_c = \frac{[NH_3]}{[H_2]^{3/2}\ [N_2]^{1/2}} = \frac{(0.412)\ mol\ L^{-1}}{(1.35)^{3/2}\ mol^{3/2}\ L^{-3/2}\ (1.15)^{1/2}\ mol^{1/2}\ L^{-1/2}}$$

$$= 2.45 \times 10^{-1}\ L\ mol^{-1}$$

The data given in the problem for equilibrium concentrations are independent of how we wish to balance the equation, but the value of K depends on which balanced equation we use. In this example, the first K expression is the **square** of the second K expression, and the first K value is equal to the **square** of the second equilibrium constant value.

Example 13.2

At 1000 K molecular iodine decomposes partially to iodine atoms, with $K_c = 3.76 \times 10^{-5}$ mol L^{-1} describing $I_2(g) \rightleftharpoons 2I(g)$. What are the equilibrium concentrations of I_2 and I, if at the start 1.00 mole of I_2 is placed in a 2.00 liter vessel?

Solution 13.2

Since the balanced equation is given, as well as its K, we know

$$K = \frac{[I(g)]^2}{[I_2(g)]} = 3.76 \times 10^{-5}\ mol\ L^{-1}$$

Also, since the data given are initial **moles**, it is easiest to set up a **moles** array, and convert to mol L^{-1} at the end. Thus we write

number of moles	$I_2(g)$	I(g)
initial	1.00	0

Before we consider the decomposition, we have only $I_2(g)$. When we let a change take place (in solving the problem), we see from the balanced equation that for every molecule of $I_2(g)$ that breaks up, 2 atoms of I(g) **must** be formed. If we say that the **change** in the number of moles (compared to the initial number of moles) is -y for I_2, then 2y moles of I(g) must be produced. Now the second line in the array looks like

number of moles	$I_2(g)$	I(g)
change	-y	2y

Note that the entries for I_2 and I are of opposite sign. I is formed from I_2, so if the change, due to reaction, for I is positive, it must be negative for I_2. The **equilibrium** number of moles for each reactant and product is just the sum of the first 2 entries in the array.

number of moles	$I_2(g)$	I(g)
equilibrium	1.00-y	2y

The equilibrium **concentrations** needed for the K expression are obtained from the relationship

$$\text{concentration} = \frac{\text{moles}}{\text{volume in liters}}$$

Thus

$$[I_2]_{equilib} = \frac{(1.00 - y)\text{ moles}}{2.00\text{ L}} = (\frac{1.00 - y}{2})\text{ mol L}^{-1}$$

$$[I_{(g)}]_{equilib} = \frac{2y\text{ moles}}{2.00\text{ L}} = y\text{ mol L}^{-1}$$

$$K = 3.76 \times 10^{-5}\text{ mol L}^{-1}$$

$$= \frac{y^2}{(\frac{1.00 - y}{2})}$$

Expanding we get

$$y^2 + (1.88 \times 10^{-5})\,y - 1.88 \times 10^{-5} = 0$$

This is a quadratic equation, i.e., of the form $ay^2 + by + c = 0$, whose solutions are

$$y = \frac{-b \pm \sqrt{b^2 - 4ac}}{2a}$$

Thus

$$y = \frac{-1.88 \times 10^{-5} \pm \sqrt{(1.88)^2 \times 10^{-10} + 4(1.88 \times 10^{-5})}}{2}$$

Since the **second** term under the square root is about 10^5 times larger than the **first** term, we can safely ignore this first term, leaving

$$y = (-1.88 \times 10^{-5} \pm 8.64 \times 10^{-3})\,/2$$

So we have 2 values of y to consider,

$$y = -4.33 \times 10^{-3} \quad \text{and} \quad y = +4.31 \times 10^{-3}$$

In the array above we had 2y equal the number of moles of I(g) at **equilibrium.** This must be **positive,** since it is meaningless to have less than zero moles of I(g) at equilibrium. So we reject as chemically unacceptable the negative root for y, and are left with

$$y = 4.31 \times 10^{-3}\text{ mol L}^{-1} = [I(g)]$$

and

$$[I_2] = \frac{1.00 - 0.00431}{2.00}\text{ mol L}^{-1} = 0.498\text{ mol L}^{-1}$$

If the problem had given initial **concentrations** then an array of **concentrations** could have been used. It makes no difference whether you base the calculation on moles or **concentration** of I(g). For example the same problem could have used

concentration (mol L^{-1})	$I_2(g)$	$I(g)$
initial	$\frac{1.00}{2}$	0
change	-y	2y
equilibruim	$\frac{1.00}{2}$ -y	2y

$$K = \frac{(2y)^2}{0.5 - y} = 3.76 \times 10^{-5} \text{ mol L}^{-1}$$

Then $2y = 4.31 \times 10^{-3}$ mol L^{-1} = $[I(g)]$

$[I_2] = 0.5 - 0.002 = 0.498$ mol L^{-1}, as before.

13.2 Heterogeneous Equilibria

Chemical equilibria involving **two or more phases** are called **heterogeneous** equilibria. For such equilibria, the concentrations of pure solids and pure liquids are **constant** and are incorporated into the equilibrium constant expression. Thus, equilibrium constants for heterogeneous equilibria do not include concentration terms for such substances.

Example 13.3

Ammonia and hydrogen sulfide gases combine in an acid-base reaction to form solid ammonium hydrogen sulfide, NH_4HS. However, the solid can also decompose into NH_3 and H_2S. When a 1.000 g sample of pure solid NH_4HS is placed in an evacuated 1.000 L vessel at 25°C, and equilibrium is established, a total pressure of 500 torr is measured. Calculate the numerical value of equilibrium constant K_p for the decomposition. How many grams of solid NH_4HS remain?

Solution 13.3

At equilibrium,

$$NH_4HS(s) \rightleftharpoons NH_3(g) + H_2S(g)$$

is established. There is a partial pressure of 250 torr due to NH_3, and 250 torr due to H_2S. In atm, these pressure are each

$$250 \text{ torr} \times \frac{1 \text{ atm}}{760 \text{ torr}} = 0.329 \text{ atm}$$

The equilibrium constant K_p is

$$K_p = p_{NH_3} \times p_{H_2S} = 0.329 \times 0.329 = 0.108 \text{ atm}^2$$

Note that NH_4HS does not appear in the equilibrium constant expression.

You may think that the second part of the problem requires you to convert K_p to K_c, and then solve for equilibrium concentrations of NH_3 and H_2S, convert to equilibrium moles of NH_3 and H_2S, convert to grams of NH_3 and H_2S, and finally obtain grams of NH_4HS. That method can certainly be followed, but there is an easier and shorter way.

Let's assume that NH_3 and H_2S are ideal gases. Then

$$n_{NH_3} = \frac{p_{NH_3}V}{RT} = \frac{0.329 \text{ atm} \times 1.000 \text{ L}}{0.0821 \text{ L atm mol}^{-1}\text{K}^{-1} \times 298 \text{ K}}$$

$$= 0.013 \text{ mol} = n_{H_2S}$$

$$\text{mass } NH_3 = 0.0134 \text{ mol } NH_3 \times \frac{17.0 \text{ g}}{1 \text{ mol}} = 0.228 \text{ g}$$

$$\text{mass } H_2S = 0.0134 \text{ mol } H_2S \times \frac{34.1 \text{ g}}{1 \text{ mol}} = 0.457 \text{ g}$$

$$\text{mass } NH_4HS = 1.000 - (0.228 + 0.457) = 0.315 \text{ g}$$

So you see that the partial pressures of NH_3 and H_2S can be used directly in this problem, without working through a formal equilibrium calculation.

13.3 The Effect of Changing Conditions on an Equilibrium

Le Chatelier's Principle

We can perturb a system at equilibrium so that, for example, the equilibrium concentrations of the products are larger than they were before we disturbed the system. **Yet the same equilibrium constant expression and the same value of K remain valid.** Thus, for example, we can increase the external pressure on an equilibrium situation (by placing the reagents in a vessel fitted with a piston), or we can add some more of one of the reactants to a system at equilibrium. The system adjusts so as to **minimize** the effect of the disturbance. But the new equilibrium state is **different** from the old one, even though K is constant. If we change the temperature, we change K, of course, and the position of the equilibrium changes accordingly.

Example 13.4

Predict the effect of i) an increase in pressure, ii) an increase in temperature, and iii) addition of hydrogen on the position of equilibrium in each of the following. Use thermodynamic data from the Appendix in the textbook as required.

a) $2H_2(g) + O_2(g) \rightleftharpoons 2H_2O(g)$

b) $C(\text{graphite}) + H_2O(g) \rightleftharpoons CO(g) + H_2(g)$

Solution 13.4

a) $2H_2(g) + O_2(g) \rightleftharpoons 2H_2O(g)$

i) There are two moles of gaseous products and three moles of gaseous reactants. An increase in pressure will shift the equilibrium to the side of **fewer** gaseous moles. Thus, the equilibrium will shift to the **right** and more H_2O will be produced.

ii) We must first calculate whether the reaction is endothermic or exothermic. From the Appendix we find

$$\Delta H° = 2\Delta H_f°(H_2O,g) - 2\Delta H_f°(H_2,g) - \Delta H_f°(O_2,g)$$
$$= 2(-241.8)kJ - 2(0) - 0$$
$$= -483.6 \text{ kJ}$$

So the reaction is exothermic. Heat is a **product** of the reaction. Thus, an increase in temperature will shift the equilibrium to the **left**.

iii) Adding the **reactant** H_2 will produce more H_2O.

b) $C(graphite) + H_2O(g) \rightleftharpoons CO(g) + H_2(g)$

i) Note that the reactant C(graphite) is a **solid**. Thus, there are more moles of gaseous products than reactants in the balanced equation, and an increase in pressure will shift the equilibrium to the **left**.

ii) From data in the Appendix we can calculate

$$\Delta H° = \Delta H_f°(CO,g) + \Delta H_f°(H_2,g) - \Delta H_f°(C,graphite) - \Delta H_f°(H_2O,g)$$
$$\Delta H° = -110.5 \text{ kJ} + 0 - 0 - (-241.8) \text{ kJ} = +131.3 \text{ kJ}$$

This reaction is endothermic. Heat can be considered to be a **reactant** here. An increase in temperature minimizes this disturbance by forming more **products**.

iii) Adding the **product** H_2 will result in the formation of more reactants.

SELF TEST

Part I True or False

Which of the following statements are true concerning the reaction?
$Ca(s) + S(s) + 2\,O_2(g) \rightleftharpoons CaSO_4(s) \quad \Delta H < 0$

1. The value of the equilibrium constant K_c is independent of temperature.
2. The units of K_c are $mol^{-2}\,L^2$.
3. The equilibrium shifts to the right if the temperature is increased.
4. The equilibrium shifts to the right if the volume is decreased.
5. The value of K_c can be increased by the correct choice of catalyst.

Part II Multiple Choice

6. For the equilibrium $A(g) + B(g) \rightleftharpoons C(g)$ the equilibrium constant is 50 L mol^{-1}. What is the equilibrium constant, in mol L^{-1}, for $C(g) \rightleftharpoons A(g) + B(g)$?
(a) 50 (b) 2 (c) 0.5 (d) 0.2 (e) 0.02
7. In the equilibrium $2SO_2(g) + O_2(g) \rightleftharpoons 2SO_3(g)$ the initial concentrations of the reactants before the reaction started were 0.1 M SO_2 and 0.02 M O_2. There was no SO_3 present initially. At equilibrium, the concentration of the SO_3 produced is (2x) M. The equilibrium concentration of SO_2 and O_2 in mol L^{-1} are, respectively:

(a) (0.1 - 2x) and (0.02 - x) (b) (0.1 - x) and (0.02 - 2x)
(c) (0.2 - 2x) and (0.04 - x) (d) (0.2 - x) and (0.04 - 2x)
(e) (0.1 - x) and (0.02 - x)

8. What is the value of the reaction quotient for the reaction

$$N_2(g) + 3H_2(g) \rightleftharpoons 2NH_3(g)$$

when $[N_2] = [H_2] = [NH_3] = 0.20$ M?
(a) 0.04 (b) 0.2 (c) 1 (d) 5 (e) 25

9. Initially, 0.84 mole of PCl_5 and 0.18 mole of PCl_3 are mixed in a 1.0-liter vessel. It is later found that 0.72 mole PCl_5 is present at equilibrium. What is K for the reaction $PCl_5(g) \rightleftharpoons PCl_3(g) + Cl_2(g)$ at this temperature?
(a) 0.0010 mol L^{-1} (b) 0.015 mol L^{-1} (c) 0.030 mol L^{-1}
(d) 0.050 mol L^{-1} (e) 0.150 mol L^{-1}

10. For the reaction $2S(g) + 3O_2(g) \rightleftharpoons 2SO_3(g)$ the equilibrium constant is
(a) $\frac{[SO_3]}{[S][O_2]}$ b) $\frac{[S]^2[O_2]^3}{[SO_3]^2}$ c) $\frac{[S][O_2]}{[SO_3]}$ d) $\frac{[SO_3]}{[O_2]}$ e) $\frac{[SO_3]^2}{[O_2]^3[S]^2}$

11. C(graphite) + $H_2O(g) \rightleftharpoons CO(g) + H_2(g)$. The production of CO and H_2 is endothermic. To shift the equilibrium towards CO and H_2
(a) increase pressure (b) decrease temperature (c) add water vapor
(d) add a catalyst (e) decrease volume of the reaction vessel

12. The value of K_p for the reaction

$$2H_2(g) + O_2(g) \rightleftharpoons 2H_2O(l)$$

is
a) $K_c(RT)^3$ b) $K_c(RT)^2$ c) $K_c(RT)$ d) $K_c(RT)^{-2}$ e) $K_c(RT)^{-3}$

13. Consider the following information:

$S(s) + O_2(g) \rightleftharpoons SO_2(g)$; K_1 is the equilibrium constant
$2SO_2(g) + O_2(g) \rightleftharpoons 2SO_3(g)$; K_2 is the equilibrium constant

The value of K for the reaction

$$2S(s) + 3O_2(g) \rightleftharpoons 2SO_3(g)$$

is
a) K_1K_2 b) $2K_1K_2$ c) $K_2/(K_1)^2$ d) $2K_1 + K_2$ e) $(K_1)^2K_2$

14. At the instant of mixing certain concentrations of chemicals A, B, C, and D together, the value of the reaction quotient Q for the process

$$A + B \rightleftharpoons C + D$$

is greater than the value of the equilibrium constant K. Which of the following statements regarding the future behavior of the reaction mixture is true?
a) No further changes to the concentrations will occur since Q already exceeds K.
b) All the reactants A and B will now be converted to products, and the concentrations of A and B will fall to zero.
c) The reverse reaction will ccur faster than the forward reaction until Q decreases to K.
d) All the products C and D will be converted to A and B, and the concentrations of C and D will fall to zero.
e) None of a-d are true.

15. A certain substance can exist in two different forms, M and N; their interconversion occurs through the reaction

$$M \rightleftharpoons N$$

for which $K_c = 5.0$ at a certain temperature. At equilibrium, the percent of the substance which exists as N is

a) 5% b) 17% c) 50% d) 83% e) 95%

Answers to Self Test

1. F; 2. T; 3. F; 4. T; 5. F; 6. e; 7. a; 8. e; 9. d; 10. e; 11. c; 12. e; 13. e; 14. c; 15. d.

CHAPTER 14

ACID-BASE EQUILIBRIA

SUMMARY REVIEW

Strong acids and bases react completely with water to give $H_3O^+(aq)$ and $[OH^-]$, respectively; these reactions have large equilibrium constant values. Only HCl, HBr, HI, HNO_3, H_2SO_4, and $HClO_4$ are **strong acids**; almost all other acids are weak. For a **weak acid,** we have the equilibrium

$$HA(aq) + H_2O(l) \rightleftharpoons H_3O^+(aq) + A^-(aq)$$

for which the **acid dissociation constant** is given by

$$K_a(HA) = \frac{[H_3O^+][A^-]}{[HA]}$$

Strong bases include OH^-, O^{2-}, H^-, and NH_2^-. Common **weak bases** include ammonia, $NH_3(aq)$, and the amines. For a **weak base,** we have the equilibrium

$$B(aq) + H_2O(l) \rightleftharpoons BH^+(aq) + OH^-(aq)$$

for which the **base dissociation** constant is given by

$$K_b(B) = \frac{[BH^+][OH^-]}{[B]}$$

For a solution of a weak acid of molar concentration C_a, $[H_3O^+]$ in the solution is given by $K_a = x^2/(C_a-x)$, where $x = [H_3O^+]$, which for $x << C_a$ gives $[H_3O^+] = \sqrt{(C_aK_a)}$.

For a solution of a weak base of molar concentration C_b, $[OH^-]$ in the solution is given by $K_b = x^2/(C_b-x)$, where $x = [OH^-]$, which for $x << C_b$ gives $[OH^-] = \sqrt{(C_bK_b)}$.

Water behaves as a very weak acid and as a very weak base. In any aqueous solution, and in pure water, $[H_3O^+][OH^-] = K_w$, the **ionic product constant of water,** which has the value 1.00×10^{-14} $mol^2\ L^{-2}$ at 25°C. In any solution H_3O^+ is in equilibrium with OH^-, but for most solutions of acids, $[H_3O^+]$ from the ionization of water is negligible compared to that from the ionization of the acid, and for most solutions of bases, $[OH^-]$ from the ionization of water is negligible compared to that from the ionization of the base.

In pure water, $[H_3O^+] = [OH^-] = 10^{-7}$ mol L^{-1} at 25°C; **acidic** solutions have $[H_3O^+] > 10^{-7}$ mol L^{-1}, **neutral** solutions have $[H_3O^+] = [OH^-] = 10^{-7}$ mol L^{-1}, and **basic** solutions have $[OH^-] > 10^{-7}$ mol L^{-1}. The **pH** of a solution (no units) is given by $pH = -\log_{10}[H_3O^+]$; pure water and neutral solutions have pH = 7.00, acidic solutions have pH < 7.00, and basic solutions have pH > 7.00.

pOH is defined in an analogous way; $pOH = -\log_{10}[OH^-]$, and for **any solution** pH + pOH = 14.00. K_a and K_b values may also be expressed on a logarithmic scale: $pK_a = -\log_{10} K_a$; $pK_b = -\log_{10} K_b$.

For any acid and its conjugate base, HA and A^-, or BH^+ and B,

$$pK_a + pK_b = -\log K_w = 14.00 \text{ (at 25°C)}$$

Anions (conjugate bases) of **strong** acids have no basic properties in water and give neutral solutions. Anions of **weak** acids are weak bases and give **basic** solutions in water; $A^- + H_2O \rightleftharpoons HA + OH^-$. Anions of polyprotic acids that contain hydrogen, e.g. HSO_4^-, $H_2PO_4^-$, or HPO_4^{2-} may be **either** acids **or** bases; if $K_a > K_b$, the solution is **acidic** and if $K_b > K_a$, the solution is **basic**.

Cations (conjugate acids) of **weak** bases are weak acids and give **acidic** solutions in water; $BH^+ + H_2O \rightleftharpoons B + H_3O^+$. **Strongly hydrated** metal ions behave as weak acids; the only common hydrated ions that **do not** behave as acids are the alkali metal ions, the alkaline earth metal ions (except Be^{2+}) and Ag^+. The acidic, neutral, or basic behavior of **salts** depends on the acid-base properties of their constituent cations and anions.

Buffer solutions contain a weak acid and a salt of the acid, or a weak base and a salt of the base; their pH's are insensitive to addition of moderate amounts of acid or base. To a good approximation the pH is given by

$$pH = pK_a + \log_{10} ([\text{base}]/[\text{acid}])$$

where [acid] and [base] are the stoichiometric concentrations (or moles) of the conjugate acid and base used to create the buffer solution.

In an **acid-base titration,** the equivalence point corresponds to the stoichiometric formation of a salt. The pH at the equivalence point depends on the nature of the salt which in turn determines the choice of a suitable indicator with which to detect the equivalence point. The titration of an acid with a base (or vice-versa) can be followed by measuring the pH during the course of the titration.

An acid that can donate more than one proton to a base is a **polyprotic acid**. In pH calculations on polyprotic acids or solutions of their salts, because of the great difference in the magnitudes of successive K_a values, successive stages of ionization can usually be treated as independent of each other.

REVIEW QUESTIONS

1. If 0.01 moles of $Al(NO_3)_3$ are dissolved to give 1 L of solution, what are the concentrations of $Al^{3+}(aq)$ and $NO_3^-(aq)$?
2. What is a strong electrolyte and what is a weak electrolyte?
3. Which of the following are strong acids and which weak acids?
   $HCl(aq)$ $HF(aq)$ $HCN(aq)$ $HClO_4(aq)$ $HBr(aq)$ $HNO_3(aq)$ $H_3PO_4(aq)$.
4. Write the expression for the acid dissociation constant, K_a, of acetic acid.
5. Write the expression for the base dissociation constant, K_b, of ammonia.
6. Why is it possible to assume that the concentration of water in a dilute solution of a weak acid or base is 55.5 mol L^{-1}?

7. Write the equilibrium expression for the self-ionization of water and show how the ion product, $[H_3O^+][OH^-]$, is derived from it.
8. At 25°C, the value of the ion product, $K_w = [H_3O^+][OH^-]$, is 10^{-14} mol^2 L^{-2}. What are the concentrations of H_3O^+ and OH^- ions (a) in pure water at 25°C (b) when $[(H_3O^+)] = 10^{-2}$ mol L^{-1}?
9. What is the concentration of OH^- ions in a 1 M solution of a strong monoprotic acid?
10. What is the definition of pH?
11. Define pK_a and pK_b.
12. What is the concentration of H_3O^+ ions in solutions with the following pH's? (a) -1 (b) +1 (c) 0
13. What are the concentrations of OH^- ions in the above solutions?
14. What is the pH of the following solutions?
(a) 0.01 M $HClO_4$ (b) 0.01 M NaOH (c) pure water.
15. Is the pH of an aqueous solution of an acid greater than or less than that of pure water?
16. What is the relationship between the K_a of a weak acid, HA, and the K_b of its conjugate base, A^-?
17. Write expressions for the base dissociation constant of ammonia, NH_3, and the acid dissociation constant of the ammonium ion, NH_4^+. If $K_b(NH_3)$ has the value 1.8×10^{-5} mol L^{-1}, what is the value of $K_a(NH_4^+)$?
18. Why is a solution of sodium chloride neutral, a solution of sodium cyanide basic, and a solution of ammonium chloride acidic?
19. Why is the pH of a solution of $AlCl_3 \cdot 6H_2O$ less than 7.0?
20. What properties must an indicator have if it is to be used to determine the pH of a solution?
21. Why must the minimum possible amount of an indicator be used when it is used to determine the pH of a solution?
22. Under what conditions is the pK_a of an indicator equal to the pH of a solution?
23. Over what pH range is an indicator effective?
24. Starting with solutions of sodium hydroxide and acetic acid, how could a buffer solution be prepared?
25. Why does a solution containing equimolar amounts of ammonia and ammonium chloride behave as a buffer solution?
26. What is the ideal pK_a of an indicator that is to be used to detect the equivalence point in the titration of acetic acid with sodium hydroxide?
27. Write acid dissociation constant expressions for every stage of ionization of the triprotic acid H_3PO_4 and name each of the anions derived from phosphoric acid.
28. Why for a polyprotic acid are there such large differences between the values of its successive acid dissociation constants?

Answers to Selected Review Questions

1. 0.01, 0.03. 3. HCl, $HClO_4$, HBr, HNO_3 are the only strong ones. 8. (a) both 10^{-7} M, (b) $[OH^-] = 10^{-12}$ M. 9. 10^{-14} M. 13. 10^{-15} M, 10^{-13} M, 10^{-14} M. 17. 5.6×10^{-10}. 24. Partially neutralize the acid, so the solution contains comparable concentrations of acetic acid and sodium acetate.

OBJECTIVES

Be able to:

1. Recognize the common strong and weak acids and bases.
2. Calculate the $[H_3O^+]$ for solutions of strong or weak acids, and the $[OH^-]$ for solutions of strong or weak bases.
3. Calculate the extent of dissociation of weak acids or bases.
4. Write the expression for K_w and use K_w to calculate the $[OH^-]$ of an acid solution and the $[H_3O^+]$ of a base solution.
5. Relate the pH scale to hydrogen (hydronium) ion concentration.
6. Calculate the pH of solutions of strong acids or strong bases, and of solutions where both are present initially.
7. Calculate the pH of solutions of weak acids or weak bases.
8. State and use the relationship between K_a for a weak acid and K_b for its conjugate base, and between K_b for a weak base and K_a for its conjugate acid.
9. Recognize the acid-base properties of certain cations and anions, write balanced equations for their reactions with water, and calculate the pH of such solutions.
10. Choose a suitable indicator for an acid-base titration.
11. Determine K_a or K_b from pH measurements.
12. Calculate the pH of buffer solutions.
13. Calculate the pH for various stages in the titration of strong acids with strong bases, strong bases with strong acids, weak bases with strong acids, and weak acids with strong bases.

PROBLEM SOLVING STRATEGIES

14.1 Acid and Base Strengths

Strong Acids and Bases

Strong acids and **strong** bases are **completely dissociated** in water; their equilibrium constants are much greater than unity. If you have forgotten what the common strong acids and bases are, refer to Chapter 5 in the text. A strong acid furnishes H_3O^+ to a solution. Normally this source of H_3O^+ is much larger than the H_3O^+ from the self-ionization of water treated in section 14.2. A strong base furnishes OH^- to a solution.

Example 14.1

Calculate the $[OH^-]$ of a solution formed from 1.00 mL of 1.00×10^{-3} M $Ca(OH)_2$ when it is diluted to 100 mL.

Solution 14.1

$Ca(OH)_2$ is a strong base. $Ca(OH)_2 \rightarrow Ca^{2+}(aq) + 2OH^-(aq)$. 1.00×10^{-3} M $Ca(OH)_2$ contains $2 \times 1.00 \times 10^{-3}$ M OH^- ions. Thus in 1.00 mL,

$$\text{mol } OH^- = 1.00 \text{ mL} \times \frac{2 \times 1.00 \times 10^{-3} \text{ mol } OH^-}{1.0 \text{ L}} \times \frac{1.0 \text{ L}}{10^3 \text{ mL}}$$

$$= 2.0 \times 10^{-6} \text{ mol}$$

This number of moles is present in the final volume (100 mL) after dilution.

$$[OH^-] = \frac{2.00 \times 10^{-6} \text{ mol } OH^-}{100 \text{ mL}} \times \frac{10^3 \text{ mL}}{1 \text{ L}} = 2.00 \times 10^{-5} \text{ M}$$

Weak Acids and Bases

A weak acid or base is only **partially dissociated** in water. For example, for the weak acid HA,

$$HA + H_2O \rightleftarrows H_3O^+ + A^-$$

$K_a = \frac{[H_3O^+][A^-]}{[HA]}$ must be obeyed. Note that in a weak acid problem, there is a large amount of **undissociated** HA at equilibrium. Each HA that does transfer a proton to H_2O results in one H_3O^+ and one A^- being formed. This stoichiometry is necessary for setting up the array of concentrations or moles.

Example 14.2

What is the $[H_3O^+]$ of a 0.0150 M HNO_2 solution? $K_a = 4.5 \times 10^{-4}$ mol L^{-1}. What percentage of HNO_2 has dissociated?

Solution 14.2

First we write the equilibrium and the K_a.

$$HNO_2 + H_2O \rightleftarrows H_3O^+ + NO_2^-$$

$$K_a = \frac{[H_3O^+][NO_2^-]}{[HNO_2]} = 4.5 \times 10^{-4} \text{ mol L}^{-1}$$

Just as for gaseous equilibrium problems, we set up an array of concentrations.

concentration (mol L^{-1})	HNO_2	H_3O^+	NO_2^-
initial	0.015	0	0
change	-y	y	y
equilibruim	0.015 - y	y	y

These **equilibrium** concentrations are then entered into the K_a expression, and y is determined.

$$4.5 \times 10^{-4} = \frac{y^2}{0.015 - y}$$

We usually try to simplify the calculation whenever possible. If we assume $y << 0.015$, then we won't have to use the quadratic formula. With that assumption,

$$y^2 = (4.5 \times 10^{-4})(1.5 \times 10^{-2})$$

$$y = \pm 2.6 \times 10^{-3}$$

We **ignore the negative root,** since $[H_3O^+]$ (which equals y) must be positive. So we are left with $y = 2.6 \times 10^{-3}$. Now we can check the assumption just made. With $y = 2.6 \times 10^{-3}$, y is about 17% of 1.5×10^{-2}. We should not have neglected y compared to 0.015; follow the suggestion in the text that states if $y > 5\%$ of a quantity from which y is to be subtracted, or to which y is to be added, then the neglecting of y is not justified. So we go back to

$$4.5 \times 10^{-4} = \frac{y^2}{0.015 - y}$$

and solve for y with the quadratic formula.

$$y^2 + (4.5 \times 10^{-4})\, y - 6.75 \times 10^{-6} = 0$$

$$y = \frac{-4.5 \times 10^{-4} \pm \sqrt{(4.5 \times 10^{-4})^2 + 4(6.75 \times 10^{-6})}}{2}$$

After rejecting the negative value for y, we have

$$y = 2.4 \times 10^{-3} \text{ M} = [H_3O^+].$$

The percentage of HNO_2 dissociated equals the **equilibrium** $[H_3O^+]$ divided by the **initial** undissociated acid concentration, times 100%.

$$\% \ HNO_2 \text{ dissociated} = \frac{2.4 \times 10^{-3} \text{ M}}{0.015 \text{ M}} \times 100\% = 16\%.$$

14.2 The Self-Ionization of Water and 14.3 The pH Scale

In pure water and in aqueous solutions, the expression $K_w = [H_3O^+][OH^-]$ must be obeyed. At 25°C K_w has the value 1.0×10^{-14} mol^2 L^{-2}. If the solution is neutral, that is, if $[H_3O^+] = [OH^-]$, as is the case for pure water, each of these concentrations is 1.0×10^{-7} M. Rather than use these negative powers of 10 to express concentration, we define $p = -\log_{10}$ of some quantity, and use $pH = -\log_{10} [H_3O^+]$ and $pOH = -\log_{10} [OH^-]$. (A summary of operations with logarithms is given in Appendix A of the text. Remember a logarithm is an exponent; for example, $\log_{10} (10^{-7}) = -7$ and $-\log_{10} (10^{-7}) = -(-7) = 7$.) For a neutral solution, pH = pOH = 7. When $[H_3O^+] > [OH^-]$, pH < 7, and the solution is **acidic.** For $[H_3O^+] < [OH^-]$, pH > 7, and the solution is **basic**. The practical pH scale extends from about -1 (for $[H_3O^+] = 10$ M) to about 15 (for $[H_3O^+] = 1.0 \times 10^{-15}$ M, $[OH^-] = 10$ M).

Example 14.3

A solution has an $[OH^-]$ of 2.0 M. What is its pH? Is this an acidic or a basic solution?

Solution 14.3

We can solve the problem two ways. Either first calculate pOH from $[OH^-]$ and then use pH = 14 - pOH or calculate $[H_3O^+]$ from $K_w/[OH^-]$ and then use pH = $-\log_{10}[H_3O^+]$. By the first method, since $[OH^-]$ = 2.0 M,

$$\begin{aligned} pOH &= -\log_{10} 2.0 = -0.30 \\ pH &= 14 - pOH = 14 - (-0.30) \\ &= 14.30 \end{aligned}$$

We get the same answer by the second method. Since $[OH^-]$ = 2.0 M and $K_w = 1.0 \times 10^{-14}\ mol^2\ L^{-2}$

$$[H_3O^+] = \frac{1.0 \times 10^{-14}\ mol^2\ L^{-2}}{2.0\ mol\ L^{-1}} = 5.0 \times 10^{-15}\ M$$

$$pH = -\log_{10} (5.0 \times 10^{-15}) = 14.30$$

This is a basic solution.

By the way, in Example 14.1 above, any OH^- from the dissociation of water is completely negligible compared to the OH^- provided by the base. We could have solved for the $[H_3O^+]$ in that solution from $K_w/[OH^-]$. Also, in Example 14.2, we have made the reasonable assumption that the overwhelming source of H_3O^+ is due to the dissociation of HNO_2. We could have solved for $[OH^-]$ in that problem from $K_w/[H_3O^+]$. You should note that most times any H_3O^+ supplied by the self ionization of H_2O can be ignored in pH calculations, compared to the H_3O^+ contributed by an acid dissociation. Such a generalization is valid when the acid contributes at least 10^{-6} M H_3O^+.

14.4 Acid-Base Properties of Salts

Some, but not all, cations and anions can act as acids or bases. See Table 14.5 in the text for a condensed list. The anions, A^-, in the list are related to their conjugate weak acids by

$$A^- + H_2O \rightleftharpoons HA + OH^-$$

In other words, the conjugate base A^- of the weak acid can act as a base, and accept a proton from water, forming HA and OH^-. The equilibrium constant describing this equilibrium is

$$K_b = \frac{K_w}{K_a} = \frac{[HA][OH^-]}{[A^-]}$$

where K_a is the **acid** dissociation constant for HA. Likewise a cation, like NH_4^+, can react as a weak acid with water. This equilibrium is described by

$$NH_4^+ + H_2O \rightleftharpoons NH_3 + H_3O^+$$

$$K_a = \frac{K_w}{K_b} = \frac{[NH_3][H_3O^+]}{[NH_4^+]}$$

where $\mathbf{K_b}$ is the base dissociation constant for NH_3. There are only 2 fundamental equilibrium constants in these problems, K_w and K_a (or K_w and K_b). Not **all** cations and anions react in this manner. Cations of **strong** bases, like Na^+, K^+, do not undergo this reaction. Also, anions of **strong** acids, like Cl^-, NO_3^- do not act in this manner. Once you can identify that the system under consideration contains a cation or anion that reacts with water, then the calculations are not difficult. After writing the equilibrium and the equilibrium constant, set up the array of concentrations or moles, and then insert the equilibrium concentrations into K.

Example 14.4

Calculate the pH of 0.50 M KF. You may use data from Chapter 14 of the textbook.

Solution 14.4

The pH is not 7, since this is the salt of a **strong** base and a **weak** acid. K^+ does not react with H_2O, but F^- is the conjugate base of the weak acid HF, and acts as a weak base; thus the reaction to consider is

$$F^- + H_2O \rightleftarrows HF + OH^-$$

Note that OH^- appears on the right side; the solution is **basic.** The equilibrium constant is

$$K_b = \frac{K_w}{K_a} = \frac{[HF][OH^-]}{[F^-]}$$

K_a is given in Table 14.1 as 3.5×10^{-4} mol L^{-1}. Then

concentration (mol L^{-1})	F^-	HF	OH^-
initial	0.50	0	0
change	-y	y	y
equillibrium	0.50 - y	y	y

$$K_b = \frac{1.0 \times 10^{-14}\ mol^2\ L^{-2}}{3.5 \times 10^{-4}\ mol\ L^{-1}} = 2.86 \times 10^{-11}\ mol\ L^{-1} = \frac{y^2}{0.50 - y}$$

Assume

$$y \ll 0.50$$

Then

$$y^2 = 1.43 \times 10^{-11}$$
$$y = \pm 3.78 \times 10^{-6}$$

Reject the negative root.

$$y = 3.78 \times 10^{-6}\ M = [OH^-]$$

Note that the assumption of $y \ll 0.50$ is valid.

$$pOH = -\log_{10}[OH^-] = -\log_{10}(3.78 \times 10^{-6})$$
$$pOH = 5.42$$
$$pH = 14 - pOH = 14 - 5.42 = 8.58$$

14.6 Buffer Solutions

A buffer solution is resistant to large changes in pH upon addition of small amounts of strong acid or strong base. A buffer is composed of two parts, either a weak acid and its salt, or a weak base and its salt. There are again two equilibria to consider. Let's take the case of the weak acid HA plus the salt of the weak acid, NaA.

$$HA + H_2O \rightleftharpoons H_3O^+ + A^- \qquad \text{Equilibrium constant here} = K_a$$

$$A^- + H_2O \rightleftharpoons HA + OH^- \qquad \text{Equilibrium constant here} = K_b = \frac{K_w}{K_a}$$

These two equilibria are not independent. We can choose **either** equilibrium to calculate the pH.

Example 14.5

Calculate the pH of a solution formed from 0.10 mole of formic acid, HCOOH, and 0.10 mole potassium formate, KCOOH, dissolved to give 250 mL of aqueous solution.

$$K_a = 1.8 \times 10^{-4} \text{ mol L}^{-1} \text{ for HCOOH}$$

Solution 14.5

One equilibrium is

$$HCOOH + H_2O \rightleftharpoons H_3O^+ + HCOO^-$$

$$K_a = \frac{[H_3O^+][HCOO^-]}{[HCOOH]} = 1.8 \times 10^{-4} \text{ mol L}^{-1}$$

In this example, let's convert moles to concentrations before setting up the array. Whether you choose to use a **moles** array or a **concentration** array is up to you, but at some point in the calculation we need expressions for the **concentrations** of all species that appear in the equilibrium constant expression.

$$\frac{0.10 \text{ mole}}{250 \text{ mL}} \times \frac{10^3 \text{ mL}}{1 \text{ L}} = 0.40 \text{ mol L}^{-1}$$

for both HCOOH and K^+HCOO^-.

concentration (mol L^{-1})	HCOOH	H_3O^+	$HCOO^-$
initial	0.40	0	0.40
change	-y	y	y
equilibrium	0.40 - y	y	0.40 + y

Note that there **must** be some H_3O^+ present at equilibrium, and we are ignoring any produced by the ionization of water. Thus the H_3O^+ must come from HCOOH, and the $[HCOO^-]$ must increase.

$$K_a = 1.8 \times 10^{-4} = \frac{(y)\ (0.40 + y)}{(0.40 - y)}$$

Assume

$$y << 0.40$$

Then

$$1.8 \times 10^{-4} = \frac{(y)\ (0.4)}{(0.4)} = y = [H_3O^+]$$

$$pH = -\log_{10}(1.8 \times 10^{-4}) = 3.74$$

If you wish, you can **memorize** equations that help in the solution of buffer problems, like the Henderson-Hasselbalch equation. However, no new insight into problem solving is provided by doing this.

14.7 Acid-Base Titrations

Titrations of a Weak Base with a Strong Acid

The situations encountered when a **weak acid** is titrated with a **strong base** are discussed in Section 14.7 of the text. Note carefully that four different types of pH calculations are required in such cases, depending on whether no base has been added, or the weak acid is in excess, or the equivalence point has been reached, or the strong base is in excess.

Here we treat the case of a **weak base** such as $NH_3(aq)$ titrated with a **strong acid** like HCl(aq). Again recognize there are four types of problems, all of which have been treated earlier in the chapter.

1. Initially the **solution is a weak base in water.** Knowing the value of K_b for the base NH_3 and its concentration, the pH of the solution can be calculated. A typical problem is solved in Example 14.8 of the text.
2. After **some** strong base has been added, but not enough to reach the equivalence point, the solution contains both the weak base NH_3 and the salt of the weak base and strong acid, NH_4Cl. The **solution is a buffer solution.** Once the concentrations of base and salt have been determined, you can apply the procedure given in Example 14.16 in the text. Note that at the **half-equivalence point**, we have a buffer solution containing **equal** concentrations of NH_3 and NH_4^+ and pH = pK_a of NH_4^+.
3. At the equivalence point the **solution contains the salt NH_4Cl.** As you have seen in Section 14.4 of the text, NH_4^+ reacts with water to form an acid solution. Example 14.12 in the text treats this case.
4. Beyond the equivalence point **excess acid is present.** The pH of the solution can be calculated once the number of moles of excess acid is known, and the volume of the solution.

Example 14.6

20.00 mL of a 0.2000 M NH_3 solution is titrated with 0.1000 M HCl. Calculate the pH of the solution
(a) before the addition of the acid, and after (b) 30.00 mL, (c) 40.00 mL, and (d) 50.00 mL of acid are added. $K_b = 1.80 \times 10^{-5}$ mol L^{-1}.

Solution 14.6

(a) Before the acid is added, there is only $NH_3(aq)$ present, and the equilibrium is

$$NH_3 + H_2O \rightleftarrows NH_4^+ + OH^-$$

equilibrium concentrations $0.2000 - y$ $\quad y \quad y \quad mol\ L^{-1}$

$$K_b = 1.80 \times 10^{-5}\ mol\ L^{-1} = \frac{y^2}{0.2000 - y}$$

Ignoring y compared to 0.2000, we find

$$y = 1.90 \times 10^{-3}$$
$$pOH = 2.72$$
$$pH = 14.00 - pOH = 11.28$$

Note that the initial number of moles of the base NH_3 present is

$$0.2000\ mol/L \times 20.00\ mL \times \frac{1\ L}{10^3\ mL} = 0.0040\ moles$$

(b) When 30.00 mL HCl are added,

$$\text{moles acid added} = \frac{0.1000\ mol}{L} \times 30.00\ mL \times \frac{1\ L}{10^3\ mL} = 0.0030$$

moles base remaining = initial moles base - moles acid added
= 0.0040 - 0.0030 = 0.0010

As well, 0.0030 moles NH_4Cl have been produced. The total volume of the solution is 0.050 L. The reaction of interest here is

$$NH_3 + H_2O \rightleftharpoons NH_4^+ + OH^-$$

$$K_b = \frac{[NH_4^+][OH^-]}{[NH_3]} = 1.80 \times 10^{-5}\ mol\ L^{-1}$$

$$[NH_3] = \frac{0.0010\ moles}{0.050\ L} = 0.020\ M$$

$$[NH_4^+] = \frac{0.0030\ moles}{0.050\ L} = 0.060\ M$$

concentration ($mol\ L^{-1}$)	NH_3	NH_4^+	OH^-
initial	0.020	0.060	0
change	-y	y	y
equilibrium	0.020 - y	0.060 + y	y

Assume

$$y \ll 0.020 \text{ and } 0.060$$

Then

$$1.80 \times 10^{-5} = \frac{(y)(0.060)}{(0.020)}$$

$$y = 5.40 \times 10^{-5}$$

$$pOH = 4.27$$

$$pH = 14.00 - pOH = 9.73$$

(c) moles acid added $= 0.1000 \ \frac{mol}{L} \times 40.00 \ mL \times \frac{1 \ L}{10^3 \ mL} = 0.0040$ mole

In part (a) we saw that 0.0040 moles base were initially present. Thus the titration has reached the equivalence point. Also 0.0040 moles of NH_4Cl have been produced. The total volume of the solution is 0.060 L. Then

$$[NH_4^+] = \frac{0.0040 \ mol}{0.060 \ L} = 0.0067 \ M$$

Since the cation NH_4^+ reacts with water, the pH at the equivalence point will not be 7. The reaction to consider is

$$NH_4^+ + H_2O \rightleftharpoons NH_3 + H_3O^+$$

$$K_a = \frac{K_w}{K_b} = \frac{1.0 \times 10^{-14} \ mol^2 \ L^{-2}}{1.8 \times 10^{-5} \ mol \ L^{-1}} = 5.56 \times 10^{-10} \ mol \ L^{-1}$$

The array of concentrations is

concentration ($mol \ L^{-1}$)	NH_4^+	NH_3	H_3O^+
initial	0.067	0	0
change	-y	y	y
equilibrium	0.067 - y	y	y

$$K = 5.56 \times 10^{-10} = \frac{(y)(y)}{0.067 - y}$$

Assume

$$y << 0.067$$

Then

$$y^2 = 6.7 \times 10^{-2} \times 5.56 \times 10^{-10} = 3.72 \times 10^{-11}$$

$$y = 6.10 \times 10^{-6} \qquad pH = 5.21$$

d) Moles acid added $= 0.100 \ \frac{mol}{L} \times 50.00 \ mL \times \frac{1 \ L}{10^3 \ mL} = 0.0050$

moles excess acid = moles acid added - moles base initially present
= 0.0050 - 0.0040 = 0.0010

Total volume is 0.070 L.

$$[H_3O^+] = \frac{0.0010 \ mol}{0.070 \ L} = 0.0143 \qquad pH = 1.84$$

SELF TEST

1. The pH of 1.0 M HCl is
   (a) -1 (b) 0 (c) 1 (d) 7 (e) 1.8×10^{-5}

2. If the pH of a solution of $Ba(OH)_2$ is 10.0 at 25°C, the $[OH^-]$ of the solution is
(a) 5×10^{-5} M (b) 1.0×10^{-4} M (c) 2×10^{-4} M
(d) 1.0×10^{-10} M (e) 2.0×10^{-10} M

3. Which of the following represents K_a for hydrofluoric acid, HF?

(a) $\frac{[H_3O^+][F^-]}{[H_2O][HF]}$ (b) $[H_3O^+][HF]$ (c) $[H_3O^+][F^-]$

(d) $\frac{[H_3O^+][HF]}{[F^-]}$ (e) $\frac{[H_3O^+][F^-]}{[HF]}$

4. The pH of a 0.3 M acetic acid solution is: (K_a = 1.8×10^{-5} mol L^{-1})
(a) 2.6 (b) 3.4 (c) 4.2 (d) 11.3 (e) 1.2

5. The pH of 0.20 M NH_3 ($K_b = 1.8 \times 10^{-5}$ mol L^{-1}) in H_2O is
(a) 2.7 (b) 2.4 (c) 12.1 (d) 11.3 (e) 10.3

6. Which one of the following solutions will be basic?
(a) 0.01 M HCl (b) 0.1 M NaCl (c) 0.1 M CH_3COONa
(d) 0.1 M CH_3COOH (e) 0.1 M NH_4Cl

7. What is the pH of a 1.0 M solution of NaCN? For HCN $K_a = 4 \times 10^{-10}$ mol L^{-1}.
(a) 2.3 (b) 3.7 (c) 7.0 (d) 10.3 (e) 11.7

8. A water solution of NH_4NO_3 is:
(a) neutral because NH_3 is a strong base and HNO_3 is a strong acid.
(b) acidic because NH_3 is a strong base and HNO_3 is a weak acid.
(c) acidic because NH_3 is a weak base and HNO_3 is a strong acid.
(d) basic because NH_3 is a strong base and HNO_3 is a weak acid.
(e) basic because NH_3 is a weak base and HNO_3 is a strong acid.

9. The indicator methyl red has a K_a of about 10^{-5} mol L^{-1}. Its "acid" form is red and its "basic" form is yellow. What is the color of a solution with pH = 5 to which a few drops of methyl red have been added?
(a) red (b) yellow (c) green (d) orange (e) colorless

10. A solution is prepared from 3.0×10^{-3} moles of a weak acid, HX, and 6.0×10^{-4} moles of NaX in a total volume of 500 mL. The pH of the solution is 4.80. The K_a for HX, in units of mol L^{-1}, is:
(a) 7.9×10^{-5} (b) 1.6×10^{-5} (c) 3.2×10^{-7} (d) 3.2×10^{-6}
(e) 8.0×10^{-5}

11. How many **moles** of solid NaF should be added to 200 mL of 0.22 M HF to produce a buffer of pH of 4.00? $K_a(HF)$ = 3.5×10^{-4} mol L^{-1}
(a) 0.15 (b) 0.77 (c) 1.5 (d) 3.5 (e) 7.7

12. In the titration of a weak base with a strong acid, the pH of the solution at the equivalence point is:
(a) acidic (b) basic (c) neutral
(d) depends on molarities of acids and bases
(e) depends on the temperature

13. What is the pH of an aqueous buffer solution prepared by combining three times as many moles of NH_4Cl as NH_3? K_b (NH_3) = 1.8×10^{-5} mol L^{-1}.
(a) 4.22 (b) 5.22 (c) 6.78 (d) 8.78 (e) 9.78

14. 24.00 mL of 0.100 M HF ($K_a = 3.5 \times 10^{-4}$ mol L^{-1}) are titrated with 0.150 M sodium hydroxide solution. Calculate the pH at the equivalence point.
(a) 5.77 (b) 5.88 (c) 7.00 (d) 8.12 (e) 8.23

15. For H_2S, $(K_a)_1 = 9.1 \times 10^{-8}$ and $(K_a)_2 = 1.3 \times 10^{-13}$ mol L^{-1}. In an 0.010 M solution of H_2S in water, the extent of dissociation of the H_2S is
a) 0.003% b) 0.3% c) 1% d) 5% e) 100%

Answers to Self Test

1. b; 2. b; 3. e; 4. a; 5. d; 6. c; 7. e; 8. c; 9. d; 10. d; 11. a; 12. a, d, e; 13. d; 14. d; 15. b.

CHAPTER 15

THE ALKALI AND ALKALINE EARTH METALS

SUMMARY REVIEW

Groups 1 and 2 of the periodic table, the **alkali metals,** Li, Na, K, Rb, Cs and Fr, with a single ns^1 valence electron, and the **alkaline earth metals,** Be, Mg, Ca, Sr, Ba and Ra, with ns^2 valence electron configurations, comprise **the s-block elements.** Calcium, sodium, potassium and magnesium are among the most abundant elements in the earth's crust. All have typical metallic properties. The alkali metals are unusual only in that their metallic bonding involves only one electron per atom, giving them rather low densities, low melting points, and a large liquid range. The alkali metals and barium have the monatomic body-centered cubic structure; most of the others have the cubic or hexagonal close-packed structures.

The ns^1 and ns^2 electrons of the group 1 and 2 metals are easily removed; they have low ionization energies and are readily oxidized to M^+ (group 1) and M^{2+} (group 2) ions, so that they are found as salts, in association with common ions such as Cl^-, SO_4^{2-} and CO_3^{2-}. Deposits of layers of salts such as **rock salt,** NaCl, **sylvite, KCl,** and **carnallite,** $KMgCl_3 \cdot 6H_2O$, resulted from the evaporation of ancient seas. Group 2 cations are commonly found as rather insoluble carbonates, such as **magnesite,** $MgCO_3$, **dolomite,** $MgCa(CO_3)_2$, and the various forms of **calcium carbonate,** $CaCO_3$ (limestone, chalk and marble) formed from the shells and skeletons of dead sea creatures. Calcium also occurs commonly as phosphate rock, $Ca_3(PO_4)_2$, as **gypsum,** $CaSO_4 \cdot 2H_2O$, and as **fluorspar,** CaF_2.

Most of the group 1 and 2 metals have characteristic line spectra and can be identified from flame tests. Preparation of the elements involves reduction of M^+ and M^{2+} ions to the metals, which requires a strong reducing agent. Electrolytic reduction is now commonly used (Chapter 17); potassium is still made by reducing KCl with sodium, and sodium was formerly made by reducing Na_2CO_3 with carbon. With the exception of a few compounds of Li and Be, all common compounds are ionic (+1 ions for the alkali metals, +2 ions for the alkaline earth metals) and the metals are good reducing agents. For example:

<u>with halogens</u> $2M + X_2 \rightarrow 2M^+X^-$; $M + X_2 \rightarrow M^{2+}(X^-)_2$

<u>with hydrogen</u> $2M + H_2 \rightarrow 2M^+H^-$; $M + H_2 \rightarrow M^{2+}(H^-)_2$

(BeH_2 is covalent). **Nitrides,** containing the nitride ion, N^{3-}, result from heating Li, Mg, Ca, Sr and Ba in nitrogen. The metals react with oxygen to give basic oxides. Normal oxides, containing the O^{2-} ion, are formed by Li, Be, Mg and Ca, while **peroxides,** containing the O_2^{2-} ion, such as Na_2O_2, and **superoxides,** containing the O_2^- ion, such as KO_2, are formed by the other metals.

Alkali Metal Compounds

The oxides and metals react with water (with increasing vigor from Li to Cs) to give solutions of the hydroxides, M^+OH^-. The solid **hydroxides** are deliquescent. Many salts are conveniently made from them by reaction with the requisite acid. **Halides, sulfates, carbonates, nitrates,** and **phosphates** are all soluble in water. Halides also result from direct oxidation of a metal with a halogen.

Potassium carbonate, K_2CO_3, potash, results from burning wood, and **sodium carbonate,** Na_2CO_3 , soda, from burning seaweed. Formerly, most Na_2CO_3 came from the Solvay process ($CaCO_3 + 2NaCl \rightarrow Na_2CO_3 + CaCl_2$). Now, large desposits of **Trona,** $Na_5(CO_3)_2(HCO_3)\cdot 2H_2O$ are used to give Na_2CO_3 on heating. **Sodium hydrogen carbonate,** $NaHCO_3$, results when CO_2 is passed into $NaOH(aq)$. **Sodium sulfate,** Na_2SO_4, and **sodium hydrogen sulfate**, $NaHSO_4$, are by-products of the manufacture of HCl(g) from sulfuric acid and sodium chloride.

Alkaline-Earth Metal Compounds

The **oxides** are most readily prepared by heating the **carbonates;** MgO, CaO, SrO and BaO all have the NaCl structure. MgO is insoluble but reacts slowly with water to give insoluble $Mg(OH)_2$. CaO reacts with water in a highly exothermic reaction to give $Ca(OH)_2$, **slaked lime,** which is sparingly soluble. $Ca(OH)_2$ is also precipitated when $OH^-(aq)$ is added to a solution of a soluble calcium salt. BeO is amphoteric.

Calcium carbonate, $CaCO_3$, is insoluble in water but dissolves in acids to give soluble $Ca(HCO_3)_2$. For example, when CO_2 is passed into **limewater,a** saturated solution of $Ca(OH)_2$, $CaCO_3(s)$ is first deposited, which is used as a test for CO_2. More CO_2 causes the $CaCO_3$ to dissolve as soluble $Ca(HCO_3)_2$. This reaction is also responsible for the formation of limestone caves; stalagmites and stalactites result from the reverse reaction ($Ca^{2+} + 2HCO_3^- \rightarrow CaCO_3 + CO_2 + H_2O$). The hardness of the water is due to Ca^{2+} and Mg^{2+} salts, such as the chloride, sulfate or hydrogen carbonate. Water can be softened by adding sodium carbonate to precipitate $CaCO_3$ and $MgCO_3$.

$MgCl_2$ and $CaCl_2$ are useful drying agents; they crystallize from water as $MgCl_2\cdot 6H_2O$ and $CaCl_2\cdot 6H_2O$. On heating, $CaCl_2\cdot 6H_2O$ dehydrates simply, while $MgCl_2\cdot 6H_2O$ reacts to give MgO, HCl and water.

Solubilities

All the salts of the singly-charged alkali metal cations, and of NH_4^+, are **soluble**. Among the salts of doubly-charged cations, such as those of group 2, salts with singly-charged anions are generally soluble (except when the ions are very small, for example F^- or OH^-), and salts with multicharged anions are **insoluble** or only **sparingly soluble** (e.g., SO_4^{2-}, CO_3^{2-} and PO_4^{3-}). Among the exceptions to these rules are the chlorides, bromides and iodides of Ag^+, Pb^{2+} and Hg_2^{2+} (insoluble or sparingly soluble), the sulfides of group 2, and $Ba(OH)_2$ (soluble).

Even salts described as "insoluble" have very small concentrations of ions in equilibrium with undissolved salt. For such heterogeneous equilibria involving an ionic compound $(A^{m+})_x(B^{n-})_y$, we have

$$A_xB_y(s) \rightleftharpoons xA^{m+}(aq) + yB^{n-}(aq)$$

for which,

$$K_c[A_xB_y(s)] = K_{sp} = [A^{m+}]^x[B^{n-}]^y$$

is called the **solubility product constant.** K_{sp} values are calculated from the experimentally measured solubilities of salts. Similarly, if the value of K_{sp} is known, the solubility of a salt may be calculated. More important uses of the solubility product constant are to predict whether a precipitate will form when two solutions are mixed, how much of a given ion remains in solution after precipitation, and salt solubility in the presence of a common ion (the common-ion effect). The solubilities of salts of weak acids (and of weak bases) are pH dependent.

REVIEW QUESTIONS

1. What groups of the periodic table constitute the alkali metals and the alkaline earth metals?
2. What are the outer (valence) shell configurations of the alkali metal and alkaline earth metal elements?
3. Why do alkali metals differ from most other metals in having rather low densities, low melting points, and a wide liquid range?
4. Why are the elements of groups 1 and 2 found in nature exclusively in the form of their cations?
5. How do the ionization energies change in going from the top to the bottom of each group?
6. Why is the extraction of salts from sea-water generally not an economically viable process?
7. What are the chemical compositions of rock salt, sylvite, carnallite, magnesite, dolomite, chalk, limestone, marble, calcite, phosphate rock and gypsum?
8. Account for the fact that both sodium metal and its compounds give a characteristic yellow flame.
9. How might you prepare a sample of potassium metal in the laboratory?
10. What is the reducing agent in electrolysis of molten NaCl to give metallic sodium?
11. What compounds result from the reaction of calcium with (a) H_2, (b) Br_2, (c) N_2, (d) O_2?
12. What are the formulas of sodium peroxide and potassium superoxide?
13. What reaction occurs between cesium and water?
14. In the above reaction, will the resulting solution be acidic, neutral, or basic?
15. What reaction occurs between lithium oxide and water?
16. How are simple salts of the alkali metals best prepared?
17. Draw diagrams to illustrate the face-centered cubic sodium chloride and primitive cubic cesium chloride unit cells.
18. Draw Lewis diagrams for the following compounds: (a) LiH, (b) BaH_2, (c) Li_3N, (d) CaC_2.

19. Write balanced equations for the reaction of the compounds in Question 18 with water.
20. Will the pH of a solution of sodium hydrogen carbonate be greater than or less than 7.00?
21. How is $NaHCO_3$ normally prepared?
22. Describe the action of baking powder in cookery.
23. Which will have the lower pH, a 0.1 M solution of $NaHCO_3$ or a 0.1 M solution of Na_2CO_3?
24. What is the formula of calcium hydrogen sulfate?
25. Classify the alkaline earth metal oxides as acidic or basic.
26. What are (a) slaked lime, (b) lime water?
27. Describe a simple test for $CO_2(g)$.
28. Write equations for the action of heat on (a) $NaHCO_3$, (b) $CaCO_3$.
29. What equilibrium is involved in the processes that lead to the formation of caves and stalactites and stalagmites in limestone areas?
30. What dissolved compounds make water hard?
31. How can the hardness of water be removed?
32. Write equations for the action of heat on (a) $CaCl_2 \cdot 6H_2O$, (b) $MgCl_2 \cdot 6H_2O$.
33. To what would you attribute the difference in the nature of the above two reactions?
34. Write the formulas for epsomite, gypsum and plaster of paris.
35. Describe a common test to detect the presence of sulfate ion in aqueous solution.
36. What is meant by the solubility of a substance in aqueous solution?
37. What can be said very generally about the solubility of alkali metal salts in aqueous solution?
38. Are there any insoluble ammonium salts?
39. Which of the following would be expected to be insoluble or sparingly soluble?
    (a) $Ca_3(PO_4)_2$, (b) $Ba(NO_3)_2$, (c) $BaSO_4$, (d) $MgCl_2$.
40. What is meant by a precipitation reaction?
41. What equilibrium exists between a solid and its saturated solution?
42. Define solubility.
43. Write the expression for the solubility product constant of
    (a) $PbSO_4$, (b) $Ca_3(PO_4)_2$, (c) $Zn(OH)_2$.
44. If the solubility of a compound MCl_2 is 0.01 mol L^{-1}, what is its solubility product constant, K_{sp}?
45. The solubility product constant, K_{sp}, of a compound CaX_2 is 4×10^{-9} mol^3 L^{-3}. What is its solubility?
46. For the compound in Question 45, equal volumes of 0.02 M $Ca(NO_3)_2$ and 0.02 M NaX were mixed. Would a precipitate be expected to form?
47. How could the solubility of $BaSO_4$ in aqueous solution be reduced?
48. Why is CaF_2 insoluble in water but soluble in dilute HCl(aq)?

Answers to Selected Review Questions

13. $2Cs(s) + 2H_2O \rightarrow 2Cs^+(aq) + 2OH^-(aq) + H_2(g)$ 14. basic 15. $O^{2-} + H_2O \rightarrow 2OH^-$ 18. $Li^+{:}H^-$ $Ba^{2+}({:}H^-)_2$ $(Li^+)_3{:}\ddot{\underset{..}{N}}{:}^{3-}$ $Ca^{2+}({}^-{:}C{\equiv}C{:}^-)$

19. $H^- + H_2O \rightarrow H_2 + OH^-$; $N^{3-} + 3H_2O \rightarrow NH_3 + 3OH^-$; $C_2^{2-} + 2H_2O \rightarrow C_2H_2 + 2OH^-$
20. >7.00, because of the reaction $HCO_3^- + H_2O \rightleftharpoons H_2CO_3 + OH^-$ 23. $NaHCO_3$

24. $Ca(HSO_4)_2$ 25. basic 33. Both salts contain hydrated cations, $Ca(H_2O)_6{}^{2+}$ and $Mg(H_2O)_6{}^{2+}$. Because of its smaller size, the hydrated Mg^{2+} ion is a weak acid relative to the hydrated Ca^{2+} ion. It reacts with Cl^- on heating to give HCl(g) and an M-OH bond. 39. (a) and (c) 43. $K_{sp}(PbSO_4) = [Pb^{2+}][SO_4{}^{2-}]$; $K_{sp}(Ca_3(PO_4)_2) = [Ca^{2+}]^3[PO_4{}^{3-}]^2$; $K_{sp}(Zn(OH)_2) = [Zn^{2+}][OH^-]^2$ 44. $K_{sp} = 4 \times 10^{-6}$ mol^3 L^{-3} 45. Solubility = 1×10^{-3} mol L^{-1} 46. Yes, since $Q > K_{sp}$.

OBJECTIVES

Be able to:

1. List the members of the alkali and alkaline earth families and describe their physical properties.
2. Give the flame colors produced by the alkali and alkaline earth metals.
3. State how the alkali and alkaline earth metals are found in nature, and how they are prepared in an elemental form.
4. Give balanced chemical reactions for the alkali and alkaline earth metals plus hydrogen, halogens and oxygen.
5. Describe physical and chemical properties of various compounds of alkali metals, such as hydroxides, halides, hydrides, carbonates, hydrogen carbonates and sulfates.
6. Describe physical and chemical properties of various compounds of alkaline earth compounds, such as oxides, hydroxides, carbonates, halides and sulfates.
7. Classify common salts and hydroxides as soluble, sparingly soluble and insoluble.
8. Write the solubility product constant, K_{sp}, and the equilibrium to which it corresponds, for a sparingly soluble salt.
9. Calculate K_{sp} from the solubility, and the solubility from K_{sp}.
10. Use the relationship between ion product and K_{sp} to determine whether a precipitate will form.
11. Calculate the solubility of a slightly soluble salt in the presence of a common ion.
12. Show how an insoluble precipitate may be dissolved by decreasing the concentration of one of the ions in the equilibrium.
13. Perform calculations involving selective precipitation.

PROBLEM SOLVING STRATEGIES

15.4 Solubility and Precipitation Reactions

Solubility Product Constant

When a slightly soluble salt is in equilibrium with its ions in solution (i.e., the solution is saturated) the equilibrium is described by an equilibrium constant K_{sp}, **the solubility product constant.** For example, for AgCl

$$AgCl(s) \rightleftarrows Ag^+(aq) + Cl^-(aq)$$

$$K_{sp} = [Ag^+(aq)][Cl^-(aq)]$$

Note several points. The salt is on the **left** side of the chemical equation. The concentration of the **solid** salt does not appear in K_{sp}. There is no undissociated AgCl **in solution,** only Ag^+ and Cl^-. Finally, there must be solid salt present for equilibrium to occur.

Example 15.1

Write the form for K_{sp} for strontium phosphate, $Sr_3(PO_4)_2$, which has a K_{sp} of 1.0×10^{-31} when concentrations are expressed in mol L^{-1}. Also give the units for K_{sp}. Strontium is in group 2.

Solution 15.1

The equilibrium between $Sr_3(PO_4)_2$ and its ions is

$$Sr_3(PO_4)_2(s) \rightleftarrows 3Sr^{2+}(aq) + 2PO_4^{3-}(aq)$$

K_{sp} only contains the concentrations of the dissolved ions, each raised to a power equal to its stoichiometric coefficient in the above equation.

$$K_{sp} = [Sr^{2+}(aq)]^3[PO_4^{3-}(aq)]^2$$

For concentrations in mol L^{-1}, K_{sp} has units of $(mol\ L^{-1})^3 \times (mol\ L^{-1})^2$, or $mol^5\ L^{-5}$.

Calculating Solubility from the Solubility Product Constant

If the solubility of a salt is given, you can calculate its K_{sp}. First, convert the solubility to mol L^{-1}, if it is given in other units. Next, write the K_{sp} equilibrium and the form for K_{sp}. Third, determine the concentrations of the ions.

If the stoichiometric coefficients in the K_{sp} expression are not all unity, the concentrations of some dissolved ions will be a **multiple** of the solubility. Finally, insert the concentrations into the K_{sp} expression and obtain a value for K_{sp}.

Example 15.2

The solubility of $Mg(OH)_2$ is 7.05×10^{-4} g/100 mL solution. What is K_{sp} for $Mg(OH)_2$?

Solution 15.2

First we convert the solubility to mol L^{-1}.

$$\text{solubility} = \frac{7.05 \times 10^{-4}\ g}{100\ mL} \times \frac{10^3\ mL}{1\ L} \times \frac{1\ mol\ Mg(OH)_2}{58.3\ g\ Mg(OH)_2} = 1.21 \times 10^{-4}\ mol\ L^{-1}$$

$$Mg(OH)_2(s) \rightleftharpoons Mg^{2+}(aq) + 2OH^-(aq)$$

$$K_{sp} = [Mg^{2+}(aq)][OH^-(aq)]^2$$

Since the solubility is 1.21×10^{-4} mol L^{-1}, for every 1.21×10^{-4} mole of $Mg(OH)_2$ that dissolves in 1 liter solution, 1.21×10^{-4} moles of $Mg^{2+}(aq)$ are produced, and $2 \times 1.21 \times 10^{-4}$ moles of $OH^-(aq)$ are obtained. The balanced equation demands that the $[OH^-]$ is twice the $[Mg^{2+}]$. Then

$$K_{sp} = (1.21 \times 10^{-4})\ (2.42 \times 10^{-4})^2 = 7.1 \times 10^{-12}\ mol^3\ L^{-3}$$

The units for K_{sp} here are $mol^3\ L^{-3}$. We had to square the $[OH^-]$ because the form for K_{sp} contained the square of $[OH^-]$.

We can also work this type of problem in reverse, and calculate the solubility from a K_{sp} value. The same steps as used in Example 15.2 can be applied.

Example 15.3

Calculate the solubility of $SrSO_4$ in mol L^{-1}; $K_{sp} = 3.2 \times 10^{-7}$ mol^2 L^{-2} for $SrSO_4$.

Solution 15.3

Let the solubility of $SrSO_4$ be S mol L^{-1}

$$SrSO_4(s) \rightleftharpoons Sr^{2+}(aq) + SO_4^{2-}(aq)$$

$$K_{sp} = [Sr(aq)^{2+}][SO_4^{2-}(aq)]$$

For S moles of $SrSO_4$ that dissolve in 1 liter solution, the balanced equation tells us that S moles of Sr^{2+} and S moles of SO_4^{2-} are produced.

$$K_{sp} = 3.2 \times 10^{-7}\ mol^2\ L^{-2} = (S)\ (S) = S^2$$

Thus

$$S = 5.6 \times 10^{-4}\ mol\ L^{-1}$$

If the problem had asked for solubility in g L^{-1}, you would have needed to convert to grams per liter at the end.

Common Ion Effect

If there is a second source of one or more of the types of ions produced in an equilibrium involving a slightly soluble salt, **the solubility of the salt is decreased.** The ion externally added is called a **common ion.** Even though the equilibrium concentration of this ion comes from two sources, the amount contributed by the slightly soluble salt is usually **negligible.** The method of attack in obtaining the solubility in the presence of a common ion is the same as outlined above for K_{sp} problems.

Example 15.4

What is the solubility of $Mg(OH)_2$ in a 0.20 M solution of KOH? $K_{sp} = 7.1 \times 10^{-12}$ mol^3 L^{-3} for $Mg(OH)_2$. Compare your result to the solubility of 1.2×10^{-4} M in pure water.

Solution 15.4

Let the solubility of $Mg(OH)_2$ here be S mol L^{-1}. The sources of the ions are

$$Mg(OH)_2 \rightleftharpoons Mg^{2+} + 2OH^-$$

$$KOH \rightarrow K^+ + OH^-$$

Let's list the effect of $Mg(OH)_2$ on the ion concentrations in 0.20 M KOH:

	$[OH^-]$	$[Mg^{2+}]$
From KOH	0.20	0
Change due to $Mg(OH)_2$	2S	S
Net	0.20 + 2S	S

Thus

$$K_{sp} = [Mg^{2+}][OH^-]^2$$

$$7.1 \times 10^{-12} \text{ mol}^3 \text{ L}^{-3} = S(0.20 + 2S)^2$$

In the $[OH^-]$ column, note that we **do not** double the 0.20, since its source, KOH, produces only one mole of OH^- when 1 mol of KOH ionizes; we do double the S since $Mg(OH)_2$ produces 2S OH^- for each S Mg^{2+}.

We can approximate 0.20 + 2S as 0.20, so

$$7.1 \times 10^{-12} = S(0.20)^2$$

$$S = 1.8 \times 10^{-10} \text{ mol L}^{-1}$$

So we see that the solubility with KOH added is consideraly less than the solubility in water, (Example 15.2). The diminished solubility can be predicted qualitatively by Le Chatelier's principle.

Selective Precipitation

When a slightly soluble salt is in **equilibrium** with its ions, the K_{sp} equation is valid. The solution is saturated. Situations can arise when the product of ion concentrations, raised to their appropriate stoichiometric coefficients, is **less than** K_{sp}. Then no **solid,** or precipitate, is present. Mathematically, we say that in this case the **ion product Q** is less than K_{sp}. Q has the same form as K_{sp}, but refers to **initial** concentrations, not **equilibrium** concentrations. If $Q = K_{sp}$, then there is an **equilibrium** between the solid and the dissolved ions. If the initial concentrations yielded a $Q > K_{sp}$, then a solid will form (often quite quickly) and continue to precipitate until $Q = K_{sp}$. Two different ions can be **separated** from each other by one of them forming a precipitate. The ion forming the precipitate is the ion for which $Q > K_{sp}$. The method of attack involves forming the Q's, calculating their values, and comparing them to the respective K_{sp}'s.

Example 15.5

The Br^- concentration in sea water is about 0.07 g L^{-1} and the Cl^- concentration is approximately 19 g L^{-1}. Which will precipitate, AgBr or AgCl, when 5×10^{-10} moles of solid $AgNO_3$ are added to a 1.00 L sample of sea water?

$K_{sp} = 5.0 \times 10^{-13}$ mol^2 L^{-2} for AgBr; $K_{sp} = 1.8 \times 10^{-10}$ mol^2 L^{-2} for AgCl

Solution 15.5

First we convert the concentrations into mol L^{-1}:

$$[Br^-(aq)] = 0.07 \text{ g L}^{-1} \times \frac{1 \text{ mol}}{79.9 \text{ g}} = 9 \times 10^{-4} \text{ mol L}^{-1}$$

$$[Cl^-(aq)] = 19 \text{ g L}^{-1} \times \frac{1 \text{ mol}}{35.5 \text{ g}} = 0.54 \text{ mol L}^{-1}$$

For AgBr, $Q = [Ag^+(aq)][Br^-(aq)] = (5 \times 10^{-10})(9 \times 10^{-4}) \text{ mol}^2\text{L}^{-2}$
$= 4.5 \times 10^{-13} \text{ mol}^2\text{L}^{-2}$

Q is less than K_{sp} for AgBr, so AgBr does **not** precipitate.

For AgCl, $Q = [Ag^+(aq)][Cl^-(aq)] = (5 \times 10^{-10})(0.54) \text{ mol}^2\text{L}^{-2}$
$= 2.7 \times 10^{-10} \text{ mol}^2\text{L}^{-2}$

This Q is greater than K_{sp} for AgCl, so AgCl precipitates.

SELF TEST

Part I True or False

1. Magnesium is the lightest of all the metals.
2. Some of the alkali metals are less dense than water.
3. The alkali metals commonly occur as chlorides.
4. Chalk is one of the forms of naturally-occurring magnesium carbonate.
5. The ions of the alkali metal family are very easily reduced to the elemental form.
6. Nitrogen reacts with lithium to form lithium nitride.
7. All the alkali metals M react with atmospheric oxygen to form the oxides M_2O.
8. The halides, sulfates, carbonates, nitrates and phosphates of the alkali metals are soluble in water.
9. The common name potash refers to potassium carbonate.
10. Sodium hydrogen carbonate is a major constituent of baking powder.
11. Calcium hydroxide is more soluble in water than sodium hydroxide.
12. Water that contains only very small concentrations of Ca^{2+} and Mg^{2+} is called hard water.
13. Barium sulfate has a very low solubility in water.
14. Insoluble metal hydroxides will not dissolve in acid.

Part II Multiple Choice

15. The flame color produced by strontium is
(a) yellow (b) lilac (c) pale green (d) red (e) blue
16. Give **all** the products of the reaction of sodium with water
(a) OH^- (b) Na_2O_2 (c) H_2 (d) Na^+ (e) O_2
17. The solubility product expression for As_2S_3 is: K_{sp} =
(a) $[As^{3+}][S^{2-}]$ (b) $[As^{3+}]^2[S^{2-}]^3$ (c) $[As^{3+}]^3[S^{2-}]^2$
(d) $[As_2][S_3]$ (e) none of these.
18. How many grams of calcium oxalate, CaC_2O_4 will dissolve in pure water to form 1.0 liters of saturated solution? The dissociation of calcium oxalate in water yields Ca^{2+} and $C_2O_4^{2-}$. K_{sp} for CaC_2O_4 is 2.5×10^{-9} $mol^2\ L^{-2}$ and the molar mass of CaC_2O_4 is 128 g mol^{-1}.
(a) 0.0064 g (b) 0.016 g (c) 0.032 g (d) 0.11 g (e) 0.26 g
19. Calculate the solubility product, in $mol^2\ L^{-2}$, for barium sulfate, whose solubility is 0.000247 g in 100 mL of water at a temperature not far from 25°C. The molar mass of $BaSO_4$ is 233 g.
(a) 1.12×10^{-10} (b) 1.06×10^{-10} (c) 1.06×10^{-8}
(d) 1.12×10^{-8} (e) 1.06×10^{-5}
20. The following gives K_{sp} data for some silver salts: AgCl, 1.8×10^{-10} $mol^2\ L^{-2}$; AgI, 8.3×10^{-17} $mol^2\ L^{-2}$; Ag_2SO_4, 1.5×10^{-5} $mol^3\ L^{-3}$; AgBr, 5.0×10^{-13} $mol^2\ L^{-2}$. Which one of the following lists the salts in order of **increasing** solubility, in moles L^{-1}, from left to right?
(a) AgCl, AgBr, AgI, Ag_2SO_4 (b) AgI, AgBr, Ag_2SO_4, AgCl
(c) Ag_2SO_4, AgBr, AgCl, AgI (d) AgI, AgBr, AgCl, Ag_2SO_4
(e) AgBr, AgCl, AgI, Ag_2SO_4
21. How many moles of AgCl will dissolve in 1 liter of 0.17 M $CaCl_2$ solution? $K_{sp} = 1.8 \times 10^{-10}$ mol L^{-1} for AgCl.
(a) 1.3×10^{-5} (b) 5.3×10^{-10} (c) 1.0×10^{-9} (d) 18 (e) 1.0×10^{-10}
22. A 1.00 L solution contains initially 1×10^{-3} moles of each of Sr^{2+}, Pb^{2+}, and Ag^+, and 1.0×10^{-2} moles of SO_4^{2-}. What precipitates are formed? K_{sp} data: $SrSO_4$, 3.2×10^{-7} $mol^2\ L^{-2}$; $PbSO_4$, 6.3×10^{-7} $mol^2\ L^{-2}$; Ag_2SO_4, 1.5×10^{-5} $mol^3\ L^{-3}$.
(a) $SrSO_4$, $PbSO_4$, Ag_2SO_4 (b) $SrSO_4$ and $PbSO_4$ only
(c) $SrSO_4$ and Ag_2SO_4 only (d) $PbSO_4$ and Ag_2SO_4 only (e) Ag_2SO_4 only

Answers to Self Test

1. F; 2. T; 3. T; 4. F; 5. F; 6. T; 7. F; 8. T; 9. T; 10. T; 11. F; 12. F; 13. T; 14. F; 15. d; 16. a,c,d; 17. b; 18. a; 19.a; 20. d; 21. b; 22. b.

CHAPTER 16

THERMODYNAMICS, ENTROPY AND FREE ENERGY

SUMMARY REVIEW

The **spontaneity** of a reaction is determined by both the change in **energy** and the change in **entropy** during a reaction; that is, by the change in **Gibbs free energy, $\Delta G = \Delta H - T\Delta S$.**

Among familiar spontaneous reactions are the expansion of a gas to fill the space available to it and the cooling of a hot object to the temperature of the environment. However, among chemical reactions there are examples of both exothermic and endothermic reactions that are spontaneous; a decrease in energy of microscopic systems is not the necessary condition for spontaneity as it is for macroscopic systems. Rather, the tendency for molecules to become more disordered must also be taken into account. In any spontaneous process **the total disorder in the system and its surroundings increases.**

On a quantitative basis the disorder of the atoms and molecules that make up a system and the dispersal of energy associated with these particles is called its **entropy, S,** which depends only on the conditions that determine the state of the system. Change in entropy ΔS, depends, therefore, only on the initial and final states of the system, and not how the change is achieved; entropy is a state function. The **second law of thermodynamics** states that in any spontaneous process the total entropy of a system and its surroundings increases, which may be alternatively stated as "the entropy of the universe increases".

At 0 K the molecules of any pure crystalline substance have a perfectly regular arrangement and have no thermal motion -- the entropy is zero. As the temperature is raised the thermal motions of the molecules increase and the entropy (state of disorder) also increases. The **standard molar entropy** of a substance, $S°$, is the entropy of 1 mole of the substance in its standard state (1 atm and 25°C) with units of J K^{-1}. In general standard entropies increase from solids to liquids to gases, and substances with large molecules usually have higher entropies than those with smaller molecules; energy is shared between more atoms and is more dispersed. The **standard entropy change** of the system undergoing a reaction, $\Delta S°$, is given by

$$\Delta S° = \sum S°(\text{products}) - \sum S°(\text{reactants})$$

To find whether or not a reaction will proceed spontaneously, the entropy change for the system and its surroundings must be calculated. The energy transferred as heat from the system to the surroundings is $-\Delta H_{system}$ and thermodynamics gives

$$\Delta S_{surroundings} = -\frac{\Delta H_{system}}{T}$$

However, for a spontaneous reaction

$$\Delta S_{total} = \Delta S_{system} + \Delta S_{surroundings} > 0 = \Delta S - \frac{\Delta H}{T}$$

Multiplying through by -T gives

$$-T\Delta S_{total} = \Delta H - T\Delta S$$

where the terms on the right side refer to the system. The quantity $-T\Delta S_{total}$ is given the symbol ΔG, so that

$$\mathbf{\Delta G = \Delta H - T\Delta S} \quad \text{at constant pressure.}$$

$G = H - TS$ is a state function of the system called the **Gibbs free energy. ΔG is negative for a spontaneous reaction.** In terms of the possible combinations of ΔH and ΔS for any reaction, we have:

	ΔH	ΔS	ΔG	
1.	-	+	-	spontaneous
2. at low T at high T	- -	- -	- +	spontaneous not spontaneous
3.	+	-	+	not spontaneous
4. at low T at high T	+ +	+ +	+ -	not spontaneous spontaneous

A reaction for which ΔG is positive in the forward direction is not spontaneous; however, ΔG is negative for the reverse reaction, so the reverse reaction will be spontaneous. **When $\Delta G = 0$, the reaction is at equilibrium.**

Standard free energy changes, $\Delta G°$, are calculated either from $\Delta H°$ and $\Delta S°$ values or from standard free energies of formation, ΔG°_f:

$$\Delta G° = \sum \Delta G^\circ_f(\text{products}) - \sum \Delta G^\circ_f(\text{reactants})$$

For a gaseous reaction, ΔG is related to Q (the reaction quotient) and K_p (the equilibrium constant) by the equation

$$\Delta G = RT \ln Q/K_p$$

and for all the reagents in their standard states, $\Delta G = \Delta G°$ and $Q = 1$ (since all concentrations and pressures are unity), so that

$$\Delta G^o = -RT \ln K_p \quad \text{or} \quad \ln K_p = \frac{-\Delta G^o}{RT}$$

The free energy change, ΔG, for a reaction is equal to the **maximum possible work** that can be obtained from the reaction. A nonspontaneous reaction can be driven in the nonspontaneous direction by coupling it to another reaction which has a negative ΔG, such that the overall ΔG is negative.

REVIEW QUESTIONS

1. Give two examples of physical changes that are spontaneous.
2. Is the following statement true or false? "Examples of spontaneous reactions include both exothermic and endothermic reactions".
3. How can the spontaneous reaction of hydrogen with oxygen to give water be reversed?
4. What change takes place at the molecular level when a gas expands spontaneously at constant temperature?
5. In any reaction, what can be said about the energy of a system and its surroundings?
6. What is the symbol for entropy? Why is entropy a state function?
7. State the second law of thermodynamics in terms of the entropy of a system and its surroundings.
8. State the first law of thermodynamics and the second law of thermodynamics in terms of the energy and entropy, respectively, of the universe.
9. Why is the entropy of a pure crystalline substance zero at 0 K?
10. What does entropy measure?
11. Define the standard molar entropy of a substance, S°. What are the units of entropy?
12. Why is the entropy of a gas generally higher than that of a solid?
13. Why do the standard entropies of the straight-chain alkanes increase from n-pentane (C_5H_{12}) to n-octane (C_8H_{18})? Why is the standard entropy of n-butane (C_4H_{10}) greater than that of n-pentane?
14. What expression relates the standard entropy change, $\Delta S°$, for a reaction to the standard entropies of the reactants and products?
15. How is the enthalpy change for a reaction at constant pressure, ΔH, related to the enthalpy change in its surroundings?
16. What quantity measures the entropy change in the surroundings, $\Delta S_{surroundings}$?
17. What equation relates the change in Gibbs free energy for a system to its enthalpy change, entropy change, and the absolute temperature?
18. What can be said about the sign of ΔG for a spontaneous reaction?
19. Which of the following statements is always true? (a) an exothermic reaction in which there is an increase in entropy is always spontaneous; (b) an endothermic reaction in which there is a decrease in entropy is always spontaneous.
20. Under what conditions is an exothermic reaction with a negative entropy change spontaneous?
21. If ΔG for a forward reaction has the value 130.9 kJ, what is the value of ΔG for the reverse reaction?
22. Define the standard free energy of formation of a substance. How is the standard free energy change for a reaction calculated from the standard free energies of formation of its reactants and products?
23. What is the relationship between ΔG for a reaction, its equilibrium constant, K_p, its reaction quotient, Q, and the temperature?

24. What is the value of ΔG for a reaction that has attained equilibrium?
25. How can the equilibrium constant for a reaction be calculated from thermodynamic quantities?
26. What is the magnitude of the equilibrium constant for a spontaneous reaction?
27. If the equilibrium constant for a reaction is less than unity, what is the sign of its standard Gibbs free energy change?
28. What is the maximum amount of work that can be obtained from a reaction?
29. Why is the maximum amount of work that can be obtained from a reaction not equal to its enthalpy change, ΔH?

OBJECTIVES

Be able to:

1. Give examples of spontaneous processes that involve a decrease in potential energy.
2. Appreciate that not all spontaneous changes are accompanied by a decrease in energy.
3. Realize that in any spontaneous process the total disorder in the system and its surroundings increases.
4. State the second law of thermodynamics in terms of the entropy of the universe and in terms of an entropy change of a system and its surroundings.
5. Use a table of standard molar entropies to calculate entropy changes in chemical reactions.
6. Predict the sign of the entropy change for a reaction involving gases.
7. Relate the entropy change in the surroundings at constant pressure and temperature to the enthalpy change of the system.
8. Define Gibbs free energy in terms of enthalpy and entropy, and use the sign of ΔG to predict spontaneity for processes at constant temperature and pressure.
9. Calculate $\Delta G°$ for a chemical reaction using $\Delta H°$ and $\Delta S°$ data, or using $\Delta G_f°$ data.
10. Calculate the temperature dependence of $\Delta G°$, assuming $\Delta H°$ and $\Delta S°$ are temperature independent.
11. Calculate equilibrium constants from $\Delta G°$, and relate the Q/K ratio to spontaneity.
12. Relate $\Delta G°$ to the maximum useful work that can be obtained from a chemical change.
13. Realize that a nonspontaneous reaction can be coupled to a spontaneous reaction, so that the net reaction is spontaneous.

PROBLEM SOLVING STRATEGIES

16.2 Entropy

Entropy is a measure of randomness or chaos. All naturally-occurring (i.e., spontaneous) changes in nature are accompanied by an **increase in the**

entropy of the universe, that is, by an increase in the entropy of the system and surroundings, taken together. To calculate an **entropy change** for a chemical reaction, consider the chemical reaction to be the **system.** In the next section we will comment on calculating an entropy change for the **surroundings.** Here are some points to consider.

(a) Write a **balanced** chemical reaction, including the correct phase (s,l,g) for each substance, and the correct allotropic form of the elements, if any are present.

(b) Since **entropy is a state function**, any set of steps, or set of balanced equations, that leads from reactants to products can be used for the calculation. The calculations in the text usually refer to 298 K and 1 atmosphere pressure, so the standard molar entropy data of Table 16.1 can be used if the reactants and products are in the table. To calculate the entropy change for the **system,** use the relation

$$\Delta S^\circ = \sum S^\circ(\text{products}) - \sum S^\circ(\text{reactants}).$$

The superscript refers to the material in its standard state (1 atmosphere). Remember that at temperatures above 0 K, **all** substances have a **positive** entropy. Thus, do not confuse the value of ΔH_f° for the most stable form of an element at 298 K and 1 atmosphere, which is **zero,** with the **nonzero** entropy of the same material at the same temperature.

(c) Calculate ΔS° for the reaction, or rearrange the ΔS° expression if you need to calculate S° for a reactant or a product, given ΔS°. Remember molar entropy has units J mol^{-1} K^{-1};entropy does not have units of energy, but energy **per degree.**

(d) As a check, if gases are present, calculate the number of moles of gaseous products minus reactants. If the number is positive, ΔS° should be positive; if negative, ΔS° should be negative.

Example 16.1

Calculate ΔS° for the combustion of one mole of liquid methanol; the products are liquid water and gaseous carbon dioxide. What would ΔS° be, for the same combustion, if the water produced were gaseous rather than liquid? Use Table 16.1 in the text.

Solution 16.1

The balanced equation is

$$CH_3OH(l) + \frac{3}{2}O_2(g) \rightarrow CO_2(g) + 2H_2O(l)$$

$$\Delta S^\circ = S^\circ(CO_2,g) + 2S^\circ(H_2O,l) - S^\circ(CH_3OH,l) - \frac{3}{2}S^\circ(O_2,g)$$

$$= 1 \text{ mol} \times 213.7 \text{ J mol}^{-1} \text{ K}^{-1} + 2 \text{ mol} \times 70.0 \text{ J mol}^{-1} \text{ K}^{-1}$$

$$- 1 \text{ mol} \times 126.8 \text{ J mol}^{-1} \text{ K}^{-1} - \frac{3}{2} \text{ mol} \times 205.0 \text{ J mol}^{-1} \text{ K}^{-1}$$

$$\Delta S^\circ = -80.6 \text{ J K}^{-1}$$

Note that $\Delta n(\text{gases}) = -1/2$ here, and ΔS° is negative. If the water were gaseous rather than liquid, the entropy of the products would be larger. The value of Δn(gases) would be +3/2, so we should expect $\Delta S°$ to be positive. The balanced equation is the same, except water is gaseous.

$$\Delta S° = S°(CO_2,g) + 2S°(H_2O,g) - S°(CH_3OH,l) - \frac{3}{2}S°(O_2,g)$$

$$= 1 \text{ mol} \times 213.7 \text{ J mol}^{-1} \text{ K}^{-1} + 2 \text{ mol} \times 188.7 \text{ J mol}^{-1} \text{ K}^{-1}$$

$$- 1 \text{ mol} \times 126.8 \text{ J mol}^{-1} \text{ K}^{-1} - \frac{3}{2} \text{ mol} \times 205.0 \text{ J mol}^{-1} \text{ K}^{-1}$$

$$\Delta S° = 156.8 \text{ J K}^{-1}$$

16.3 Gibbs Free Energy

Calculating Standard Free Energy Changes

Since the second law of thermodynamics requires a knowledge of **the sign of ΔS for the universe** to predict the direction of spontaneous change, **we must calculate both the ΔS for a system and ΔS for its surroundings.** It would be advantageous if we could **ignore** any calculations associated with the surroundings, and only deal with changes in properties of a **system. The Gibbs free energy, G, was introduced for that purpose, since the change in G** or ΔG, for a system at constant temperature and pressure actually contains ΔS for the **universe.** In other words, for a change carried out at constant presure and temperature, as the text shows, $\Delta G_{system} = -T\Delta S_{universe}$. So we really are not "ignoring" the surroundings; we are merely performing calculations for a set of conditions where the entropy change of the surroundings has a simple mathematical form.

There are two methods for calculating $\Delta G°$'s for chemical transformations. They both involve the same first step.

(a) You must **balance** the chemical reaction and include the correct phase (s,l,g) for each substance; also you need to know the proper allotropic forms of all **elements** used in the reaction.

(b) Depending on the data given in the problem, or the table of thermodynamic data available, use $\Delta G° = \Delta H° - T\Delta S°$ and go to step (c), or use $\Delta G° = \Delta G_f°$(products) - $\Delta G_f°$(reactants) and go to step d.

(c) Since you already know how to calculate $\Delta H°$ from Chapter 6 and ΔS from this chapter, merely substitute $\Delta H°$ and $\Delta S°$ into the $\Delta G°$ relation. **Be careful that temperature is expressed in Kelvins and S in kJ K^{-1} if $\Delta H°$ is in kJ.**

(d) Use Table 16.2 in the text, which contains standard Gibbs free energies of formation of substances. Remember, just as for **enthalpy** considerations, the **Gibbs free energy of formation of the most stable form of an element at 298 K and 1 atmosphere is defined to be zero.**

A negative $\Delta G°$ signifies that the reaction, at the given temperature (usually 298 K) and 1 atmosphere is **spontaneous;** that is, the reaction **may** proceed on its own in the direction written. However, the **rate** or speed of the reaction cannot be predicted from the sign and magnitude of $\Delta G°$.

Example 16.2

Calculate $\Delta G°$ at 298 K for the formation of 1 mole of $CO_2(g)$ from its elements in their standard states, using ΔH_f° and S° data.

Solution 16.2

The balanced chemical equation is

$$C(graphite) + O_2(g) \rightarrow CO_2(g) \qquad \Delta G° = \Delta H° - T\Delta S°$$

From Table 6.2 in the text

$$\Delta H° = \Delta H_f^\circ(CO_2,g) - \Delta H_f^\circ(C,graphite) - \Delta H_f^\circ(O_2,g)$$
$$= 1 \text{ mol} \times (-393.5 \text{ kJ mol}^{-1}) - 0 - 0$$
$$= -393.5 \text{ kJ}$$

Note that the enthalpies of formation for C(graphite) and $O_2(g)$ are zero. From Table 16.1 in the text

$$\Delta S° = S°(CO_2,g) - S°(C,graphite) - S°(O_2,g)$$
$$= 1 \text{ mol} \times 213.7 \text{ J mol}^{-1} \text{ K}^{-1} - 1 \text{ mol} \times 5.8 \text{ J mol}^{-1} \text{ K}^{-1}$$
$$- 1 \text{ mol} \times 205.0 \text{ J mol}^{-1} \text{ K}^{-1}$$
$$= +2.9 \text{ J K}^{-1}$$

Note that the entropies of the elements carbon and oxygen are not zero.

$$\Delta G° = \Delta H° - T\Delta S°$$

$$\Delta G° = -393.5 \text{ kJ} - 298 \text{ K} \times 2.9 \text{ J K}^{-1} \times \frac{10^{-3} \text{ kJ}}{1 \text{ J}}$$
$$= -394.4 \text{ kJ}$$

The reaction is spontaneous.

Example 16.3

Calculate $\Delta G°$ at 298 K for the reaction of Example 16.2, using only ΔG_f° data.

Solution 16.3

The desired reaction has already been balanced.

$$\Delta G° = \sum \Delta G_f^\circ(\text{products}) - \sum \Delta G_f^\circ(\text{reactants})$$

From Table 16.2 in the text we find a value of -394.4 kJ mol^{-1} for ΔG_f° for $CO_2(g)$. Carbon(graphite) and $O_2(g)$ are not listed in the table, since they are the most stable forms of the elements carbon and oxygen, respectively, at 298 K and 1 atmosphere. They have $\Delta G_f^\circ = 0$. Thus

$$\Delta G° = 1 \text{ mol} \times -394.4 \text{ kJ mol}^{-1} - 0 - 0$$
$$= -394.4 \text{ kJ}$$

as before, in Example 16.2. It makes no difference whether you use ΔG_f° data, or ΔH_f° and S° data, to calculate $\Delta G°$.

Gibbs Free Energy and Equilibrium Constant

As we have seen in Chapter 13, a comparison between the **reaction quotient** Q and **the equilibrium constant K** for a chemical reaction can tell us if a set of reactants with given concentrations and/or pressures will proceed in the direction of forming more **products,** in the direction of forming more **reactants,** or is at **dynamic equilibrium**. In that chapter you were given K's without being told how they were obtained. In the current chapter we find that there is a relationship between the Gibbs free energy change for a reaction and the Q/K ratio, in the form

$$\Delta G = RT \ln Q/K_p$$

In Chapter 16 you have also seen $\Delta G° = -RT \ln K_p$. ΔG for a reaction can be calculated if Q and K_p are known. The initial conditions allow us to calculate Q. K_p can be found using Table 16.2. So when you used the Q/K ratio to predict the direction of spontaneity in Chapter 13, you were using the result developed here in Chapter 16, namely that the sign of ΔG, at constant temperature and pressure, can be used to predict the direction of spontaneous change.

Example 16.4 in the text shows how to calculate K from $\Delta G°$, with use of $\Delta G° = -RT \ln K$. Be sure to convert R to $kJ\ mol^{-1}\ K^{-1}$ if $\Delta G°$ is in $kJ\ mol^{-1}$. Also don't forget to express temperature in Kelvins. Once you know K_p, and calculate Q (which has the same **form** as K_p, but contains **initial** concentrations and/or pressures), the sign and magnitude of ΔG follow directly. Remember, when the constant pressure is **not** 1 atmosphere, the sign of **ΔG**, and not **$\Delta G°$**, is a criterion for spontaneity. The sign of **$\Delta G°$** is a criterion for spontaneity only when one considers constant temperature and constant **1 atmosphere** pressure transformations.

Example 16.4

The water gas reaction is $H_2(g) + CO_2(g) \rightarrow H_2O(g) + CO(g)$.
The initial pressures for the four gases are 2.0 atm, 3.0 atm, 4.0 atm and 5.0 atm, respectively. Calculate ΔG for the water gas reaction for these conditions. Do reactants spontaneously form products under these conditions?

Solution 16.4

From the balanced equation we can write the form of Q.

$$Q = \frac{p_{H_2O} \times p_{CO}}{p_{H_2} \times p_{CO_2}} = \frac{4.0 \times 5.0}{2.0 \times 3.0} = 3.33$$

From $\Delta G° = \sum \Delta G_f°(\text{products}) - \sum \Delta G_f(\text{reactants})$ we calculate $\Delta G°$.

$$\Delta G° = \Delta G_f°(H_2O,g) + \Delta G_f°(CO,g) - \Delta G_f°(H_2,g) - \Delta G_f°(CO_2,g)$$

$$= 1\ mol \times (-228.6\ kJ\ mol^{-1}) + 1\ mol \times (-137.2\ kJ\ mol^{-1}) - 0$$

$$- 1\ mol \times (-394.4\ kJ\ mol^{-1})$$

$$= 28.6\ kJ$$

Then, since $\Delta G° = -RT \ln K_p$

$$\ln K_p = \frac{\Delta G°}{-RT} = \frac{28.6 \text{ kJ}}{8.31 \text{ J mol}^{-1} \text{ K}^{-1} \times 298 \text{ K} \times \frac{1 \text{ kJ}}{10^3 \text{ J}}} = -11.55$$

$$K_p = 9.6 \times 10^{-6}$$

Finally, from $\Delta G = RT \ln Q/K_p$, we have

$$\Delta G = 8.31 \text{ J mol}^{-1} \text{ K}^{-1} \times 298 \text{ K} \ln(\frac{3.33}{9.6 \times 10^{-6}}) \times \frac{1 \text{ kJ}}{10^3 \text{ J}} = 366 \text{ kJ mol}^{-1}$$

This positive ΔG signifies that the reaction will proceed in the direction of forming more **reactants.** The reaction is **not** spontaneous in the direction written.

SELF TEST

1. The second law of thermodynamics states that when contributions from the system and the surroundings are summed together, then for any spontaneous reaction
   a) energy is always conserved b) energy always increases
   c) entropy is always conserved d) entropy always increases
   e) both energy and entropy are always conserved
2. Which of the following substances have zero entropy at 298 K and 1 atm?
   (a) $O_2(g)$ (b) C(graphite) (c) $Br_2(g)$ (d) C(diamond) (e) none of these
3. Which of the following has the largest molar entropy at 298 K and 1 atm? (Predict without using any data).
   (a) $H_2O(g)$ (b) $Br_2(l)$ (c) C(graphite) (d) Ag(s) (e) $H_2O(l)$
4. Which of the following changes has (have) $\Delta S°_{system} = +$? (Predict without using any data).
   (a) $H_2O(l, 0°C) \rightarrow H_2O(s, 0°C)$
   (b) $CH_4(g, 25°C) + 2O_2(g, 25°C) \rightarrow 2H_2O(l, 25°C) + CO_2(g, 25°C)$
   (c) $H_2O(s, 0°C) \rightarrow H_2O(g, 100°C)$
   (d) $Br_2(l, 25°C) \rightarrow Br_2(g, 25°C)$
   (e) $CaO(s, 25°C) + CO_2(g, 25°C) \rightarrow CaCO_3(s, 25°C)$
5. Using Table 16.1 in the text, determine the entropy change $\Delta S°$ for the following reaction at 25°C.
   $$C_6H_6(l) + \frac{15}{2} O_2(g) \rightarrow 3H_2O(l) + 6CO_2(g)$$
   (a) 138.6 J K^{-1} (b) +217.5 J K^{-1} (c) 1320 J K^{-1}
   (d) -217.5 J K^{-1} (e) -138.6 J K^{-1}
6. Which of the following sets of signs for $\Delta H°$ and $\Delta S°$ are consistent with a spontaneous change at 1 atm and the appropriate constant temperature?
   (a) $\Delta H° = -$, $\Delta S° = -$, high T (b) $\Delta H° = +$, $\Delta S° = +$, low T
   (c) $\Delta H° = +$, $\Delta S° = -$, high T (d) $\Delta H° = -$, $\Delta S° = -$, low T
   (e) $\Delta H° = -$, $\Delta S° = +$, low T
7. Which of the following substances have zero Gibbs free energy of formation?
   (a) $O_2(g)$ (b) C(graphite) (c) $Br_2(g)$ (d) C(diamond) (e) none of these.

8. Using data from Tables 6.1 and 16.1 of the text, calculate $\Delta G°$ for the formation of one mole of gaseous benzene from the most stable form of its constituent elements in their standard states at 25°C:
(a) 129.7 kJ (b) 124.7 kJ (c) 46990 kJ (d) 4020 kJ (e) 86.8 kJ

9. At approximately what temperature do you expect the following reaction to be spontaneous at 1 atm?

$$N_2O_4(g) \rightleftharpoons 2NO_2(g)$$

For this reaction, $\Delta S°$ and $\Delta H°$ at 25°C are 176.6 J K^{-1} and 58.0 kJ, respectively.
(a) 300 K (b) 310 K (c) 0.3 K (d) 330 K (e) 30 K

10. Calculate ΔG, in kJ, for the following reaction at 25°C:

$$CaCO_3(s) \rightleftharpoons CaO(s) + CO_2(g)$$

The CO_2 pressure is 10^{-3} atm. You may use data from Chapter 16 of the text.
(a) 123.5 (b) 148.0 (c) 113.8 (d) 138.3 (e) -17.1

11. Which of the following is (are) true?
(a) coupled reactions violate the second law of thermodynamics
(b) coupled reactions often occur in biochemical systems
(c) it is impossible to convert ADP to ATP in living systems
(d) it is impossible to convert ATP to ADP in living systems
(e) the net ΔG for a set of coupled reactions must be negative for a spontaneous change.

12. Using data in Chapter 16, calculate the equilibrium constant K_p (in units of atm^{-1}) at 298 K for the reaction

$$2NO_2(g) \rightleftharpoons N_2O_4(g)$$

(a) 0.14 (b) 1.00 (c) 6.9 (d) 69 (e) 1.07×10^{10}

13. For the reaction $NH_3(g) \rightleftharpoons 1/2N_2(g) + 3/2H_2(g)$, determine the direction of spontaneous change when a mixture containing 4.00 atm N_2, 4.00 atm H_2 and 2.00 atm NH_3 is allowed to react at 25°C.
(a) reaction proceeds to right (b) reaction proceeds to left
(c) reaction is at equilibrium

14. The value of the reaction quotient Q for the mixture given in question 13 is
(a) 0.016 (b) 0.125 (c) 8 (d) 16 (e) 64

<u>Answers to Self Test</u>

1. d; 2. e; 3. a; 4. c,d; 5. d; 6. d,e; 7. a,b; 8. a; 9. d; 10. c; 11. b,e; 12. c; 13. b; 14. c.

CHAPTER 17

ELECTROCHEMISTRY

SUMMARY REVIEW

In **electrolysis,** electrical energy is used to drive nonspontaneous reactions; in **electrochemical cells,** spontaneous reactions are used to generate electrical energy.

An **electrolytic cell** has two electrodes connected to a DC power supply and immersed in an electrolyte (molten salt or aqueous solution); oxidation-reduction (electron transfer) reactions occur at the electrode surfaces and may involve ions in the electrolyte, the solvent, or the electrodes themselves. In electrolysis of a **molten salt** using inert electrodes, positive ions accept electrons from the cathode and are reduced. Simultaneously, negative ions transfer electrons to the anode and are oxidized. Within the electrolyte the movement of the ions carries the current by ionic conduction; within the electrodes and the external circuit, the movement of electrons occurs by metallic conduction.

Many metals, especially very reactive metals, such as those of groups 1 and 2, have cations that are difficult to reduce chemically, and therefore are made by electrolysis of their molten salts. For example, molten NaCl may be electrolyzed using inert electrodes to give sodium metal and chlorine gas. In the electrolysis of a molten mixture of Al_2O_3 and Na_3AlF_6 -- the **Hall process** for preparing aluminum -- the stainless steel cathode is inert but the carbon anodes are not. The complex electrode reactions may be summarized approximately as:

cathode $\quad 4 \times (Al^{3+} + 3e^- \rightarrow Al)$ $\quad$ reduction

anode $\quad 3 \times (C(s) + 2O^{2-} \rightarrow CO_2(g) + 4e^-)$ $\quad$ oxidation

$$4Al^{3+} + 6O^{2-} + 3C \rightarrow 4Al + 3CO_2$$

In the electrolysis of **aqueous solutions,** the electrolyte ions, the electrodes, or some of the water may be selectively oxidized or reduced; with inert electrodes, water may be oxidized and/or reduced in preference to any of the ions in solution:

Cations	Anions
Less easy to reduce than water Na^+, K^+, Mg^{2+}, Ca^{2+}, Al^{3+}	Less easy to oxidize than water F^-, SO_4^{2-}, NO_3^-, ClO_4^-, CO_3^{2-}, PO_4^{3-}
More easy to reduce than water Cu^{2+}, Ag^+, H_3O^+	More easy to oxidize than water Cl^- (borderline), Br^-, I^-

Electrolysis of aqueous solutions using non-inert electrodes is employed in **electrorefining** and **electroplating.**

Quantitatively, the charge on 1 mole of electrons (the **Faraday, F)** is 96 485 coulombs (C): **1 F = 96 485 C.** A current of **A** amps flowing for **t** secs generates **A·t** coulombs of charge. By first converting information about the current used in an electrolysis, and its duration, into moles of electrons, calculations involving oxidation-reduction half-reactions are handled using the normal rules of stoichiometry.

In setting up an **electrochemical cell** in order to utilize the spontaneous flow of electrons to produce a usable electric current, two **half-cells** are constructed. The two solutions usually are connected by a porous barrier, or salt bridge, so that ions can move from one half-cell to the other without the solutions mixing, and the two electrodes are connected by a conducting wire.

The voltage of a single half-cell cannot be measured; it must always be combined with another half-cell to obtain a current (voltage). However, by constructing cells where one half-cell contains a hydrogen electrode, and taking the potential of a **standard hydrogen electrode** to be **zero,** a table of **standard reduction potentials** can be established. These values are for half-cells in which all ions have concentrations of 1 M and all gases have pressures of 1 atm. Substances with very negative E°_{red} values, e.g., $Li^+(aq)$, are very weak oxidizing agents; those with large positive values, e.g., $F_2(g)$ are very strong oxidizing agents. E°_{red} values are used to predict standard cell voltages and the direction of spontaneous oxidation-reduction that occurs when any two half-cells are joined. For the half-cell that is a reduction, the half-cell potential is E°_{red} and for the half-cell that is an oxidation, the half-cell potential E°_{ox} is $-E^\circ_{red}$ The cell potential, E°_{cell} results from adding E°_{red} and E°_{ox}. If the result is a **positive** voltage, the cell reaction occurs **as written;** if the voltage obtained is **negative,** it is the **reverse** reaction that is spontaneous.

The cell voltage measures the tendency of the cell reaction to proceed, and is directly related to the Gibbs free energy, ΔG, by the relationship

$$\Delta G = -nFE_{cell}$$

where F is the Faraday constant and n is the number of moles of electrons transferred in the reaction. When the cell reaction has achieved equilibrium, ΔG and E_{cell} are both zero.

For **non-standard conditions** the cell voltage is obtained using the **Nernst equation**

$$E_{cell} = E^\circ_{cell} - \frac{RT}{nF} \ln Q \quad \text{or} \quad E_{cell} = E^\circ_{cell} - \frac{0.026}{n} \ln Q \ (25°C)$$

where Q is the reaction quotient and **n** is the number of moles of electrons transferred in the reaction. As the cell reaction proceeds, the cell voltage decreases to **zero** and the reaction achieves equilibrium: $E_{cell} = 0$ and $Q = K_{eq}$, so that

$$E_{cell} = E^\circ_{cell} - \frac{0.026}{n} \ln K_{eq} = 0 \quad \text{or} \quad \ln K_{eq} = nE^\circ_{cell}/0.026 \; (25^\circ C)$$

This enables the equilibrium constant for the oxidation-reduction reaction to be calculated from the standard cell voltage.

Batteries are electrochemical cells that have been developed for practical use. **Primary** cells deliver an almost constant voltage until they attain equilibrium and are dead. **Secondary cells,** such as the lead storage battery, can be "recharged" by passing a current through them. In **fuel cells** the reactants are constantly replaced as they are used up.

Some metals are protected from the electrochemical process called **corrosion** by a protective oxide layer; for others an oxide layer is porous and permits moisture, oxygen, and water to reach the metal surface, which allows a cell to be set up with one point on the metal surface acting as the anode ($M(s) \rightarrow M^{n+} + ne^-$) and another point as the cathode, where $O_2(g)$ is reduced in the presence of water ($O_2(g) + 2H_2O(l) + 4e^- \rightarrow 4OH^-(aq)$). If the sum of E_{ox} for the metal and E_{red} for O_2 is positive, the metal corrodes, and the rate of corrosion is enhanced if the water contains dissolved salts. Corrosion is limited by coating the metal with a less easily oxidized metal, by painting it, or alloying it with another metal. In cathodic protection, the metal is connected to a more easily oxidized metal which corrodes preferentially.

REVIEW QUESTIONS

1. Define the term electrolysis.
2. In the electrolysis of molten NaCl, how are electrons carried through the electrolyte, and how are they carried through the electrodes and the external circuit?
3. In terms of their capacities for supplying or accepting electrons, define anode and cathode.
4. Why are positively charged ions called cations and negatively charged ions called anions?
5. Is the reaction that occurs at a cathode a reduction or an oxidiation? Is the reaction at an anode a reduction or an oxidation?
6. What reaction occurs at the cathode, and what reaction occurs at the anode, during the electrolysis of molten $MgCl_2$.
7. Using electrolysis, how can potassium metal be prepared?
8. What molten mixture is electrolyzed commercially to make aluminum? Why is it important to utilize it rather than say molten Al_2O_3?
9. In the electrolytic production of aluminum (Hall process), what equations approximately express the reactions that occur at the anode and at the cathode?
10. Can you suggest a way in which small amounts of aluminum could be made in the laboratory without using electrolysis?
11. What charge corresponds to that on one mole of electrons, given that the charge on one electron is 1.60206×10^{-19} C?
12. How many coulombs of electricity are produced when a current of 5.00 A is passed for 10.00 hrs? What is the charge expressed in faradays?

13. What electrode reactions account for the production of hydrogen at the cathode and oxygen at the anode when an aqueous solution of Na_2SO_4 is electrolyzed?
14. In practice, why is it impossible to electrolyze very pure water?
15. Write an equation expressing the reaction that occurs when a piece of zinc is dipped into a solution of $CuSO_4(aq)$. Write equations for the oxidation and reduction reactions.
16. What experimental set-up is necessary in order to produce an electric current from an oxidation-reduction reaction, such as that in question 15?
17. Using the convention for writing electrochemical cells, depict the cell containing a zinc electrode immersed in a solution of zinc sulfate and a copper electrode immersed in a solution of copper sulfate. Write equations for the half-reactions.
18. Why is it necessary to separate the two half-cells of an electrochemical cell with a porous barrier or salt bridge?
19. How is a standard hydrogen electrode constructed? What arbitrary assumption is made about the potential (voltage) of the standard hydrogen half-cell?
20. How is the potential of a cell operating under standard conditions calculated?
21. How can you tell from their standard reduction potentials that $H_2O_2(aq)$ is a stronger oxidizing agent than dichromate ion, $Cr_2O_7^{2-}$, in acidic solution under standards conditions?
22. What is the significance of a negative cell potential?
23. What is the standard oxidation potential for the reaction $Ca(s) \rightarrow Ca^{2+}(aq) + 2e^-$?
24. Under what conditions is the potential of a cell zero?
25. What relationship relates $\Delta G°$ for a cell reaction to its $E°_{cell}$ value and the number of electrons transferred in the cell reaction?
26. Write the general form of the Nernst equation -- the quantitative relationship between the cell potential, the standard cell potential and the concentrations of the reagents.
27. How is the solubility product constant of AgCl calculated from standard reduction potentials?
28. What is the essential difference between a primary cell, a secondary cell, and a fuel cell?
29. What reactions occur in the Leclanché or dry cell?
30. What reactions occur at the electrodes of the lead-storage battery when it is recharged?
31. What reactions occur at the electrodes of a hydrogen-oxygen fuel cell?
32. Why is iron galvanized to protect it from corrosion?
33. Why is a layer of tin not a very satisfactory way to protect iron from rusting?
34. Would copper be a satisfactory metal to be joined to iron to protect it from corrosion?

<u>Answers to Selected Review Questions</u>

12. 1.80×10^5 C; 1.87 F. 22. Reverse reaction is spontaneous. 24. When

cell reaction is at equilibrium. 34. No, since Fe is more easily oxidized than Cu.

OBJECTIVES

Be able to:

1. Define the terms electrolysis, anode, cathode, electrolyte.
2. Describe the electrolysis of molten sodium chloride in terms of reactions taking place at the electrodes; draw a cell illustrating this electrolysis.
3. Give examples of metals that are prepared by electrolysis of molten salts.
4. Perform calculations involving current flow and the time required to produce a given quantity of product in an electrolysis.
5. Give examples of electrolysis of aqueous solutions in which water undergoes oxidation and/or reduction.
6. Describe how chlorine and sodium hydroxide are made industrially.
7. Give examples involving electrolysis with non-inert electrodes and describe the electroplating process.
8. Construct an electrochemical cell from half-cells and a salt bridge, use a cell diagram, and write the balanced reaction corresponding to a cell diagram in terms of half-reactions.
9. Use a table of standard reduction potentials to calculate the potential of an electrochemical cell under standard conditions.
10. Use cell potential values to predict which way a reaction proceeds.
11. Calculate $\Delta G°$ values and equilibrium constants from electrochemical cell voltages.
12. Use the Nernst equation to calculate cell potentials for non-standard conditions.
13. Obtain equilibrium constants using the Nernst equation.
14. Explain how pH can be measured using an electrochemical cell.
15. Describe two types of batteries: primary and secondary cells.
16. Describe the corrosion process and how it can be inhibited.

PROBLEM SOLVING STRATEGIES

17.1 Electrolysis

In an electrolysis, **reduction takes place at the cathode,** and **oxidation occurs at the anode.** It is important to be able to decide what are the products in an electrolysis. First, in a **molten salt**, the cation of the salt is reduced, and the anion is oxidized, if the electrodes are inert. Second, in an **aqueous solution**, not only is a cation a candidate for reduction, but, as well, the **solvent** may be reduced. Whether a cation or the solvent is reduced depends on the relative ease of reduction of these two species. In the same way, either an anion or the solvent can be oxidized, depending on which has the greater oxidation tendency. Third, there is the possibility, in molten salt or aqueous solution electrolysis, that the **electrodes** undergo reaction. It is a good idea to learn the summary of electrode reactions given at the end of Section 17.1 of the text. This lists some of the common cations and anions and their tendencies to be reduced or oxidized, relative to the solvent water. It also contains some examples of reactive electrodes. So in order to predict the products of electrolysis, you need to consider three possibilities: the ions can undergo reaction; the solvent can undergo reaction; the electrode can undergo reaction.

Example 17.1

Give the products at the cathode and anode for the following electrolyses: (a) molten sodium fluoride, using inert electrodes; (b) aqueous copper nitrate, using inert electrodes.

Solution 17.1

(a) In a molten salt, with the electrodes inert, the cation is reduced at the cathode. Thus the process

$$2Na^+ + 2e^- \rightarrow 2Na(l)$$

occurs at the cathode. This is a reduction. At the anode, F^- is oxidized to F_2 gas.

$$2F^- \rightarrow F_2(g) + 2e^-$$

The overall cell reaction is

$$2Na^+ + 2F^- \rightarrow 2Na(l) + F_2(g)$$

Both reduction and oxidation must involve the same number of electrons. We could have written these as **one** electron changes. Then the overall reaction would have been

$$Na^+ + F^- \rightarrow Na(l) + 1/2F_2(g)$$

(b) When we consider reduction at the cathode, either Cu^{2+} or H_2O can be reduced. If we know that Cu^{2+} has a much greater tendency for reduction that has H_2O, then we will be able to write

$$2Cu^{2+} + 4e^- \rightarrow 2Cu$$

Thus, even though the electrode is inert (i.e., it is not decomposed by the electrolysis), copper is plated on the electrode. At the anode, oxidation takes place. We know that NO_3^- contains nitrogen in the highest possible (+5) oxidation state, so the solvent H_2O is more readily oxidized. The electrode reaction is

$$2H_2O \rightarrow O_2 + 4H^+ + 4e^-$$

The overall cell reaction is the sum of the 2 half-reactions.

$$2Cu^{2+} + 2H_2O \rightarrow 2Cu + O_2 + 4H^+$$

After you know the products of an electrolysis, you can perform calculations on how much a product is formed under given conditions. You need to realize that when 96,500 coulombs flow in an electrolysis, one mole of electrons is consumed at the cathode and produced at the anode. Steps to follow in electrolysis calculations are these:

1. Knowing the products of the electrolysis, write the half reactions and the overall reaction.
2. Determine how many moles, n, of electrons are involved, using $Q = nF$, where F is the Faraday constant, 96 500 C mol^{-1}. Remember charge in coulombs equals amperes times seconds, i.e., $Q = it$ ("Quit").
3. Determine how many moles of product are formed from the n moles of electrons. For example, from $2Cl^- \rightarrow Cl_2 + 2e^-$, we know that **one** mole of Cl_2 is produced in an electrolysis involving **two** moles of electrons.
4. If needed, convert moles of product into grams (or volume, if a gaseous product).

Example 17.2

A current of 5.00 A is passed for exactly 2.00 hours through two electrolysis cells joined in series. One contains Zn^{2+}, the other Fe^{3+}. Assume that the only reduction is that of the ion to the metal. How many grams of each metal will be deposited on each of the two cathodes?

Solution 17.2

At one cathode, $Zn^{2+} + 2e^- \rightarrow Zn$ takes place, while at the other, $Fe^{3+} + 3e^- \rightarrow Fe$ occurs. The number of coulombs Q that pass through each of the cells (since the cells are joined in **series**, the same current flows through both) is

$$Q = 5.00 \text{ A} \times \frac{1 \text{ C}}{1 \text{ A s}} \times 2.00 \text{ h} \times \frac{60 \text{ min}}{1 \text{ h}} \times \frac{60 \text{ s}}{1 \text{ min}} = 3.60 \times 10^4 \text{ C}$$

We divide by the Faraday constant to convert this current flow into an equivalent number of moles of electrons.

$$n = \frac{Q}{F} = \frac{3.60 \times 10^4 \text{ C}}{9.65 \times 10^4 \text{ C mol}^{-1}} = 0.373 \text{ mol electrons}$$

Since 2 moles of electrons are required to produce 1 mole of Zn, which has a mass of 65.4 g,

$$\text{gm of Zn produced} = 0.373 \text{ mol electrons} \times \frac{1 \text{ mol Zn}}{2 \text{ mol electrons}} \times \frac{65.4 \text{ g}}{1 \text{ mol}}$$

$$= 12.2 \text{ g Zn}$$

And since 3 moles of electrons are needed to yield 1 mole of Fe, which has a mass of 55.8,

$$\text{gm of Fe produced} = 0.373 \text{ mol electrons} \times \frac{1 \text{ mol Fe}}{3 \text{ mol electrons}} \times \frac{55.8 \text{ g}}{1 \text{ mol}}$$

$$= 6.94 \text{ g Fe}$$

17.2 Electrochemical Cells

Cell Potential and Standard Reduction Potential

The overall cell voltage **must be positive** in an electrochemical cell. From what we have seen in Chapter 16, this is the same as the statement that ΔG is **negative** in an electrochemical cell. Under standard conditions, the cell voltage is called E°_{cell} . Two half-reactions take place in the cell, a reduction and an oxidation. No electrons appear in the balanced cell reaction since the electrons supplied by the oxidation half-reaction are used up in the reduction half-reaction. It is convenient for tabulation purposes to consider all half-reactions as **reductions**. The tendency for a reduction half-reaction to occur, under standard conditions, is measured by the standard reduction potential, E°_{red}, for that half-reaction. The tendency for the same half-reaction to occur as an **oxidation**, under standard conditions, is called E°_{ox}, where $E^\circ_{ox} = -E^\circ_{red}$. Use E°_{red} when the half-reaction is written as a reduction, and use E°_{ox} when it is written as an oxidation. The cell voltage, under standard conditions, equals the standard **reduction** potential for the half-reaction that is a **reduction** plus the standard **oxidation** potential for the half-reaction that is an **oxidation**, or $E^\circ_{cell} = E^\circ_{red} + E^\circ_{ox}$. Often a shorthand notation is used for an electrochemical cell; it is called a **cell diagram**. In such a diagram, materials on the left are involved in an **oxidation**, while those substances on the **right** take part in a **reduction**. To calculate a cell voltage from such a diagram, under standard conditions, follow these steps.

1. Write the 2 half-reactions, and combine them to get the overall cell reaction.
2. Consult a table of E°_{red}'s and find E°_{red} for the reduction half-reaction.
3. For the oxidation half-reaction required, read the table from right to left. Convert the E°_{red} listed to an E°_{ox}, by $E^\circ_{ox} = -E^\circ_{red}$.
4. Calculate E°_{cell} using $E^\circ_{cell} = E^\circ_{red} + E^\circ_{ox}$.

One of the many uses of standard cell potentials is to predict whether or not a proposed reaction (composed of 2 half-reactions) will have the possibility of occurring. If we use the 2 half-reactions to calculate an E°_{cell}, and it is positive, we know the reaction can happen. Just use the steps outlined above.

Example 17.3

Which of the halides ions X^- can be oxidized to the corresponding halogen X_2 with Fe^{3+}, producing Fe^{2+}, under standard conditions?

Solution 17.3

There are 4 oxidations to consider,

$2F^- \rightarrow F_2 + 2e^-$, $2Cl^- \rightarrow Cl_2 + 2e^-$, $2Br^- \rightarrow Br_2 + 2e^-$, $2I^- \rightarrow I_2 + 2e^-$

and one reduction

$$Fe^{3+} + e^- \rightarrow Fe^{2+}$$

Thus there are 4 cell reactions for which we need to calculate E°_{cell} values.

$$2F^- + 2Fe^{3+} \rightleftarrows F_2 + 2Fe^{2+} \qquad 2Cl^- + 2Fe^{3+} \rightleftarrows Cl_2 + 2Fe^{2+}$$

$$2Br^- + 2Fe^{3+} \rightleftarrows Br_2 + 2Fe^{2+} \qquad 2I^- + 2Fe^{3+} \rightleftarrows I_2 + 2Fe^{2+}$$

From Table 17.1 we find $E^\circ_{red} = 0.77$ V for Fe^{3+} being reduced to Fe^{2+}. For the 4 **oxidation** half reactions, we find in the order listed above, $E^\circ_{ox} = -2.87$, -1.36, -1.09, -0.54 V. Using E°_{cell} $E^\circ_{red} + E^\circ_{ox}$, we get a positive voltage only for the last reaction,

$$2I^- + 2Fe^{3+} \rightleftarrows I_2 + 2Fe^{2+} \qquad E^\circ = 0.77 - 0.54 = 0.23 \text{ V}$$

So only I^-, among the halide ions, can be oxidized to the halogen by Fe^{3+}. Remember from earlier chapters that the order of **halogens** as **oxidizing agents** (the order of ease of reduction of the halogens) is: $F_2 > Cl_2 > Br_2 > I_2$. The calculation tells us that Fe^{3+} is a better oxidizing agent than is I_2. Thus, Fe^{3+} is reduced and I^- is oxidized.

Nonstandard Conditions: Nernst Equation

The voltage developed by an electrochemical cell depends on the concentrations of any ions present in the reaction, and on the pressures of any gases taking part. Under nonstandard conditions, the cell voltage is called E_{cell}. It is related to E°_{cell} by the Nernst equation,

$$E_{cell} = E^\circ_{cell} - \frac{0.0592}{n} \log Q$$

To calculate E_{cell}, first calculate E°_{cell}, according to the steps outlined earlier. Second, since you already have the balanced cell reaction, you can form Q, the reaction quotient, and calculate its value. Remember Q refers to **initial** concentrations. The number of moles of electrons, n, in the Nernst equation is obtained from either of the 2 half-reactions that you combined to get the balanced cell reaction. When $E_{cell} = 0$, the tendencies for reduction and oxidation are exactly balanced, and $\Delta G = 0$. **The system is at equilibrium, and is described by an equilibrium constant K.** This has the same **form** as Q, but Q = K only at equilibrium when $E_{cell} = 0$. Thus we can write the Nernst equation, with $E_{cell} = 0$, set Q = K, and determine K from E°_{cell}. Note carefully that $E_{cell} \neq E^\circ_{cell}$. Also realize that the same symbol Q has been used earlier in the chapter to represent charge.

Example 17.4

For the cell

$$Fe|Fe^{2+}\ (1.0 \times 10^{-2}\ M)||Pb^{2+}(2.0 \times 10^{-1}\ M)|Pb$$

write the cell reaction and determine E_{cell}. Would the cell, as diagrammed, actually behave as an electrochemical cell?

Solution 17.4

First, write the half-reactions and the overall cell reaction.

Left, oxidation $\quad Fe \rightarrow Fe^{2+} + 2e^-$

Right, reduction $Pb^{2+} + 2e^- \rightarrow Pb$

$$Fe + Pb^{2+} \rightleftarrows Fe^{2+} + Pb$$

Next, calculate E°_{cell} from E°_{red} data in Table 17.1.

$$E^\circ_{cell} = E^\circ_{red} + E^\circ_{ox} = -0.12 + (+0.44) = +0.32 \text{ V}$$

Finally, we write the Nernst equation with the appropriate form for Q, substitute the given concentrations, insert n from the half reactions, and obtain E_{cell}.

$$E_{cell} = E^\circ_{cell} - \frac{0.0592}{n} \log \frac{[Fe^{2+}]}{[Pb^{2+}]}$$

$$= 0.32 - \frac{0.0592}{2} \log \frac{1.0 \times 10^{-2}}{2.0 \times 10^{-1}}$$

$$= 0.32 + 0.04 = 0.36 \text{ V}$$

Since E_{cell} is positive, the cell diagram does correspond to an electrochemical cell, and operates as diagrammed.

Equilibrium Constants from the Nernst Equation

At **equilibrium,** the potential of an electrochemical cell is zero, and the reaction quotient Q equals the equilibrium constant K for the cell reaction. Thus we write

$$0 = E^\circ_{cell} - \frac{0.0592}{n} \log K$$

Likewise $\Delta G = 0$ at equilibrium, but not ΔG°.

Example 17.5

Write the form of K, calculate its value, and calculate ΔG°, in kJ, for

$$Cr_2O_7^{2-} + 3Cu + 14H^+ \rightleftarrows 2Cr^{3+} + 3Cu^{2+} + 7H_2O$$

Solution 17.5

By inspection we can write the form of K.

$$K = \frac{[Cr^{3+}]^2[Cu^{2+}]^3}{[Cr_2O_7^{2-}][H^+]^{14}}$$

Note that the concentration terms for Cu and H_2O are omitted. To calculate K, first we break the overall reaction into two balanced half-reactions.

oxidation $3Cu \rightarrow 3Cu^{2+} + 6e^-$

reduction $Cr_2O_7^{2-} + 14H^+ + 6e^- \rightarrow 2Cr^{3+} + 7H_2O$

Next we calculate E°_{cell} using Table 17.1. We find E°_{red} = 1.33 V for the $Cr_2O_7^{2-}$ half-reaction. However, we do not find an entry for $3Cu^{2+} + 6e^- \rightarrow 3Cu$. Instead we find $Cu^{2+} + 2e^- \rightarrow Cu$. But the tendency for Cu^{2+} to form Cu does not depend on how we choose to state the balanced half-reaction. So the E°_{ox} we desire is

$$E^\circ_{ox} = -E^\circ_{red} = -(0.34\ V)$$

Then

$$E^\circ_{cell} = E^\circ_{red} + E^\circ_{ox} = 1.33 - 0.34 = 0.99\ V$$

From the Nernst equation, with $E_{cell} = 0$,

$$0 = E^\circ - \frac{0.0592}{6} \log K = 0.99 - 0.00987 \log K$$

$$\log K = 100.30 \qquad K = 2.0 \times 10^{100}$$

Such a large value for K signifies that the position of equilibrium lies very far to the right. In other words, the reaction is essentially complete. Finally, to calculate ΔG° for the reaction we use

$$\Delta G^\circ = -nFE^\circ_{cell}$$

$$\Delta G^\circ = -6 \times 96500\ C \times 0.99\ V \times \frac{1\ kJ}{1000\ J} \times \frac{1\ J}{1\ V\ C}$$

$$\Delta G^\circ = -573\ kJ$$

SELF TEST

Part I True or False

1. The electrolysis of aqueous sodium chloride is the principal method of industrial production of chlorine and sodium hydroxide.
2. In the electrolytric refining of copper, impure copper is used as the cathode.
3. The main industrial source of aluminum is the electrolysis of pure cryolite, Na_3AlF_6.
4. In metallic conduction the metal cations are the charge carriers.
5. Tables of half-cell electrode potentials are usually given for reduction half-reactions, since reductions are easier to carry out than oxidations.
6. Fluorine is more easily reduced than H^+.
7. The lead storage battery is an example of a secondary cell.
8. The oxidation of Fe to Fe^{2+} is one of the steps in the complex process of rusting.

Part II Multiple Choice (You may use Table 17.1 from the text for these problems).

9. In the electroslysis of molten KCl, chlorine gas would be produced
   (a) by oxidation at the anode (b) by oxidation at the cathode
   (c) by reduction at the anode (d) by reduction at the cathode
   (e) in negligible quantities, and instead oxygen gas would be produced.
10. The number of coulombs required to deposit 26.0 grams of chromium (molar mass Cr = 52.0 g) by electrolysis of an aqueous solution of $CrCl_3$ is (96 500 Coulombs = 1 Faraday)
    (a) 4.83×10^4 (b) 9.65×10^4 (c) 3.22×10^4
    (d) 1.45×10^5 (e) 9.65×10^5

11. In electrolysis, decide which grouping of the following statements is true.
   (1) When a water solution of NaCl is electrolyzed, the solution become basic.
   (2) The anode is the negative electrode.
   (3) In the electrolysis of water, O_2 may be formed at the anode, but never at the cathode.
   (4) In the electrolysis of water, the same gases are formed as are obtained from boiling water in a beaker in the laboratory.
   (5) High purity sodium is readily and directly obtained (say for use in the laboratory) by electrolysis of water solutions of NaCl, using platinum electrodes.
   (a) (1),(2),(3),(4), and (5) are all true.
   (b) Only (3) and (4) are true.
   (c) Only (1) and (3) are true.
   (d) Only (1) and (5) are true.
   (e) None of the statements are true.
12. What weight of copper is produced by the passage of 1.600 amps for 1.000 hour through a Cu^{2+} solution? Molar mass of copper is 63.54 g. (96 500 Coulombs = 1 Faraday)
   (a) 0.948 g (b) 1.896 g (c) 3.792 g (d) 0.0316 g (e) 0.0632 g
13. The purpose of a salt bridge in a galvanic cell is
   (a) to provide a pathway so the reducing agent can reach the oxidizing agent.
   (b) to provide a connection between the cathode and anode solutions so positive ions can gather around the anode.
   (c) to provide a pathway for the current carriers in solutions.
   (d) to increase the resistance to the flow of current so that the cell maintains its potèntial.
   (e) none of the above.
14. The standard hydrogen electrode potential is zero volts at 25°C because;
   (a) There is no potential difference between the electrode and the solution.
   (b) It has been defined that way.
   (c) It has been measured accurately as zero with respect to many electrodes.
   (d) The hydrogen ion acquires electrons from a platinum electrode.
   (e) Hydrogen gas, H_2, is the naturally-occurring form of the element.
15. Using a cell diagram, the galvanic cell in which the following reaction occurs, $Sn + 2Ag^+ \rightleftharpoons Sn^{2+} + 2Ag$, may be represented as
   (a) $Ag|Ag^+||Sn^{2+}|Sn$ (b) $Sn|Sn^{2+}||Ag^+|Ag$ (c) $Ag|Sn^{2+}||Ag^{2+}|Sn$
   (d) $Sn|Ag^+||Sn^{2+}|Ag$ (e) none of the above
16. The tendency, (measured in volts) under standard conditions, for the half-reaction $2Ni \rightarrow 2Ni^{2+} + 4e^-$ to occur is
   (a) -0.25 (b) +0.25 (c) -0.50 (d) +0.50 (e) -1.00
17. E° for the cell $Fe|Fe^{2+}||Pb^{2+}|Pb$ is:
   (a) -0.57 V (b) 0.57 V (c) 0.31 V (d) -0.31 V (e) 0.155 V
18. Which of the following will be oxidized by Cu^{2+} at 25°C under standard conditions?
   (a) Ni (b) Ag (c) Al (d) Cr^{3+} (e) F^-

19. For the overall cell reaction: $Au + 3K^+ \rightleftharpoons Au^{3+} + 3K$, the standard cell potential is -4.43 V. The negative sign of E° indicates that;
(a) gold, Au, is a stronger reducing agent than K^+
(b) the oxidizing power of K^+ is much greater than that of Au^{3+}
(c) Au^{3+} is a stronger oxidizing agent than K^+
(d) the reaction proceeds on its own in the direction written
(e) none of the above

20. The standard reduction potential at 25°C for $Zn^{2+} + 2e^- \rightarrow Zn$ is -0.76 V. Calculate ΔG°, in kJ, for

$$Zn(s) + 2H^+(aq) \rightleftharpoons Zn^{2+}(aq) + H_2(g)$$

The faraday constant is 96,500 C
(a) 146.7 (b) -146.7 (c) 73.3 (d) -73.3 (e) none of these

21. At 25°C, the cell potential (in volts) for the reaction $Cu + 2Ag^+ (0.01\ M) \rightleftharpoons Ag + Cu^{2+} (0.01\ M)$ is
(a) $E° - 0.059$ (b) $E° - \frac{0.059}{2}$ (c) $E° + \frac{0.059}{2}$
(d) $E° + 0.059$ (e) $E° - 0.059 \log 2$

22. For the cell whose reaction is $2H^+ + Cu \rightleftharpoons H_2(g) + Cu^{2+}$ calculate the cell potential at 25°C when $[Cu^{2+}] = 10^{-4}$ M, $p(H_2) = 10^{-3}$ atm, $[H^+] = 10^{-2}$ M.
(a) -0.82 V (b) -0.53 V (c) -0.34 V (d) -0.25 V (e) +0.19 V

Answers to Self Test

1. T; 2. F; 3. F; 4. F; 5. F; 6. T; 7. T; 8. T; 9. a; 10. d; 11. c; 12. b; 13. c; 14. b; 15. b; 16. b; 17. c; 18. a,c; 19. c; 20. b; 21. a; 22. d.

CHAPTER 18

FURTHER CHEMISTRY OF NITROGEN AND OXYGEN

SUMMARY REVIEW

Nitrogen is essential to life but N_2 in the atmosphere is very unreactive. Conversion of N_2 to readily-utilized compounds, such as NH_3 and substances containing NH_4^+ or NO_3^-, is called **nitrogen fixation.** Nature accomplishes this with catalysts such as the enzyme nitrogenase; industrially the most important process is the Haber process using N_2 and H_2. Nitrogen oxides are formed by the reaction of N_2 and O_2 in lightning discharges.

Nitrogen monoxide, NO (nitric oxide), is manufactured by reacting NH_3 with O_2 using a hot platinum-rhodium wire gauze as catalyst. In the laboratory it is made by reducing **dilute** $HNO_3(aq)$ with Cu. It is a major atmospheric pollutant from automobile exhaust and coal- and oil-fired electricity generating plants. NO(g) reacts with $O_2(g)$ to give **nitrogen dioxide,** NO_2, a red-brown gas, which exists in equilibrium with its dimer, N_2O_4, **dinitrogen tetroxide,** both in the gas phase and when condensed to a deep red-brown liquid at 21°C. Below -11°C, a white solid consisting entirely of N_2O_4 molecules is formed. In the laboratory NO_2 is prepared by reducing **concentrated** $HNO_3(aq)$ with Cu, or by heating Pb(II) nitrate.

Molecules such as NO and NO_2 contain an odd number of valence electrons and are known as **free radicals.** Unlike other free radicals, which are very reactive, NO is rather inert; NO molecules do not combine to form N_2O_2 except at low temperatures. In writing its Lewis structures, the odd electron may be placed either on nitrogen, $(:\dot{N}{=}\ddot{O}:)$ or on oxygen, $(^{-}:\ddot{N}{=}\dot{O}:^{+})$, or, alternatively, the odd electron may be shared between nitrogen and oxygen, $(^{1/2-}:\dot{N}{\dot{=}}\dot{O}:^{1/2+})$. The latter is consistent with the observed bond length of 115 pm, the length expected for a bond order of $2\frac{1}{2}$. The molecule NO_2 is bent with a bond angle of 134°. Two NO_2 molecules combine to give N_2O_4 while an NO_2 and an NO molecule combine to give N_2O_3, **dinitrogen trioxide,** a deep blue liquid which decomposes back to NO_2 and NO below room temperature. N_2O_3, N_2O_4, and N_2O_2 all have long and rather weak N-N bonds.

When $NH_4NO_3(s)$ is gently heated, it melts and decomposes to give N_2O, **dinitrogen monoxide,** (nitrous oxide) and water. The anhydride of nitric acid, **dinitrogen pentaoxide,** N_2O_5, is obtained by dehydrating HNO_3 with P_4O_{10}. In the gas phase it is covalent; in the solid it is ionic, $NO_2^+NO_3^-$.

Nitric acid, HNO_3, an important industrial chemical, is made industrially as a 68% aqueous solution. NH_3 and O_2 are reacted to give NO, which is then further oxidized to NO_2; the NO_2 is dissolved in water to give HNO_3 and NO which is recycled:

$$3NO_2(g) + H_2O(l) \rightarrow 2HNO_3(aq) + NO(g)$$

This is a **disproportionation** reaction in which N(IV) gives N(V) and N(II). Nitric acid is a strong acid. Most nitrate salts are soluble in water. NH_4NO_3, $NaNO_3$, and KNO_3 are important fertilizers. **Nitration** of organic compounds is achieved with a mixture of nitric and sulfuric acids

$$HNO_3 + 2H_2SO_4 \rightarrow NO_2^+ + H_3O^+ + 2HSO_4^-$$

in which the nitronium ion, NO_2^+, is the important nitrating agent, e.g.,

$$C_6H_5CH_3 + 3NO_2^+ \rightarrow C_6H_2(NO_2)_3CH_3 + 3H^+$$

toluene 2,4,6-trinitrotoluene (TNT)

Pure (100%) acid is obtained by removing the excess water from 68% HNO_3(aq) with sulfuric acid and distilling off the volatile HNO_3, or by simply heating a nitrate salt with H_2SO_4. All the NO bonds of the NO_3^- ion have the same length and a bond order of 1 1/3. HNO_3 and NO_3^- are species with nitrogen in the +5 oxidation state and are good oxidizing agents. They may be reduced to any of the other possible oxidation states of N (+4, +3, +2, +1, 0 and -3) depending on the concentration, temperature, and nature of the reducing agent. Nitric acid reacts with all of the metals except Au, Pt, Rh and Ir; it oxidizes many metals that are not oxidized by H_3O^+.

Nitrous acid, HNO_2, is a weak acid in water and is made by dissolving N_2O_3 , or a mixture of NO and NO_2, in water. Only dilute solutions can be obtained. On heating they decompose to NO, NO_2 and water. Alkali metal nitrites result from reacting a mixture of NO and NO_2 with NaOH(aq), and from heating alkali metal nitrates above their melting points. Nitrous acid and nitrites are good reducing agents since they are readily oxidized to nitric acid and nitrates.

Hydrazine, $H_2N\text{-}NH_2$, and **hydroxylamine,** $H_2N\text{-}OH$, can be regarded as derived from NH_3 by replacing an H atom by a $-NH_2$ or -OH group, respectively. N_2H_4 can be prepared by oxidizing ammonia with a solution of sodium hypochlorite ($2NH_3 + OCl^- \rightarrow N_2H_4 + H_2O + Cl^-$). It is a colorless liquid, a weak base, and a strong reducing agent. **Hydroxylamine** results from the reduction of nitric acid with tin or SO_2. It is a weak base and a strong reducing agent. **Hydrazoic acid,** HN_3 (aq), is a liquid that boils at 37°C. It is a weak acid in water and dangerously explosive. Its salts are the **azides** containing the N_3^- ion. Heavy metal azides are used as detonators.

Ozone, O_3, is an allotrope of oxygen prepared by passing an electric discharge through oxygen. In the upper atmosphere it is formed by the action of ultraviolet light on O_2 molecules. It is a very powerful oxidizing agent. Isoelectronic with NO_2^-, it has an angular AX_2E structure, and bond order 1½.

Hydrogen peroxide, H_2O_2 , has an angular AX_2E_2 geometry at each O atom, with free rotation around the O-O bond, except in the solid state where it is non-planar. It is a colorless liquid. Due to the low energy of the O-O bond it readily forms two very reactive OH radicals. H_2O_2 is a stronger oxidizing agent than NO_3^- or MnO_4^-. The pure liquid is strongly associated by hydrogen bonding. On heating it decomposes explosively to water and O_2. Aqueous solutions are more stable but their decomposition is catalyzed by many substances, including Fe^{2+} and MnO_2. It is used as a bleaching agent.

Sodium peroxide, Na_2O_2, and barium peroxide, BaO_2, containing the **peroxide ion,** $^{-}$:Ö-Ö:$^{-}$, result from heating the metals in excess O_2. Aqueous solutions of H_2O_2 result from treating BaO_2 with H_2SO_4(aq) and filtering off the insoluble $BaSO_4$ formed. Heating potassium, rubidium, or cesium in a limited amount of air or oxygen gives the normal oxides containing the O^{2-} ion. With excess oxygen they form the **superoxides,** KO_2 , RbO_2 and CsO_2 , containing the **superoxide ion,** O_2^- , which is a stable free radical ion.

REVIEW QUESTIONS

1. Draw the electron configurations of nitrogen and oxygen. What are the valences of nitrogen and oxygen?
2. What is nitrogen fixation? Give two examples of nitrogen fixation in nature.
3. What is the principal industrial process for the fixation of N_2 from the air?
4. Write formulas for the oxides of nitrogen having nitrogen in the +5, +4, +3, +2 and +1 oxidation states. Name each oxide.
5. How is nitric oxide, NO, made commercially from ammonia? How is NO made conveniently in the laboratory?
6. How does NO originate in the atmosphere?
7. Which oxide of nitrogen results from the reaction of NO with O_2?
8. What is a free radical? What free radicals would result from breaking the carbon-carbon bond in an ethane molecule, the O-O bond in a hydrogen peroxide molecule, and the N-N bond in an N_2O_4 molecule?
9. Draw three possible electronic structures for NO. Which of the structures is most consistent with the observed N-O bond length in NO?
10. For nitrogen dioxide, what are the molecular species found in (a) the gas, (b) the liquid, (c) the solid?
11. How could the equilibrium concentrations of NO_2 and N_2O_4 in the gas phase be found experimentally?
12. Why does the concentration of N_2O_4 molecules in a gaseous N_2O_4-NO_2 mixture increase with decreasing temperature and increase with increasing pressure?
13. How is NO_2 conveniently made in the laboratory?
14. Draw the possible Lewis structures for NO_2.
15. What are the bond orders of the NO bonds in (a) NO_2, (b) N_2O_4?
16. Draw the possible Lewis structures for N_2O_4, N_2O_3 and N_2O_2.
17. What is the reaction when a sample of ammonium nitrate is gently heated?
18. Draw two resonance structure for N_2O.
19. What are the expected geometric shapes of NO_2 and N_2O?
20. How is dinitrogen pentaoxide prepared from nitric acid?
21. Draw Lewis structures for the structure of N_2O_5 in (a) the gas phase, (b) the solid.
22. What is the shape of the nitronium ion, NO_2^+?
23. Give examples of two molecules that are isoelectronic with NO_2^+.
24. By what method is NO_2 normally prepared in solution?
25. How is nitric acid prepared commercially?
26. Why is the reaction of NO_2 with water described as a **disproportionation** reaction?

27. How can a sample of pure 100% nitric acid be prepared?
28. Draw two Lewis structures for nitric acid; what is the expected geometry and what are the bond orders of the N-O bonds?
29. Draw the possible Lewis structures for the nitrate ion, NO_3^-; what is the bond order of the NO bonds?
30. Why is copper oxidized to Cu^{2+} by nitric acid but is unreactive to HCl(aq)?
31. Draw the Lewis structures of nitrous acid.
32. What reactions occur when (a) potassium nitrate, (b) lead(II) nitrate, are heated?
33. What reaction might be expected to occur between dilute aqueous solutions of nitrous acid and bromine?
34. What is the structural relationship between ammonia, hydrazine, and hydroxylamine?
35. What ions are obtained when hydrazine is protonated?
36. How is hydroxylamine prepared from nitric acid?
37. Draw Lewis structures for the weak acid hydrazoic acid, and its anion, the azide ion.
38. Name two molecules with which the azide ion is isoelectronic.
39. Draw the two resonance structures for ozone and describe its molecular shape.
40. How is ozone formed in the upper atmosphere?
41. Write a balanced equation to illustrate the reaction between ozone and an acidic solution of potassium iodide.
42. How is the molecular structure of hydrogen peroxide related to that of water?
43. What salt can be reacted with sulfuric acid to give an aqueous solution of hydrogen peroxide?
44. Write the half-reaction for hydrogen peroxide behaving as an oxidizing agent in aqueous acidic solution.
45. Write the formulas and draw the Lewis structures for sodium oxide, sodium peroxide, and potassium superoxide. What are the bond orders of the O-O bonds in oxygen, superoxide ion, and peroxide ion?

Answers to Selected Questions

22. Linear.

OBJECTIVES

Be able to:

1. List the oxides of nitrogen, give the oxidation state for nitrogen in each, and list the preparation and some chemical and physical properties for each.
2. Draw Lewis structures for the oxides of nitrogen, including the free radicals NO and NO_2.
3. Describe the preparation, properties, and uses of nitric acid.
4. Draw Lewis structures for nitric acid and the nitrate ion.
5. Describe the preparation and properties of nitrous acid, and draw its Lewis structure.

6. Draw Lewis structures and give chemical properties of hydrazine, hydroxylamine, and hydrazoic acid.
7. List some physical and chemical properties of ozone.
8. Give the preparation and physical and chemical properties of hydrogen peroxide.
9. Give examples of peroxides and superoxides, and draw Lewis structures for the peroxide and superoxide ions.

PROBLEM SOLVING STRATEGIES

18.1 Oxides of Nitrogen

Lewis Diagrams for Oxides of Nitrogen and Free Radicals

For some oxides of nitrogen, and for other molecules, there is an odd number of valence electrons. The rules for drawing Lewis structures cannot be exactly obeyed in these cases, since we cannot arrange the electrons **in pairs** around all the atoms. So a Lewis structure for an odd electron species, called a free radical, will **not** have all electrons paired. You begin by drawing a Lewis structure as described in Chapter 4. However, after joining all the bonded atoms with single bonds, distribute the remaining electrons, forming multiple bonds if necessary, and apply the octet rule **preferentially** for the more electronegative atom(s) from the list C, N, O, F. Then calculate a formal charge for all atoms in the structure. Next repeat the process, but allow a more electropositive atom(s) to have the octet(s). Recalculate formal charges for this structure. Remember that **any** single Lewis structure for **any** molecule emphasizes a **localized** electron pair point of view. When we need to describe the bonding by using other than the language of single bond, double bond or triple bond, or need to represent the equivalency of several bonds, we use the concept of resonance. Therefore, the electron structures of free radicals may involve resonance. Please also note that it is sometimes possible for the octet rule to be obeyed for all the atoms in the free radical (see the NO_2 example in the text); however the bond orders may not be integers, even in the individual contributing resonance structures.

Example 18.1

Draw the Lewis structure(s) for dinitrogen monoxide, N_2O.

Solution 18.1

Since there is a total of 16 valence electrons, this is not an example of a free radical. You should know that N is the central atom in N_2O. So after joining the central N to N and O by single bonds, using 4 valence electrons, we have

$$N - N - O$$

By increasing these bonds to double bonds, and obeying the octet rule, we have

$$^{-}\underset{\cdot\cdot}{\ddot{N}} = \overset{+}{N} = \underset{\cdot\cdot}{\ddot{O}}$$

We next try to reduce formal charge, if possible, by writing the nitrogen-nitrogen bond as a triple bond, and the nitrogen-oxygen bond as a single bond. With formal charges, this looks like

$$:N \equiv \overset{+}{N} - \ddot{\underset{\cdot\cdot}{O}}:^{-}$$

We have not reduced the formal charge, but we do have a new Lewis structure. We represent N_2O as a resonance hybrid of two non-equivalent structures.

$$^{-}\ddot{\underset{\cdot\cdot}{N}} = \overset{+}{N} = \ddot{\underset{\cdot\cdot}{O}} \longleftrightarrow \quad :N \equiv \overset{+}{N} - \ddot{\underset{\cdot\cdot}{O}}:^{-}$$

Example 18.2

Draw the Lewis structure for NO, nitrogen monoxide.

Solution 18.2

The molecule contains 11 electrons, so it is a free radical. When we join N and O by a single bond and then arrange the 9 remaining electrons so that O follows the octet rule, we get

$$\ddot{\underset{\cdot}{N}} = \ddot{\underset{\cdot\cdot}{O}}$$

There are no formal charges associated with this structure. If we keep a double bond between N and O,but let N obey the octet rule, we have

$$^{-}\ddot{\underset{\cdot\cdot}{N}} = \ddot{\underset{\cdot}{O}}^{+}$$

The electron structure for NO is better represented by the first Lewis structure, since there are no formal charges. If we include **both** structures, then we can depict the bonding as

$$\ddot{\underset{\cdot}{N}} = \ddot{\underset{\cdot\cdot}{O}} \longleftrightarrow \quad ^{-}\ddot{\underset{\cdot\cdot}{N}} = \ddot{\underset{\cdot}{O}}^{+}$$

Finally we can show the electron distribution by having the odd electron neither on nitrogen nor on oxygen, but in the internuclear region.

$$^{1/2-}\ddot{\underset{\cdot}{N}} \overset{\cdot}{=} \ddot{\underset{\cdot}{O}}^{1/2+}$$

Note that this is quite a departure from Lewis's electron **pair** idea. It has the advantage of suggesting that the bonding is somewhat greater than a double bond, which is consistent with the observed bond length.

Example 18.3

Draw the Lewis structure for ethyl radical, C_2H_5.

Solution 18.4

This is a 13 electron free radical. A single bonded structure requires 6 single bonds, and 12 electrons.

```
    H   H
    |   |
H---C---C
    |   |
    H   H
```

We put the remaining electron on the right carbon.

```
      H   H
      |   |
  H---C---C·
      |   |
      H   H
```

Note that the right carbon does not obey the octet rule. Also all the formal charges are zero. There are no important resonance structures here.

18.2 Oxoacids of Nitrogen and
18.3 Hydrazine, Hydroxylamine, and Hydrazoic Acid

Oxidation State for Nitrogen-Containing Molecules

Different nitrogen compounds exhibit a wide range of oxidation numbers, from -3 to +5. By assigning oxidation numbers to nitrogen in compounds, not only can we balance redox reactions and identify oxidizing and reducing agents, but also we can systematize the chemistry of nitrogen compounds. For example, HNO_3, NO_3^- and N_2O_5 all have nitrogen in the +5 oxidation state; these are all known as nitrogen(V) compounds. All of them can behave as oxidizing agents, and be reduced to lower oxidation state nitrogen-containing substances. Review oxidation number assignments in Chapter 8 if necessary.

Example 18.4

Hydrazoic acid, HN_3, can be slowly oxidized to nitrogen gas by I_2. Give a balanced equation for this reaction. Is HN_3 an oxidizing agent or a reducing agent here?

Solution 18.4

Since H always has an oxidation number of +1, except for elemental H_2 and metallic hydrides, there is an oxidation number of -1 to distribute among 3 nitrogens in HN_3. Thus each nitrogen has an oxidation number of -1/3. Molecular nitrogen has zero oxidation number. Thus nitrogen is being oxidized in this reaction, and therefore iodine must be reduced. I^- the the obvious choice for the product. Thus the 2 half-reactions are

$$2HN_3 \rightarrow 3N_2 + 2H^+ + 2e^-$$

$$2e^- + I_2 \rightarrow 2I^-$$

and the balanced equation is the sum of these.

$$2HN_3 + I_2 \rightarrow 3N_2 + 2H^+ + 2I^-$$

Since N is being oxidized, HN_3 is acting as a reducing agent. It is the species that transfers electrons to I_2, allowing I_2 to be reduced to $2I^-$.

SELF TEST

Part I True or False

1. All nitrogen oxides are colorless.
2. All nitrogen oxides are gases at -50°C.
3. The oxides of nitrogen play an important role in photochemical smog.

4. Hydrogen peroxide is a linear molecule.
5. Hydrazine reacts with N_2O_4 in an exothermic reaction.
6. Nitrous acid is a weak acid and a poor reducing agent.
7. Hydrazoic acid is a linear molecule.
8. The conversion of ozone to molecule oxygen is exothermic.
9. Nitrogen monoxide is very soluble in water.
10. A chlorine atom is a free radical.
11. Concentrated NO_2 dimerizes to colorless N_2O_4.
12. N_2O_3 is the anhydride of nitric acid.
13. Concentrated nitric acid decomposes to N_2O upon exposure to light.
14. Concentrated nitric acid can oxidize carbon to CO_2.
15. The nitronium ion is NO_2^+.
16. Concentrated solutions of nitrous acid are stable.
17. Hydroxylamine is an oxidizing agent.
18. Ozone has an equilateral triangular structure.
19. Hydrogen peroxide, when heated, decomposes to O_2 and water.
20. Hydrogen peroxide can react as an oxidizing agent and as a reducing agent.
21. When sodium is heated in excess air, Na_2O is formed.

Part II Multiple Choice

22. Which **two** of the following compounds have nitrogen in the same oxidation state?
(a) N_2O_3 (b) NH_3 (c) NH_2OH (d) HNO_2 (e) HN_3
23. From the following list, choose all the oxides of nitrogen that are not linear.
(a) NO (b) NO_2 (c) N_2O (d) N_2O_4 (e) N_2O_3
24. The bond order of the N-O bond in the nitrate ion is
(a) 1/3 (b) 2/3 (c) 1 (d) 1 1/3 (e) 1 2/3
25. When copper reacts with concentrated nitric acid, the nitrogen-containing reduction product has nitrogen in which oxidation state?
(a) 4 (b) 3 (c) 2 (d) 1 (e) -3
26. For the balanced equation describing the reaction of hydrazine and hydrogen peroxide, forming nitrogen and water, when all the stoichiometric coefficients are the smallest possible integers, the coefficient of H_2O is
(a) 2 (b) 4 (c) 6 (d) 8 (e) 10
27. In the presence of excess oxygen, potassium forms the oxide
(a) KOOK (b) K_2O (c) OK (d) KO_2 (e) K_2O_3
28. Which two of the following Lewis structures, with formal charges omitted, are the resonance structures for N_2O?
(a) :N≡N=Ö: (b) :N̈=N=Ö: (c) :N̈-N≡O (d) :N̈=N̈-Ö (e) :N≡N-Ö: (with an additional lone pair below O)

Answers to Self Test

1. F; 2. F; 3. T; 4. F; 5. T; 6. F; 7. F; 8. T; 9. F; 10. T; 11. T; 12. F; 13. F; 14. T; 15. T; 16. F; 17. F; 18. F; 19. T; 20. T; 21. F; 22, a,d; 23. b,d,e; 24. d; 25. a; 26. b; 27. d; 28. b,e.

CHAPTER 19

RATES OF CHEMICAL REACTIONS

SUMMARY REVIEW

The rates of reactions vary from extremely slow to very fast; usually reaction rate increases with increased temperature. Reactions take place most readily in the gas or liquid phases. In many reactions, molecules must collide before a reaction can occur.

Reaction rate is the change $\Delta[\ ]$ in concentration of any reactant or product divided by the time interval Δt. Experimentally, the choice of reactant or product concentration to follow with time is a matter of convenience. The slope of the plot of concentration versus time at any given time gives the **instantaneous reaction rate** at that time.

For the general reaction $aA + bB + cC + \ldots \rightarrow$ products, the experimental rate information usually can be expressed empirically in a rate law of the form

$$\text{Rate} = k[A]^x [B]^y [C]^z \ldots$$

where k is the **rate constant,** x, y, and z, are the **orders of reaction** with respect to A, B, and C, and $x + y + z + \ldots$ is the **overall order.** The experimentally determined values of x, y, z, ... are not necessarily the same as a, b, c, ... in the balanced equation; rate laws must be determined experimentally. If there is only one reactant, A, then for **rate = k[A],** a plot of rate versus [A] gives a straight line with slope k; for rate = $k[A]^2$, a plot of rate versus $[A]^2$ gives a straight line with slope k. For reactions which have more complex rate laws, the **initial-rate method** or the **integrated rate law method** is used. For a **first order** reaction with the rate law, rate = k[A], the integrated rate law has the form

$$\ln [A]_t = \ln [A]_0 - kt$$

where $[A]_0$ is the concentration of A at time t = 0, and $[A]_t$ is the concentration of A at time t = t. Thus, a plot of ln $[A]_t$ versus t is a straight line with slope -k. For a **second order** reaction with the rate law, rate = $k[A]^2$, the integrated rate law has the form

$$\frac{1}{[A]_t} = \frac{1}{[A]_0} + kt$$

and a plot of $1/[A]_t$ versus time gives a straight line with slope k.

The **Arrhenius equation** has the form, $k = Ae^{-E_a/RT}$. If every molecular collision in a bimolecular reaction led to reaction, the rate constant k would be the same as the molecular collision frequency Z, and all reactions would be complete almost instantaneously. The fact that almost all reactions are much slower is accounted for by two factors: 1) colliding molecules must have an appropriate **steric orientation**, and 2) colliding molecules must have some minimum energy, E_a, the **activation energy** of the reaction, for reaction to occur. The fraction of collisions with energy equal to or greater than E_a is $e^{-E_a/RT}$, and if the fraction of molecules with an appropriate steric orientation is p, then $k = pZe^{-E_a/RT}$. Since pZ is approximately constant with change in temperature in comparison to $e^{-E_a/RT}$, we can write, $k = Ae^{-E_a/RT}$, or, in logarithmic form, $\ln k = \ln A - E_a/RT$. Thus, a plot of ln k versus 1/T is a straight line with the slope $-E_a/R$. The energy maximum along the reaction path corresponds to the energy of the **activated complex** or **transition state;** the **activation energy,** E_a, is the difference between the transition state energy and the initial energy of the reactants. For two temperatures T_1 and T_2 with rate constants k_1 and k_2, respectively

$$\ln (k_2/k_1) = (E_a/R)(1/T_1 - 1/T_2)$$

In calculations involving the Arrhenius equation, the appropriate value for the gas constant R is 8.314 J K^{-1} mol^{-1}.

For **first order** reactions, the half-life, $t_{1/2} = (\ln\ 2)/k_1 = 0.693/k_1$, but for second order reactions $t_{1/2}$ is not independent of the initial concentration, and is given by $t_{1/2} = 1/(k_2\ [A_0])$.

A balanced equation gives no information about the **reaction mechanism,** the detailed way in which reactants are converted to products. The simplest (elementary) reactions occur in a **single step. Molecularity** is the number of molecules that take part in the single step. **Bimolecular reactions** are the most common and must follow a second order rate law. **Unimolecular reactions** involve the decomposition or rearrangement of single molecules and follow a first order rate law.

Multistep reaction mechanisms involve more than a single step; each step is almost always unimolecular or bimolecular. The steps occur with different rates and, in the cases considered, one step is much slower than any of the others and is the **rate-determining** or **rate-limiting** step that determines the overall reaction time and the rate law for the reaction. When the rate-limiting step involves intermediates that do not appear as reactants or products of the reaction, the rate law predicted for it is changed to the conventional form by making use of the equilibrium constants for the steps which precede the slow one in the mechanism. If for a reaction only one mechanism that is consistent with the experimental rate law has been proposed, it may be considered to be the mechanism for the reaction. If several different proposed mechanisms lead to the observed rate law, additional information regarding the nature of intermediates is needed before any one mechanism can be adopted.

When a reaction is at equilibrium, every individual step in the mechanism must also be at equilibrium. If we combine the equilibrium expressions for the individual steps, to eliminate all the concentration terms for intermediates, we are left with the normal equilibrium constant expression; the equilibrium constant expression can always be written from the balanced equation and is independent of any proposed mechanism.

A catalyst is a substance which increases the rate at which equilibrium is attained but is not consumed in a reaction. Normally a catalyst changes the mechanism of the reaction, so that the altered mechanism has a lower activation energy than the uncatalyzed reaction. The catalyst takes part in the reaction but is then regenerated; it does not appear in the balanced equation for the reaction, but it can appear in the rate law. Catalysts may also improve the steric factor for a reaction.

A catalyst in the same phase as the other reactants is a **homogeneous** catalyst; a catalyst in a different phase is **heterogeneous.** Heterogeneous catalysis is often called **surface catalysis** because the catalyst is often a solid and the reactants are gases. Important industrial processes that use heterogeneous catalysts are the Ostwald, Haber and Contact processes. **Enzymes** are biochemical catalysts.

Chain reactions, such as that between $H_2(g)$ and $Cl_2(g)$, consist of three types of steps: **chain initiation, chain propagation,** and **chain termination.** The initiation step is normally relatively slow and gives a highly reactive intermediate (often a free radical). This reacts rapidly with a reactant molecule to give a product molecule and another molecule or reactive intermediate; these propagation steps sustain the reaction until it is terminated by removal of the intermediate. Chain reactions are important in **explosions** and **polymerization** reactions.

REVIEW QUESTIONS

1. What is meant by the rate of a reaction? Does the rate at which a reaction takes place affect the final position of equilibrium?
2. What does an increase in temperature generally do to the rate of a reaction?
3 For the reaction $N_2(g) + 3H_2(g) \rightarrow 2NH_3(g)$, what is the relationship between the rate measured in terms of the disappearance of H_2 and that measured in terms of the appearance of ammonia?
4. How is the instantaneous rate of a reaction obtained from a plot of the concentration of one of its components plotted against time?
5. Write a general form for a rate law.
6. If, for the reaction between NO(g) and $O_3(g)$, the rate law is rate = $k[NO][O_3]$, what is (a) the order with respect to NO, (b) the order with respect to O_3, (c) the overall order? What will be the change in rate of the reaction if we simultaneously double the concentration of NO and halve the concentration of O_3?
7. The rate law for the reaction of NO(g) with $Cl_2(g)$ to give NOCl(g) is first order in Cl_2 and second order in NO. What happens to the rate of reaction if the concentrations of both reactants are doubled?
8. If the initial rate of a reaction increases by a factor of eight when the initial concentration of a reagent is doubled, what is the order of the reaction with respect to that reagent?
9. For the decomposition of N_2O_5, we have the concentrations of N_2O_5 at various times. How can it be shown that the reaction is first order? If in an experiment the concentration of N_2O_5 was 0.01 mol L^{-1} initially,

0.005 mol L^{-1} after 15 mins and 0.0025 mol L^{-1} after 30 mins, what is the value of the rate constant?

10. What is the steric or orientation effect?
11. Give two reasons why every collision between an NO molecule and an O_3 molecule does not lead to the formation of NO_2 and O_2.
12. What is the activation energy, E_a, of a reaction? What is the source of this energy?
13. What is the transition state?
14. If a reaction takes place in a single step, what will be the relationship between the balanced equation for the reaction and the rate law?
15. If a reaction takes place in several steps, what determines the observed reaction rate?
16. What equation relates the rate constant k′ at a temperature T′ to the rate constant k" at another temperature T"?
17. What is meant by the term reaction mechanism?
18. Write the mechanism for the reaction of NO with O_3, and show that it is consistent with the rate law, rate = $k[O_3][NO]$.
19. Write the mechanism for the reaction of H_2 and I_2 to give HI and show that it is consistent with the rate law, rate = $k[H_2][I_2]$.
20. Write the mechanism for the reaction of gaseous NO, O_2 and H_2O, to give nitric acid and show that it is consistent with the rate law, rate = $k[NO_2]^2[H_2O]$.
21. Does the equilibrium constant expression for a reaction depend on its mechanism?
22. How does a catalyst affect the stoichiometry of a reaction? How does it affect the mechanism?
23. Does a catalyst raise or lower the activation energy of a reaction?
24. What is a homogeneous catalyst? What is a heterogeneous catalyst?
25. What is an enzyme?

Answers to Selected Review Questions

3. Hydrogen is used up 3/2 times as fast as NH_3 is formed. 6. 1, 1, 2; it is unchanged. 7. rate = $k[Cl_2][NO]^2$ -- rate increases 8 fold. 8. 3. 9. k = 0.046 min^{-1}

OBJECTIVES

1. Express the rate of a reaction as a change in concentration with time, using the stoichiometry of a balanced chemical equation.
2. Write a general expression for a rate law, relating rate, rate constant, and concentrations of reactants.
3. Define the term order of a reaction, and use the initial rate method to determine whether a given reaction is first or second order.
4. Use the integrated rate law method to show whether a given reaction is first or second order.
5. Use the half-life expression for first order reactions.
6. Draw plots of the variation of potential energy with reaction pathway (reaction coordinate), for both endothermic and exothermic reactions; indicate ΔH and the activation energy, E_a, on the plots.

7. Use the Arrhenius equation in calculations relating rate constants at two temperatures to the activation energy.
8. Define the term molecularity and realize that a complex reaction can involve many elementary steps or processes.
9. Define the term mechanism.
10. Use a given mechanism for a reaction with a given rate-determining step to predict the rate law for that reaction, and decide whether the predicted rate law is consistent with the experimental rate law.
11. Appreciate that an equilibrium constant can be written for any chemical reaction, even without a knowledge of the reaction mechanism.
12. Define the term catalyst and explain how a catalyst can increase a reaction rate.
13. Distinguish between homogeneous and heterogeneous catalysis.
14. Write the three types of steps involved in chain reactions.

PROBLEM SOLVING STRATEGIES

19.2 Rate Laws and Reaction Order

Rates of chemical reactions are observed experimentally. An experimentally-determined rate law equates a rate (expressed in concentration/time) to a rate constant times concentrations of some or all of the reactants, raised to powers that are small whole numbers or fractions. Each power is called the **order** of the reaction with respect to that reactant. Only after we have experimentally determined the order, can we suggest a detailed set of steps (a **mechanism**) that may account for the observed rate law.

Initial-Rate Method

In the **initial-rate method**, the rate is measured very early in a reaction, before much product is formed, and before the reactant concentrations have changed appreciably from their initial values. All reactant concentrations except one are kept constant, and the dependence of the rate on that one concentration is determined. Next, a different reactant concentration is varied, with all the rest held constant, and the order with respect to that reactant found. The overall reaction order equals the sum of the individual reaction orders.

Example 19.1

Nitrogen monoxide reacts with hydrogen, forming elemental nitrogen and water.

$$2NO + 2H_2 \rightarrow N_2 + 2H_2O$$

In a set of initial rate experiments, the following results were obtained.

Experiment	[NO], mol L^{-1}	$[H_2]$, mol L^{-1}	rate, mol L^{-1} s^{-1}
1	4.2×10^{-2}	1.8×10^{-3}	2.20×10^{-4}
2	4.2×10^{-2}	3.6×10^{-3}	4.40×10^{-4}
3	1.26×10^{-1}	1.8×10^{-3}	1.98×10^{-3}

What is the order of the reaction with respect to H_2 and NO? What is the rate law? What is the numerical value of the rate constant, and what are its units?

Solution 19.1

The rate law is expected to be of the form

$$\text{rate} = k[NO]^x[H_2]^y$$

since NO and H_2 are the reactants; note that x and y need not be equal to the stoichiometric coefficients in the balanced equation. To determine y, the order with respect to H_2, we consider 2 experiments where [NO] is fixed, and examine how the rate changes as only $[H_2]$ changes. Let's use the first two results.

$$\frac{\text{rate}_2}{\text{rate}_1} = \frac{k[NO]_2^x[H_2]_2^y}{k[NO]_1^x[H_2]_1^y}$$

$$\frac{(4.4 \times 10^{-4})}{(2.2 \times 10^{-4})} = \frac{(3.6 \times 10^{-3})^y}{(1.8 \times 10^{-3})^y} = \left(\frac{3.6 \times 10^{-3}}{1.8 \times 10^{-3}}\right)^y$$

$$2 = 2^y \; ; \quad y = 1$$

So the order is 1 with respect to H_2. Next we want to keep $[H_2]$ constant, and vary [NO]. We use the first and last results.

$$\frac{\text{rate}_3}{\text{rate}_1} = \frac{k[NO]_3^x[H_2]_3^y}{k[NO]_1^x[H_2]_1^y}$$

$$\frac{1.98 \times 10^{-3}}{2.2 \times 10^{-4}} = \frac{(1.26 \times 10^{-1})^x}{(4.2 \times 10^{-2})^x} = \left(\frac{1.26 \times 10^{-1}}{4.2 \times 10^{-2}}\right)^x$$

$$9 = 3^x \; ; \quad x = 2$$

Thus the order with respect to NO is 2. The rate law is rate $= k[NO]^2[H_2]$. To get the value of the rate constant, we use the rate law and substitute the results for any one experiment. Using the first experiment, we obtain

$$k = \frac{\text{rate}}{[NO]^2[H_2]} = \frac{2.2 \times 10^{-4} \text{ mol } L^{-1} \, s^{-1}}{(4.2 \times 10^{-2})^2 \text{ mol}^2 \, L^{-2} \, (1.8 \times 10^{-3}) \text{ mol } L^{-1}}$$

$$= 69.3 \text{ mol}^{-2} \, L^2 \, s^{-1}$$

In general, kinetics experiments will not have the concentration ratios equal to an **exact** multiple of the original concentrations. Even so, the method of attack is the same. For example, when we found that x was equal to 2, we could have determined the order using logarithms.

$$9 = 3^x$$

$$\log 9 = x \log 3$$

$$0.954 = (x)(0.477)$$

$$x = 2$$

Half-Life

The **half-life** of a reaction is the time needed for one-half of a substance to be used up. For a **first order reaction**, the half-life is constant. Thus after n half-lives, the fraction of a substance **remaining** is $(1/2)^n$; the fraction **used up** is $1 - (1/2)^n$. For a second order reaction, the half-life is not constant, but depends on the initial concentration and the extent of reaction.

Example 19.2

The half-life of the first order gas reaction $A \rightarrow 2B$ is 30 minutes. Enough A is placed in a vessel so that the pressure is 400 mm Hg. After two hours, what is the **total** pressure of gases in the vessel?

Solution 19.2

Since 2 hours corresponds to 4 half-lives, only $(1/2)^4 = 1/16$ of the original moles of A remains. Thus the partial pressure exerted by A is 1/16 x 400 mm Hg = 25 mm Hg, where we have used the proportionality between pressure and molar concentration that is valid at constant temperature. As well, 15/16 of the original moles of A have produced moles of B = 2 x 15/16 = 30/16 of the original moles of A. So the partial pressure of B is

$$30/16 \times 400 \text{ mm Hg} = 750 \text{ mm Hg}$$

The total pressure is 25 + 750 = 775 mm Hg. You should note that after many half-lives, the total pressure would approach 800 mm Hg.

19.3 Factors Influencing Reaction Rates

Temperature and Reaction Rate: The Arrhenius Equation

The Arrhenius equation shows the relationship between the rate constant k, the absolute temperature T, and the activation energy E_a.

$$k = Ae^{-E_a/RT}$$

If we specify two temperatures, T_1 and T_2, and their corresponding rate constants k_1 and k_2, and subtract the $\ln k_1$ expression from that for $\ln k_2$, we get (assuming A is temperature independent)

$$\ln \frac{k_2}{k_1} = \frac{E_a}{R}\left(\frac{1}{T_1} - \frac{1}{T_2}\right)$$

When you use this form of the Arrhenius equation, four of the five quantities k_1, k_2, T_1, T_2, E_a must be known. Be sure to realize that E_a will have the same units as the gas constant R. Sometimes you must apply the equation **twice**; once to solve for E_a from the given data, and a second time to get an unknown rate constant or temperature. Here is an example.

Example 19.3

A reaction is carried out at two temperatures, with concentrations kept constant, and the following rates determined.

Temperature, °C	Rate, mol L^{-1} s^{-1}
20	1.5
30	2.4

What is the rate at 50°C, for the same concentrations?

Solution 19.3

Remember **rate** equals **rate constant** times some function of concentration. We assume the temperature dependence of the **rate** (which is what is asked for) equals the temperature dependence of the **rate constant**.

$$\frac{\text{rate at } T_2}{\text{rate at } T_1} = \frac{k_2 \times \text{function of concentration}}{k_1 \times \text{function of concentration}}$$

Thus, knowing the ratio of **rates** allows us to use the Arrhenius equation, which deals with the ratio of **rate constants**.

$$\ln \frac{k_2}{k_1} = \frac{E_a}{R}\left(\frac{1}{T_1} - \frac{1}{T_2}\right)$$

First we convert °C to the Kelvin scale.

$$20°C = 20 + 273 = 293 \text{ K}$$
$$30°C = 30 + 273 = 303 \text{ K}$$
$$50°C = 50 + 273 = 323 \text{ K}$$

You can see from the form of the Arrhenius equation that, in order to calculate a rate constant for 323 K, we must know E_a. We solve for E_a with the given data for 293 K and 303 K.

$$\ln \frac{2.4}{1.5} = \frac{E_a}{8.31 \text{ J K}^{-1} \text{ mol}^{-1}} \left(\frac{1}{293 \text{ K}} - \frac{1}{303 \text{ K}}\right)$$

$$E_a = 3.55 \times 10^4 \text{ J mol}^{-1}$$

With an E_a, we use the Arrhenius equation a second time, and calculate the value of the ratio k at 323 K to k at 293 K.

$$\ln \frac{k_2}{k_1} = \frac{3.55 \times 10^4 \text{ J mol}^{-1}}{8.31 \text{ J K}^{-1} \text{ mol}^{-1}} \left(\frac{1}{293 \text{ K}} - \frac{1}{323 \text{ K}}\right)$$

$$\ln \frac{k_2}{k_1} = 1.32$$

This is also the value of the ln of the **ratio** of the rates at these two temperatures.

$$\ln\left(\frac{\text{rate at 323 K}}{\text{rate at 293 K}}\right) = 1.32 \qquad \frac{\text{rate at 323 K}}{\text{rate at 293 K}} = 3.74$$

Since we know the rate at 293 K is 1.5 mol L^{-1} s^{-1},

$$\text{rate at 323 K} = 3.74 \times 1.5 \text{ mol } L^{-1} s^{-1} = 5.6 \text{ mol } L^{-1} s^{-1}$$

19.4 Reaction Mechanisms

A reaction mechanism is a set of elementary steps that we write in an attempt to account for the experimentally determined orders of reaction. Each elementary step usually involves only one or two molecules, since it is rare at the molecular level for 3 molecules to come together. There are some multistep reactions that have one elementary step much slower than all the others. This step is called the **rate determining step.** Its rate will give the rate of the overall reaction. The **rate for an elementary step can be written by inspection.** For example, for the **elementary step**

$$A + B \xrightarrow{k_1} \text{Products}$$

rate = $k_1[A][B]$. This is an example of a **bimolecular** elementary step. If this step is rate determining, then the mechanism containing this step predicts the rate of the overall reaction to be second order, that is, first order in A and first order in B. If this **predicted** rate law is the same as the **experimentally measured** rate law for the reaction, we say the mechanism is **consistent** with the experimental results. If A and B are reaction intermediates, such as free radicals, or if A and B have not been monitored in the experiment, we can still make a comparison between prediction and experiment. This will be possible if there is some chemical equilibrium relating concentrations not measured to ones that were determined in the kinetic study. Here are 3 steps to follow in testing a mechanism.

1. Write the predicted rate law of the overall reaction as equal to the rate law for the rate **determining step.** This is done by inspection, since orders equal coefficients of reactants in this step.
2. If the substances whose concentrations appear in the predicted rate law do not appear in the experimental rate law, try to relate them to substances that do appear by use of equilibrium constant expressions for earlier reaction steps.
3. Compare the predicted and experimental rate laws. If they do not agree, the mechanism is not valid. If they do agree, the mechanism may or may not be correct -- at least, the mechanism is **consistent** with the experimental results.

Example 19.4

The reaction

$$2NO(g) + Cl_2(g) \rightarrow 2NOCl(g)$$

follows the rate law

$$\text{rate} = k[NO]^2[Cl_2]$$

Are either or both of the following mechanisms consistent with the observed behavior?

Mechanism 1

$$NO(g) + Cl_2(g) \underset{<}{\overset{K_1}{\longrightarrow}} NOCl_2(g) \qquad \text{fast equilibrium}$$

$$NO(g) + NOCl_2(g) \xrightarrow{k_2} 2NOCl(g) \qquad \text{slow}$$

Mechanism 2

$$NO(g) + Cl_2 \xrightarrow{k_1} NOCl(g) + Cl(g) \qquad \text{slow}$$

$$NO(g) + Cl(g) \xrightarrow{k_2} NOCl(g) \qquad \text{fast}$$

Solution 19.4

Mechanism 1

We write the overall predicted rate law from the slow step, the rate determining step.

$$\text{rate} = k_2 [NO][NOCl_2]$$

But we cannot compare this predicted rate law with the experimental rate law, since the latter does not contain $[NOCl_2]$. So we form an equilibrium constant expression for the first step, since the mechanism supposes this step to be an equilibrium. Thus

$$K_1 = \frac{[NOCl_2]}{[NO][Cl_2]}, \qquad [NOCl_2] = K_1[NO][Cl_2]$$

If we substitute this concentration into our original predicted rate law, we find

$$\text{rate} = k_2[NO]K_1[NO][Cl_2] = k_2K_1[NO]^2[Cl_2]$$

So if we let k_2K_1 equal a constant k, we have a **consistent** mechanism.

Mechanism 2

Again we write the predicted rate law from the rate law for the slow step.

$$\text{rate} = k_2[NO][Cl_2]$$

Obviously this rate law is not consistent with the experimental facts, since the orders differ.

SELF TEST

1. Which two of the following are correct expressions for the rate of the reaction, $A + B \rightarrow C$, which is first order in A and zeroth order in B?
   (a) $\Delta[A]/\Delta t = k[B]$ (b) $-\Delta[B]/\Delta t = k[A]$ (c) $-\Delta[C]/\Delta t = k[A]$
   (d) $-\Delta[A]/\Delta t = k[C]$ (e) $-\Delta[A]/\Delta t = k[A]$
2. For a reaction of A and B to form C, the following initial rate data were obtained from three experiments:

Experiments	[A]	[B]	rate of formation of C
1	0.60 M	0.15 M	6.3×10^{-3} M min^{-1}
2	0.20 M	0.60 M	2.8×10^{-3} M min^{-1}
3	0.20 M	0.15 M	7.0×10^{-4} M min^{-1}

The rate law for the reaction is:

(a) $\frac{\Delta[C]}{\Delta t} = k[A][B]$ (b) $\frac{\Delta[C]}{\Delta t} = k[A]^2[B]$ (c) $\frac{\Delta[C]}{\Delta t} = k[A][B]^2$

(d) $\frac{\Delta[A]}{\Delta t} = k[A]^3[B]^2$ (e) $\frac{\Delta[C]}{\Delta t} = k[A]^2[B]^2$

3. The numerical value of the rate constant k for question 2 is
(a) 2.3×10^{-2} (b) 1.8×10^{-1} (c) 1.3×10^{-2}
(d) 1.2×10^{-1} (e) 7.8×10^{-1}

4. For the gaseous reaction $A + 2B \rightarrow C$ whose rate law is rate = $k[A]^2[B]$, by what factor would the rate increase if the concentration of both A and B were doubled?
(a) 2 (b) 3 (c) 4 (d) 6 (e) 8

5. The second order reaction $2B \rightarrow$ products obeys the rate law rate = $k[B]^2$. Which graph shows how the rate varies with time?

(a)

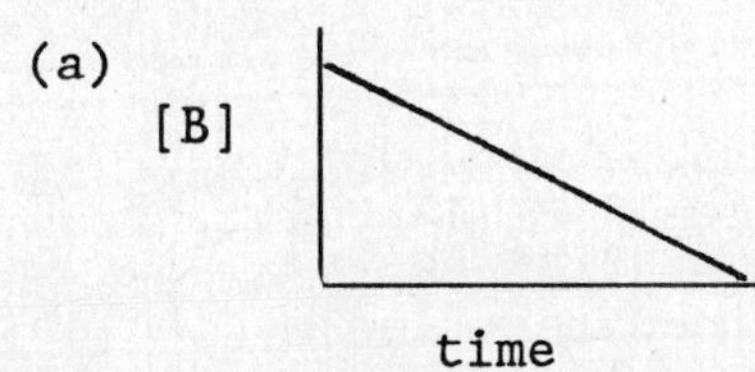

(b)

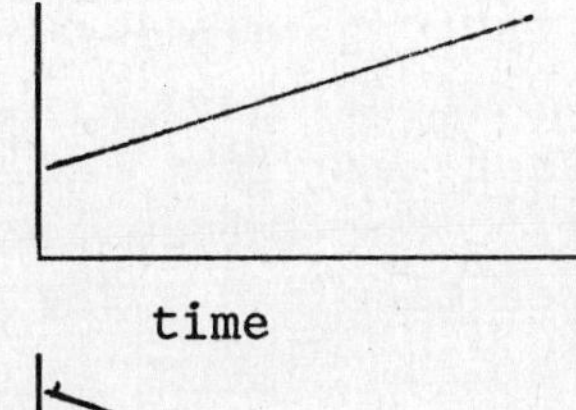

(c)

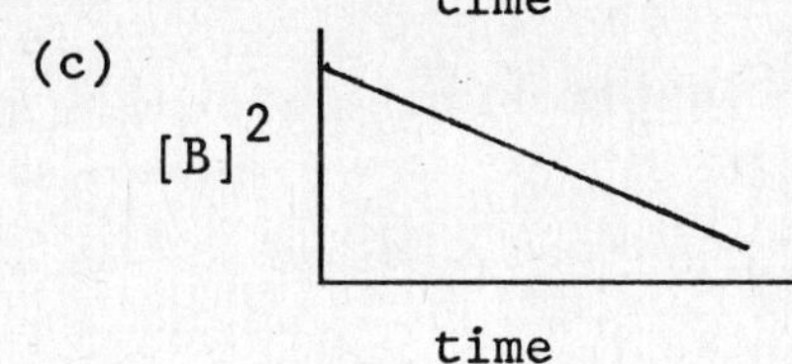

(d)

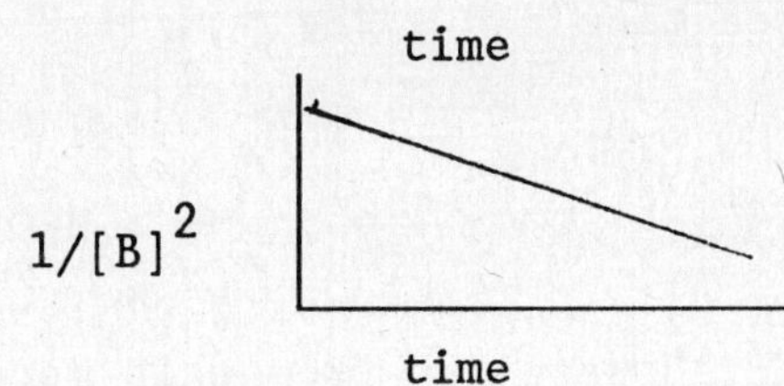

(e)

1/[B]

time

6. The reaction $X \rightarrow Y + Z$ is first order. The half-life is 10.5 min at 25°C. The rate constant for the reaction is:
(a) 3.3×10^{-4} min^{-1} (b) 6.6×10^{-2} min^{-1} (c) 10.5 min^{-1}
(d) 7.27 min (e) 15.2 min^{-1}

7. In the reaction $A + B \rightarrow C$ the activation energy, E_a in in kJ mol^{-1} for the forward reaction is

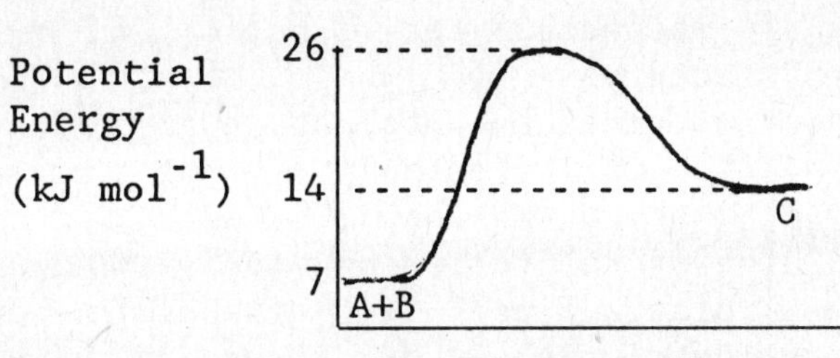

(a) +7 (b) +12 (c) -7 (d) -12 (e) +19

8. The reaction of hexaaquochromium (III) ion, $Cr(OH_2)_6^{3+}$, and thiocyanate ion, SCN^-, to form a complex ion,

$$Cr(OH_2)_6^{3+} + SCN^- \rightarrow Cr(OH_2)_5NCS^{2+} + H_2O$$

is governed by the rate law

$$rate = k[Cr(OH_2)_6^{3+}][SCN^-]$$

The value of k is 2.0×10^{-6} L mol^{-1} s^{-1} at 14°C and 2.2×10^{-5} L mol^{-1} s^{-1} at 30°C. The value of E_a in kJ mol^{-1} is

(a) 54.9 (b) 10.8 (c) 108 (d) 28.9 (e) 289

9. The overall reaction $2ICl + H_2 \rightarrow 2HCl + I_2$ has an observed rate equal to $k[ICl][H_2]$. A proposed mechanism is:

$$ICl + H_2 \rightarrow HI + HCl \quad (1)$$
$$ICl + HI \rightarrow HCl + I_2 \quad (2)$$

In order for the mechanism to be consistent with experiment, the relative rates of the individual steps (1) and (2) must be

(a) (1) slow, (2) fast
(b) (1) fast, (2) slow
(c) (1) slow, (2) slow
(d) (1) fast, (2) fast
(e) insufficient information; the temperature must be known.

10. For the reaction $2NO + 2H_2 \rightarrow N_2 + 2H_2O$ a proposed reaction mechanism is

Step 1 $NO + H_2 \underset{}{\overset{K_1}{\rightleftharpoons}} NOH_2$ (fast equilibrium)

Step 2 $NOH_2 + NO \xrightarrow{k_2} N_2 + H_2O_2$ (slow)

Step 3 $H_2O_2 + H_2 \xrightarrow{k_3} 2H_2O$ (fast)

The predicted rate law from this mechanism is

(a) rate = $k[NO]$ (b) rate = $k[NO][H_2]$ (c) rate = $k\frac{[H_2]}{[NO]}$

(d) rate = $k[NO]^2[H_2]$ (e) rate = $k[NO][H_2]^2$

11. The rate law for a reaction is a clue to

(a) the activation energy of the reaction
(b) the equilibrium constant of the reaction
(c) the mechanism of reaction
(d) the temperature coefficient of the reaction rate
(e) the stoichiometry of the reaction

12. The addition of a catalyst in a chemical reaction

(a) increases the concentration of products at equilibrium
(b) increases the fraction of reactant molecules with a given kinetic energy
(c) provides an alternate path with a different activation energy
(d) lowers the enthalpy change in the overall reaction
(e) none of the above.

13. Which of the following involve(s) heterogeneous catalysis?

(a) platinum used in the Haber process of NH_3 production
(b) catalytic converters used in the exhaust system of automobiles
(c) hydrogenation of ethylene in the presence of finely divided nickel
(d) the decomposition of ozone in the upper atmosphere by nitrogen monoxide
(e) the cis-trans isomerization of gaseous 2-butene in the presence of iodine atoms.

14. In the chain reaction between H_2 and Cl_2, light is important in which step?
(a) chain initiation (b) chain propagation (c) chain termination
(d) thermal decomposition (e) boiling point elevation

15. If for a reaction the activation energy is 80 kJ mol^{-1} the factor $A = 8 \times 10^8$ and the temperature is 47°C, the rate constant k equals
(a) 9×10^{-81} (b) 7×10^{-5} (c) 8×10^8 (d) 4×10^8 (e) 9×10^{21}

16. The half-life of a reaction is 30 seconds. The fraction, expressed as a decimal, of the original reactant which remains after the passage of 120 seconds is
(a) zero (b) 0.0312 (c) 0.0625 (d) 0.125 (e) 0.250

17. The rate constant k for a first-order reaction is 0.40 min^{-1}. The fraction, expressed as a decimal, of the original reactant which remains after 2.0 minutes of reaction is
(a) 0.45 (b) 0.62 (c) 0.80 (d) 2.2
(e) a number which cannot be computed.

Answers to Self Test

1. b,e; 2. b; 3. d; 4. e; 5. e; 6. b; 7. e; 8. c; 9. a; 10. d; 11. c; 12. c; 13. a,b,c; 14. a; 15. b; 16. c; 17. a.

CHAPTER 20

THE NOBLE GASES: MORE CHEMISTRY OF THE HALOGENS

SUMMARY REVIEW

Until 1962, it was thought that group 8 elements formed no compounds. This supposed inertness led to the idea of the stable octet. Nevertheless, because elements such as phosphorus, sulfur and chlorine exceed the octet in their higher oxidation states, a few chemists thought that group 8 elements beyond the second period could perhaps form compounds. For expansion of the valence shell of an atom beyond the s and three p orbitals, states have to be formed in which electrons have been promoted to d orbitals. Only the most electronegative ligands, such as O, F and Cl can induce this; compounds of the nonmetals in higher oxidation states are only formed with highly electronegative ligands. Compounds formed between two different halogens are known as interhalogen compounds. With the exception of SF_6, which does not react with water, covalent nonmetal halides react with water to give an oxide or oxoacid and a hydrogen halide. Many behave as Lewis acids; halide ion transfer reactions give compounds such as $K^+BrF_4^-$.

The structures of higher oxidation state nonmetal halides usually are based on the arrangements of **five** and **six** electron pairs in the valence shell of the central nonmetal atom. Six **electron pairs** are arranged at the corners of an **octahedron** and **five electron pairs** at the corners of a **trigonal bipyramid.** The corners of an octahedron are equivalent; those of the trigonal bipyramid divide into two equivalent **axial** corners and three equivalent **equatorial** corners. In an AX_5 molecule,the colinear axial bonds are longer than the equatorial bonds.

AX_5 molecules are **trigonal bipyramidal,** AX_4E are **disphenoidal,** AX_3E_2 are **T-shaped** and AX_2E_3 are **linear.** AX_6 molecules are **octahedral,** AX_5E are **square pyramidal** and AX_4E_2 have **square planar** geometry. Nonbonded electron pairs are arranged so as to minimize their repulsions with other electron pairs. In AX_5E molecules, the square pyramid distorts so that the XAX angles are less than 90°, and in AX_4E_2 molecules the lone pairs are **trans.** In AX_4E, AX_3E_2 and AX_2E_3 molecules, lone pairs are equatorial, rather than axial. Bond angles in AX_4E and AX_3E_2 molecules are smaller than the ideal 90° and 120° angles.

Oxoacids of chlorine are HOCl (hypochlorous acid), HOClO (chlorous acid), $HOClO_2$ (chloric acid), and $HOClO_3$ (perchloric acid). Acid strength increases as m increases in the general formula $HOClO_m$. HOCl results from reacting $Cl_2(g)$ with liquid water, ($Cl_2 + 2H_2O \rightarrow HOCl + Cl^- + H_3O^+$); the equilibrium is shifted to the right by precipitating Cl^- as insoluble AgCl(s). Hypochlorites, such as NaOCl, result from reacting $Cl_2(g)$ with NaOH(aq), which is done commercially by electrolysis of cold NaCl(aq). Hypochlorites are strong oxidizing agents. NaOCl(aq) decomposes slowly to give oxygen ($2OCl^- \rightarrow O_2 + 2Cl^-$) and disproportionates to chlorate ion and Cl^- ion ($3OCl^- \rightarrow 2Cl^- + ClO_3^-$). Chlorates are prepared by passing Cl_2 into hot alkali solutions ($3Cl_2$

$+ 6OH^- \rightarrow 5Cl^- + ClO_3^- + 3H_2O$); this is achieved industrially by electrolysis of hot KCl(aq). ClO_3^- is a pyramidal AX_3E species. Chloric acid and chlorates are strong oxidizing agents. In the presence of a catalyst (MnO_2), $KClO_3$ decomposes on heating to give O_2, ($2KClO_3(s) \rightarrow 2KCl(s) + 3O_2(g)$); without a catalyst, disproportionation to KCl and perchlorate occurs, ($4KClO_3(s) \rightarrow 3KClO_4(s) + KCl(s)$). Perchlorate ion, ClO_4^-, is a tetrahedral AX_4 molecule. Anhydrous $HClO_4$ is a very powerful oxidizing agent prepared by distilling a mixture of $KClO_4$ and H_2SO_4. Chlorous acid, $HClO_2$, is known only in aqueous solution; its salts are the chlorites, containing the ClO_2^- ion.

The oxides of chlorine include Cl_2O, Cl_2O_3, ClO_2, Cl_2O_6 and Cl_2O_7, all of which are unstable and explosive. ClO_2 is a stable free radical, with less tendency to dimerize than NO_2. It is made on a large scale by reducing $NaClO_3(aq)$ with $SO_2(g)$ in acidic solution ($2ClO_3^- + SO_2 + H_2SO_4 \rightarrow 2ClO_2 + 2HSO_4^-$).

The **oxoacids of bromine,** HOBr, $HOBrO_2$ and $HOBrO_3$, resemble the corresponding oxoacids of chlorine. **Iodine** forms hypoiodous acid, HOI, iodic acid, $HOIO_2$, and the periodic acids $HOIO_3$ and $(HO)_5IO$. Hypoiodite ion, IO^-, results from reaction of I_2 with $OH^-(aq)$ ($I_2 + 2OH^- \rightarrow I^- + IO^- + H_2O$), but rapidly disproportionates to iodate ion and iodide ion, ($3IO^- \rightarrow IO_3^- + 2I^-$). Iodic acid, HIO_3, is made by oxidizing iodine with nitric acid.

All the **noble gases** occur in the atmosphere as monatomic gases. Argon is the most abundant. Helium is also found in association with radioactive minerals and in natural gas. Their uses depend on their inertness.

The formulas of possible oxides and fluorides of the noble gases are predicted by extrapolation from related isoelectronic compounds of the elements of groups 5, 6 and 7. XeF_2 and XeF_4 are formed by heating Xe and F_2 together, and XeF_6 results from the reaction of xenon with excess fluorine under pressure. XeO_3 results from reacting XeF_4 or XeF_6 with water. The only known compound of Kr is KrF_2. The structures of all of the noble gas compounds are those predicted by VSEPR theory.

REVIEW QUESTIONS

1. What are the oxidation states of P and S in PF_5, SF_6, PF_3, SF_4, SCl_2, P_4 and S_8?
2. Why are the highest oxidation states of elements observed only in compounds with highly electronegative ligands?
3. Write valence state electron configurations for Cl(I), Cl(III), Cl(V) and I(VII).
4. Classify each of the following molecules and ions in terms of the AX_nE_m nomenclature: SF_6, PF_5, ClF, ClF_3, ClF_5, IF_7, XeF_2, XeF_4, XeF_6, IF_2^+, ICl_4^-.
5. What are the expected arrangements of five and six valence shell electron pairs?
6. Why do the unshared pairs of AX_4E, AX_3E_2 and AX_2E_3 molecules occupy equatorial rather than axial positions?

7. Describe the shapes of AX_5, AX_4E, AX_3E_2 and AX_2E_3 molecules. Give examples of each.
8. Describe the shapes of AX_6, AX_5E and AX_4E_2 molecules. Give examples of each.
9. What is the structure of PCl_5 in the gas phase, and in the solid?
10. What ionic compounds are formed between KF and BrF_3, ICl_3 and $AlCl_3$? Which species are behaving as Lewis acids and which as Lewis bases?
11. Explain why iodine, which is insoluble in water, dissolves in NaI(aq).
12. Draw Lewis structures for the oxoacids of chlorine corresponding to each of the oxidation states +1, +3, +5 and +7.
13. Give two reasons why the acidities of oxoacids of the general formula $ClO_m(OH)$ might be expected to increase from m = 0 to m = 3.
14. Write a balanced equation for the reaction of chlorine with water.
15. Write balanced equations for the reaction of $Cl_2(g)$ with (a) cold potassium hydroxide solution, (b) **hot** potassium hydroxide solution.
16. What are the products obtained by electrolysis of (a) molten sodium chloride, (b) cold aqueous sodium chloride, (c) hot aqueous sodium chloride, with vigorous stirring of the electrolyte?
17. What is a disproportionation reaction? Give an example.
18. Write balanced equations for (a) the formation of $O_2(g)$ from aqueous sodium hypochlorite solution, (b) the disproportionation of aqueous sodium hypochlorite to sodium chloride and sodium chlorate.
19. What reactions occur when potassium chlorate is heated (a) alone, (b) together with the catalyst MnO_2?
20. How could a sample of anhydrous perchloric acid be prepared?
21. Draw Lewis structures for perchloric acid and perchlorate ion.
22. Draw Lewis structures for the following oxides: Cl_2O, Cl_2O_3, ClO_2 and Cl_2O_7.
23. Why is ClO_2 described as a free radical?
24. What are the anhydrides of hypochlorous acid and perchloric acid?
25. Draw Lewis structures for metaperiodic acid and paraperiodic acid and deduce the expected geometric shapes.
26. Why is helium found in uranium minerals and in pockets of natural gas?
27. Why does helium form no known compounds?
28. What are the formulas of the known fluorides of Kr and Xe?
29. Classify XeF_2 and XeF_4 in terms of the AX_nE_m nomenclature and predict their geometries.
30. Is XeF_6 expected to be a regular octahedral molecule?
31. What oxides of xenon have been prepared and what are their shapes?
32. Would you expect the oxides of xenon to be good oxidizing agents?

Answers to Selected Review Questions

1. +5, +6, +3, +4, +2, 0, 0.
4. AX_6, AX_5, AXE_3, AX_3E_2, AX_5E, AX_7, AX_2E_3, AX_4E_2, AX_6E, AX_2E_2, AX_4E_2.
10. $K^+BrF_4^-$, $ICl_2^+AlCl_4^-$. BrF_3 and $AlCl_3$ are Lewis acids; F^- and ICl_3 are Lewis bases.
11. $I_2(s) + I^-(aq) \rightleftharpoons I_3^-(aq)$
32. Yes.

OBJECTIVES

Be able to:

1. Give examples of interhalogen compounds, and assign oxidation numbers to each of the atoms in these compounds.
2. Give typical products of the reactions of nonmetallic halides with water.
3. Give examples of polyhalide ions.
4. Draw the Lewis structures and predict the geometries of typical interhalogen molecules and ions.
5. Name, prepare and list reactions and uses of the halogen oxoacids, their salts, and the halogen oxides.
6. Arrange the chlorine oxoacids in order of increasing acid strength, and rationalize the order.
7. List the occurrence and uses for the noble gases.
8. Show how the fluorides and oxides of xenon can be prepared, and give some of their typical reactions.
9. Draw the Lewis structures and predict the shapes of typical noble gas compounds.

PROBLEM SOLVING STRATEGIES

20.1 Halides of Nonmetals in their Higher Oxidation States

You have already seen in Chapter 9 how to draw Lewis structures and predict shapes for various types of AX_nE_m molecules and ions. Review that material if necessary. In the present chapter some higher oxidation state non metal halides and other halogen compounds are discussed. Halogen compounds span the range from -1 oxidation state halide ions, like Cl^-, all the way to +7 oxidation state for perbromic acid, $HBrO_4$. The assignment of oxidation states has been treated in Chapter 8. The Lewis structures and shapes of some halogen compounds and ions namely BrF_3, BrF_4^-, SCl_4, and ClO_4^- have been presented in an earlier example (9.2) in this Guide. Let's continue with these molecules and ions.

Example 20.1

Assign oxidation numbers (O.N.'s) to all halogen atoms in (a) BrF_3 (b) BrF_4^- (c) SCl_4 (d) ClO_4^-.

Solution 20.1

(a) Fluorine has an O.N. of -1 in all compounds (other than in F_2, where its O.N. is zero). Since the oxidation numbers add to zero for a neutral molecule, the O.N. of Br in BrF_3 is +3.

(b) The sum of oxidation numbers must equal the charge on any ionic species. With an O.N. of -1 for each fluorine, we can write

$$4 \times \text{O.N. (F)} + \text{O.N. (Br)} = -1$$

$$\text{O.N. (Br)} = -1 + 4 = +3$$

(c) Halogen atoms other than fluorine exhibit O.N.'s of -1 except in the free halogens X_2 and when they are combined with a more electronegative atom such as F, O, or a more electronegative halogen. Since chlorine is more electronegative than sulfur, Cl in SCl_4 has an O.N. of -1. Then S has an O.N. of +4 (which you should recall from Chapter 8 is a typical oxidation state for sulfur.

(d) Oxygen is more electronegative than chlorine; its O.N. equals -2 except in O_2 and in peroxides, and in OF_2. Thus for ClO_4^-,

$$\text{O.N. (Cl)} + 4 \times \text{O.N. (O)} = -1$$

$$\text{O.N. (Cl)} = 8 - 1 = 7$$

This is the highest oxidation state exhibited by halogen atoms.

Example 20.2

Classify the interhalogen ions as AX_nE_m: a) ClF_2^+ (b) ClF_2^-. Which ion has the larger F-Cl-F bond angle?

Solution 20.2

Note at the start that Cl is the central atom for both ions, and that there are two fluorine atoms bonded to each Cl. Thus, both ions are AX_2E_m. We need to draw Lewis structures to find values for m.

(a) This ion has 7 + (2 x 7) - 1 = 20 valence electrons. Since fluorine obeys the octet rule, we place 8 electrons around each F, using one bonding pair to Cl for each fluorine. Then we have accounted for 16 valence electrons with the following partial structure:

$$:\ddot{\underset{..}{F}}-Cl-\ddot{\underset{..}{F}}:$$

The remaining four electrons are arranged in pairs about chlorine. The final Lewis structure is

$$:\ddot{\underset{..}{F}}-\overset{+}{\ddot{\underset{..}{Cl}}}-\ddot{\underset{..}{F}}:$$

Note we have placed a formal charge of +1 on Cl. This ion is AX_2E_2; it is bent. We predict the F-Cl-F angle to be slightly less than 109.5°.

(b) There are 7 + (2 x 7) + 1 = 22 valence electrons for ClF_2^-. Again we account for 16 electrons by drawing

$$:\ddot{\underset{..}{F}}-Cl-\ddot{\underset{..}{F}}:$$

The remaining 6 electrons we place, as nonbonding pairs, around chlorine.

$$:\ddot{\underset{..}{F}}-\overset{-}{\dot{\ddot{\underset{..}{Cl}}}}-\ddot{\underset{..}{F}}:$$

The formal charge of -1 is placed on chlorine. ClF_2^- is AX_2E_3. Recall that the 3 nonbonding pairs of an AX_2E_3 species are equatorial, with the two Cl-F bonds axial. Then the F-Cl-F bond angle is 180°.

The larger bond angle is predicted for ClF_2^-.

20.3 The Noble Gases

The noble gases of greater mass than helium have 8 valence electrons. When they form compounds, like xenon with fluorine and oxygen, the octet rule is violated. The geometries of the **fluorides** XeF_2 and XeF_4 can be predicted, since they are AX_2E_3 and AX_4E_2, respectively. XeF_6 has 7 electron pairs around xenon, and is described as a distorted octahedron. For the **oxides**, XeO_3 and XeO_4, we can use the rules of Chapter 9, since these molecules minimize formal charge by forming multiple bonds.

Example 20.3

Draw the Lewis structure and predict the geometry for XeO_3.

Solution 20.3

Xenon has 8 valence electrons, and each oxygen contributes 6, for a total of 26 electrons. Oxygen must obey the octet rule. If we form **single** O-Xe bonds, we get

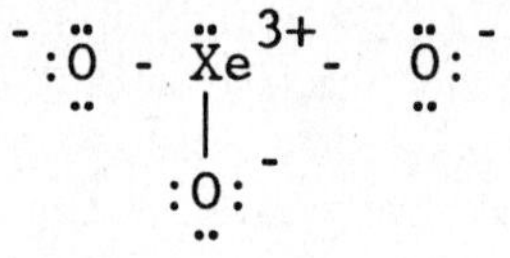

```
 -  ..    ..  3+    ..  -
   :O  -  Xe   -    O:
    ..    |         ..
          |  -
         :O:
          ..
```

We reduce formal charge by using 3 **double** bonds.

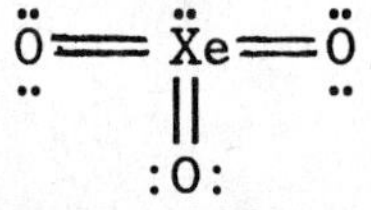

```
 ..     ..     ..
 O ==== Xe ==== O
 ..     ||     ..
       :O:
```

The formal charges now are zero. By counting each double bond as a single bond we classify the molecule as AX_3E, which is trigonal pyramidal, like ammonia.

Example 20.4

YF_4 has a dipole moment, but ZF_4 does not. Identify Y and Z from the following: Xe, K, N, S. Give the shapes for YF_4 and ZF_4.

Solution 20.4

Potassium does not form 4 bonds; neither does neutral nitrogen. So let's draw the Lewis structures for XeF_4 and SF_4, predict their geometries, and vectorially add their bond moments.

XeF_4: Xenon has 8 valence electrons, and each fluorine has 7, for a total of 36. By forming 4 single F-Xe bonds, we have

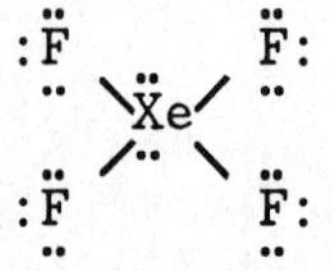

```
  ..          ..
 :F           F:
  ..  \ ..  / ..
        Xe
  ..  / ..  \ ..
 :F           F:
  ..          ..
```

This is an AX_4E_2 molecule, which is square planar. The 4 bond moments ($Xe^{\delta+}$ - $F^{\delta-}$) vectorially add to give a zero molecular dipole moment. This is the same geometry as for ICl_4^-, Figure 20.4 in the text.

SF_4: This molecule has the same structure as SCl_4 discussed in Example 9.2, part c in this Guide. It is AX_4E, and has a squashed tetrahedral geometry. The 4 bond moments ($S^{\delta+}$ - $F^{\delta-}$) vectorially add to give a nonzero molecular dipole moment.

So Y is S and Z is Xe.

SELF TEST

1. The oxidation state of I in ICl_4^- is
   (a) -1 (b) +1 (c) +3 (d) +5 (e) +7
2. Give the product(s) of the reaction of $PCl_5(s)$ with water
   (a) $H_3PO_3(aq)$ (b) $HCl(aq)$ (c) $PH_3(g)$ (d) $Cl_2(g)$ (e) $H_3PO_4(aq)$
3. The range of oxidation states in chlorine compounds is
   (a) 0 to 8 (b) -2 to +6 (c) -3 to +5 (d) -1 to +7 (e) -4 to +4
4. BrF_5 has a shape described as
   (a) trigonal bipyramid (b) pentagonal (c) octahedral
   (d) squashed tetrahedron (e) square pyramidal
5. Which of the following is(are) example(s) of AX_2E_3 molecule(s)?
   (a) H_2O (b) ICl_2^+ (c) XeF_2 (d) ClO_2 (e) ICl_2^-
6. Which of the following is(are) strong acids in water?
   (a) $HClO_4$ (b) $HClO$ (c) $HClO_3$ (d) $HClO_2$ (e) BrO_3H
7. Which of the following is(are) **not** strong oxidizing agents?
   (a) Cl^- (b) ClO_4^- (c) ClO_3^- (d) $HClO$ (e) $HClO_2$
8. The most abundant of the noble gases is
   (a) He (b) Ne (c) Ar (d) Kr (e) Xe
9. The products of the reaction of $XeF_6(s)$ with water are
   (a) XeF_2 (b) O_2 (c) XeO_3 (d) HOOH (e) HF
10. In an AX_5 type molecule, which of the following are true:
    (a) all the bond angles are equal
    (b) there are three equivalent axial bonds in the same plane
    (c) there are two axial bonds shorter than the equatorial bonds
    (d) the bond angles between the equatorial bonds and the axial bonds are 90°
    (e) all the bond lengths are equal
11. The maximum possible valence for xenon, if all electron promotions occur, is
    (a) 0 (b) 2 (c) 4 (d) 6 (e) 8
12. In water, hypochlorous acid reacts to yield
    (a) Cl_2 (b) Cl^- (c) O_2 (d) Cl_2^+ (e) OH^-
13. The reaction of Cl_2 with hot aqueous solutions of hydroxide yields
    (a) Cl^- (b) ClO^- (c) ClO_2^- (d) ClO_3^- (e) ClO_4^-
14. XeF_4 reacts with H_2O to produce
    (a) XeO_3 (b) Xe (c) F_2 (d) HF (e) O_2
15. Which of the following molecules do <u>not</u> react with H_2O?
    (a) PCl_3 (b) PCl_5 (c) SF_4 (d) SF_6 (e) ClF_5
16. When heated in the presence of the catalyst MnO_2, the substance $KClO_3$ decomposes, yielding
    (a) KCl (b) KClO (c) $KClO_3$ (d) $KClO_4$ (e) O_2
17. Which of the following oxides of xenon have been prepared?
    (a) XeO_2 (b) XeO_3 (c) XeO_4 (d) XeO_5 (e) XeO_6

Answers to Self Test

1. c; 2. b,e; 3. d; 4. e; 5. c,e; 6. a,c,e; 7. a; 8. c; 9. c,e; 10. d; 11. e; 12. b,c; 13. a,d; 14. a,b,d,e; 15. d; 16. a,e; 17. b,c.

CHAPTER 21

THE TRANSITION METALS

SUMMARY REVIEW

The first series of transition metals includes the ten elements from scandium, Sc to zinc, Zn, in which the 3d subshell is progressively filled with electrons from $3d^1$ to $3d^{10}$.

The 3d and 4s energy levels are similar in energy; for the neutral atoms 3d > 4s but for their ions 4s > 3d. The ionization energies are similar to those for K and Ca and increase from Sc to Zn. All three valence electrons may be removed from Sc ($4s^2 3d^1$) to give the Sc^{3+} ion with a d^0 electron configuration. With increasing core charge, it becomes increasingly difficult to ionize electrons and ultimately only a limited number of valence electrons are used, even for covalent bonding. The maximum oxidation states are Sc +3, Ti +4, V +5, Cr +6, Mn +7, in all of which all the valence electrons are utilized; iron has a maximum of +6, although the commonest oxidation state is +3, and the commonest oxidation state for Co, Ni, Cu and Zn is +2. In their oxidation states up to +3, the metals form cations; in their higher oxidation states, Ti, V, Cr and Mn form covalent compounds and behave more like nonmetals than metals. In comparison to group 1 and 2 metals, transition metals are hard, dense and have high melting points; these properties are related to the number of valence electrons available for metallic bonding, which reaches a maximum at V and Cr.

Many of the metals are important industrially and as alloys. Commonly, the oxides are reduced with coke, H_2, or electrolytically. Important ores include: **rutile,** TiO_2; **ilmenite,** $FeTiO_3$; ores containing the VO_4^{3-} , **vanadate,** ion and its condensed forms; **chromite,** $FeCr_2O_4$; **pyrolusite,** MnO_2; **hematite,** Fe_2O_3; magnetite, Fe_3O_4; CoS; NiS; **chalcocite,** Cu_2S; **chalcopyrite,** $CuFe_2$; **cuprite,** Cu_2O; **malachite,** $Cu_2(CO_3)(OH)_2$; and **sphalerite,** ZnS. Titanium is made by heating TiO_2 with C and Cl_2, to give $TiCl_4$, which is then reduced with Mg. **V, Cr** and **Mn** are often prepared as alloys with iron (ferrovanadium, ferrochrome and ferromanganese).

The number of valence shell electrons for the neutral atoms, their principal oxidation state, and the colors of their aqueous ions are given in the table on page 257.

Species in oxidation states lower than the most stable are readily oxidized; those in higher oxidation states are readily reduced. The M^{2+} and M^{3+} cations are strongly hydrated in aqueous solution, most often as $M(H_2O)_6^{n+}$. Lower oxidation state oxides and hydroxides are basic; higher oxidation state oxides and hydroxides are acidic. $Cr(OH)_3$, $Ni(OH)_2$ and $Zn(OH)_2$ are amphoteric.

Small cations with high core charges readily form **complexes** and the transition metals form many **coordination compounds.** The number of bonds formed between donor **ligands** and a metal atom is its coordination number. A metal cation behaves as a Lewis acid and the ligands as Lewis bases. Common **monodentate** ligands include H_2O, NH_3, halide ions, OH^- and CN^-. **Bidentate** ligands include oxalate ion, $C_2O_4^{2-}$, and ethylenediamine, $H_2NCH_2CH_2NH_2$, (en). Ethylenediaminetetraacetate ion, $^{2-}(O_2CCH_2)_2N(CH_2)_2N(CH_2CO_2)_2{}^{2-}$, **($EDTA^{4-}$)**, is a common **polydentate ion** which forms very stable **six-coordinated** complexes. Complexes containing monodentate ligands are generally less stable than chelates containing polydentate ligands.

Many coordination compounds and ions have some unused nonbonding d electrons, which are largely "inside" the electron pairs associated with the metal-ligand bonds. The d-shell may be incomplete and therefore nonspherical, which may distort the geometry predicted by the VSEPR model. The d electrons are also responsible for the absorption of light, and thus the colors of transition metal complexes. In a free metal atom the five d orbitals all have the same energy, but in a coordination complex the interaction between the ligands and the d orbitals causes the d orbitals to have different energies; transitions between these energies accounts for the colors. In an octahedral AX_6 arrangement of ligands, electrons in the d_{xy}, d_{yz}, and d_{xz} orbitals have a lower energy than electrons in the $d_{x^2-y^2}$ and d_{z^2} orbitals; the d orbital energies are split into two sets where the energy difference is Δ_o, the value of which can be obtained from the energy of the light quantum absorbed when an electron is promoted between the two energy levels. The magnitude of Δ_o depends on the strength of interaction between the ligands and the d orbitals. The value of Δ_o increases for common ligands in the order

$$Cl^- < F^- < H_2O < NH_3 < en < CN^-$$

which is called the **spectrochemical series;** the ligands on the left are called **weak-field** ligands, and those on the right are called **strong-field** ligands.

Transition metal complexes that have unpaired d electrons are paramagnetic; those with all of their electrons paired, including most of the compounds of the main group elements, are **diamagnetic.** Experimentally, the number of unpaired d electrons can be determined by measuring the force by which they are attracted in a magnetic field. Complexes with a maximum number of unpaired electrons are high-spin complexes; those with fewer unpaired electrons are **low-spin** complexes.

Many AX_4 type complexes are **square planar** in shape, rather than **tetrahedral,** and some AX_6 type complexes are **elongated octahedral,** rather than **regular octahedral.** Such distortions occur when there is a strong interaction between the d shell and the ligands.

Formation constants, K_f, of complex ions are related to the equilibrium constant for the reaction between ligand and $M^{n+}(aq)$, but do not include solvent concentrations. For example, for:

$$Co(H_2O)_6^{2+}(aq) + 6NH_3(aq) \rightleftarrows Co(NH_3)_6^{2+}(aq) + 6H_2O(l)$$

$$K_f = [Co(NH_3)_6^{2+}]/[Co(H_2O)_6^{2+}]\ [NH_3]^6$$

Element s+d Electrons	Sc 3	Ti 4	V 5	Cr 6	Mn 7	Fe 8	Co 9	Ni 10	Cu 11	Zn 12
Oxidation State										
+1									Cu^{+}	
+2		$TiCl_2$ purple	V^{2+}	Cr^{2+} p.pink	**Mn^{2+}** p.green	Fe^{2+} p.pink	**Co^{2+}** green	**Ni^{2+}** green	**Cu^{2+}** blue	Zn^{2+} cls
+3	**Sc^{3+}** cls	Ti^{3+} violet	V^{3+} green	**Cr^{3+}** violet	MnO(OH)	**Fe^{3+}** yellow-red	Co^{3+}			
+4		**$TiCl_4$**	VO^{2+} blue							
+5			**VO_4^{3-}** cls							
+6				CrO_4^{2-} yellow $Cr_2O_7^{2-}$ orange	MnO_4^{2-} green					
+7					MnO_4^{-} purple					

Note: cls = colorless
p = pale

Note = Principal oxidation states are in bold type

Cu, Ag and **Au** are the **coinage metals.** The most stable oxidation state of Cu is +2, that of Ag is +1, while Au has stable +1 and +3 oxidation states. The metals are unreactive, gold especially so. They occur as the metals in nature. Like Cu, Ag requires an oxidizing medium stronger than $H_3O^+(aq)$ to dissolve it; it dissolves in dilute $HNO_3(aq)$ but not in dilute HCl(aq) or $H_2SO_4(aq)$. $AgNO_3$ is the commonest salt. Addition of NaOH(aq) to $AgNO_3(aq)$ gives a precipitate of very slightly soluble dark-brown Ag_2O. Of the halides only AgF is appreciably soluble. AgCl and AgBr (but not AgI) dissolve in $NH_3(aq)$, due to formation of the $Ag(NH_3)_2{}^+(aq)$ ion. In general, an insoluble salt dissolves in a solution of a complexing agent if K_f of the expected complex is sufficiently large. AgCl, AgBr and AgI decompose to Ag and halogen photochemically. Gold is very inert and dissolves only in a mixture of concentrated hydrochloric and nitric acids to give auric acid, $HAuCl_4$, which crystallizes as $(H_3O^+)(AuCl_4{}^-)\cdot 3H_2O$. On heating, this dehydrates to give $AuCl_3$, and then decomposes to AuCl and finally the metal. $HAuCl_4(aq)$ gives a precipitate of $Au(OH)_3$ with $OH^-(aq)$, which gives Au_2O_3 on heating, before decomposing to Au and O_2 at 150°C. Mercury, in the same group as Cd and Zn, readily forms Hg^{2+} and mercury(II) complexes. It also forms the $Hg_2{}^{2+}$ ion containing an Hg-Hg bond. Hg is oxidized by concentrated $HNO_3(aq)$ or concentrated $H_2SO_4(aq)$ to Hg^{2+} (mercuric) salts. $HgCl_2$ with linear AX_2 molecules is volatile. **Mercury** reacts with an $Hg^{2+}(aq)$ solution to give the $Hg_2{}^{2+}$ ion. With $Cl^-(aq)$, $Hg_2{}^{2+}$ gives a precipitate of linear covalent Cl-Hg-Hg-Cl. Organometallic **dimethyl mercury** $(CH_3)_2Hg$, results from the reaction of CH_3Cl with Na/Hg amalgam. It reacts with $HgCl_2$ to give linear covalent CH_3-Hg-Cl. CH_3HgCl and $(CH_3Hg)_2SO_4$ are ionic and contain the methylmercury cation, CH_3Hg^+, which is water soluble.

REVIEW QUESTIONS

1. Locate the positions of the three series of transition metals in the periodic table.
2. What is the maximum capacity for electrons of the n = 3 shell?
3. Write the chemical symbols and electron configurations of the ten elements in the first series of transition metals.
4. What are the highest oxidation states for each element of the first transition series? Why are the highest possible oxidation states not observed for the transition metals beyond manganese?
5. In terms of the formation of covalent molecules or ions, what is the most obvious difference between the chemistry of transition elements in high oxidation states and in low oxidation states?
6. What are the more important oxidation states of each of the elements in the first transition series? What is the most stable oxidation state of each?
7. In the fourth period, the melting points of the metals increase from potassium to vanadium and the tendency is then to decrease again from chromium to zinc. Why?
8. What model is used to explain properties such as hardness, melting points and other physical properties of metals?
9. With which other element would it be most useful to compare the chemistry of scandium?

10. How is titanium prepared from TiO_2? What are the principal ores of titanium? Why is titanium an important space-age metal?
11. Why are vanadium, chromium, manganese, cobalt and nickel most often manufactured as their alloys with iron?
12. How is nickel obtained commercially from a mixture of the sulfide ores of nickel, copper and iron(II)?
13. How is zinc prepared from its principal ore?
14. Could nickel be plated onto iron?
15. What are the common oxidation states of titanium and which is the most stable? How can $TiCl_3$ and $TiCl_2$ be obtained from $TiCl_4$?
16. What is the geometric arrangement of the water molecules coordinated to the violet $Ti^{3+}(aq)$ ion?
17. Why is $Ti^{2+}(aq)$ not found in aqueous solution?
18. What are the principal oxidation states of vanadium?
19. What species are formed when V_2O_5 is dissolved in basic solution?
20. With what other element is it useful to compare the chemistry of vanadium(V)?
21. What ions are formed by successive reduction of an acidic solution of vanadium(V) with zinc?
22. What are the principal oxidation states of chromium? Are chromium(VI) compounds ionic or covalent?
23. What main group element does Cr(VI) most resemble?
24. What ion results from addition of acid to a solution of potassium chromate?
25. How is Cr(VI) oxide prepared? What is its formula and structure?
26. Write a balanced equation for the reaction of ferrous sulfate with potassium dichromate in acid solution, to give ferric sulfate and chromium sulfate.
27. What are the products of the oxidation of ethanol and 3-methyl-2-butanol with CrO_3?
28. What ion is formed by Cr(III) in aqueous solution? What ion results from reduction of $Cr_2(SO_4)_3$ with zinc in aqueous acid?
29. What are the principal oxidation states of manganese?
30. What unusual property has the compound normally formulated as MnO_2? Why is MnO_2 called a "non-stoichiometric" compound?
31. Write an equation for the reaction of MnO_2 with HCl(aq) to give $Cl_2(g)$ and manganese(II) chloride.
32. Write an equation for the reaction of MnO_2 with potassium hydroxide and oxygen, to give potassium manganate, K_2MnO_4.
33. By what methods can potassium permanganate, $KMnO_4$, be prepared from potassium manganate, K_2MnO_4?
34. Write a balanced equation for the reduction of potassium permanganate by oxalic acid in acidic solution, to give manganese(II) sulfate and carbon dioxide.
35. Write a balanced equation for the reaction of potassium permanganate with potassium iodide in KOH solution, to give potassium iodate (KIO_3) and MnO_2.
36. What are the common oxidation states of (a) iron, (b) cobalt, (c) nickel?
37. Describe what happens when (a) NaOH(aq), (b) $NH_3(aq)$, are added to a solution containing $Ni^{2+}(aq)$.
38. Why does $Zn(OH)_2$ dissolve in excess NaOH(aq)?
39. Give two examples of inorganic complexes.

40. What complex ions does Ni^{2+} form with (a) water, (b) ammonia, (c) ethylenediamine?
41. What is a ligand? Give two examples of each of (a) a monodentate ligand, (b) a bidentate ligand, (c) a polydentate ligand.
42. What is a chelate?
43. Why does $[Co(NH_3)_4Cl_2]Cl$ exist in the form of two geometric isomers?
44. How many donor atoms are there in the chelating agent ethylenediaminetetraacetate ion?
45. Why is carbon monoxide poisonous to humans?
46. Write expressions for the formation constants of (a) $Ag(CN)_2^-$, (b) $[Mn(en)_3]^{2+}$, (c) $Zn(EDTA)^{2-}$.
47. If you wished to remove $Zn^{2+}(aq)$ ions from solution, which of the following complexing agents would prove effective? (a) NH_3, (b) CN^-, (c) $C_2O_4^{2-}$, (d) ethylenediamine.
48. Why are the shapes of some transition metal complexes distorted from the shapes predicted by the VSEPR model?
49. What is the spectrochemical series of ligands?
50. Among the ligands discussed in this chapter, give an example of a strong-field ligand and an example of a weak-field ligand.
51. How are the energy levels of the five d orbitals split by (a) an octahedral arrangement, (b) a tetrahedral arrangement, and (c) a square planar arrangement of ligands?
52. To what is the property of paramagnetism attributed?
53. What are the principal oxidation states of Ag, Au and Hg?
54. What complexes are formed between silver and gold and CN^- ion?

Answers to Selected Review Questions

2. 18. 9. aluminum. 14. $Ni^{2+}(aq) + Fe(s) \rightarrow Ni(s) + Fe^{2+}(aq)$ $E^\circ_{cell} = +0.19$ V, yes. 20. P(V). 23. S(VI). 26. $6FeSO_4 + K_2Cr_2O_7 + 7H_2SO_4 \rightarrow 3Fe_2(SO_4)_3 + Cr_2(SO_4)_3 + K_2SO_4$. 27. CH_3CHO and $H_3C\text{-}\overset{O}{\overset{\|}{C}}\text{-}CH(CH_3)_2$. 28. $Cr(H_2O)_6{}^{3+}(aq)$, $Cr(H_2O)_6{}^{2+}(aq)$. 34. $2KMnO_4 + 3H_2SO_4 + 5H_2C_2O_4 \rightarrow 2MnSO_4 + 10CO_2 + K_2SO_4 + 8H_2O$. 35. $2KMnO_4 + KI + H_2O \rightarrow 2MnO_2 + 2KOH + KIO_3$. 47. CN^-.

OBJECTIVES

Be able to:

1. List the common characteristics of the transition metals, and some of their practical applications.
2. Give the general features of the electron configurations and the trends in ionization energies for the transition metals.
3. Sketch the trend in metallic radii, show the variations in melting point and density, and list the common oxidation states of the first series of transition metals.
4. List the preparation, properties, uses, and important compounds of the first series of transition metals.
5. Define the terms complex, complex ion, ligand, coordination number and coordination compound.

6. State the most common coordination numbers found in complex ions, and the geometries of these ions.
7. Draw cis and trans isomers for complexes of the type MX_4Y_2.
8. Recognize the role of nonbonding electrons in VSEPR predictions of the geometries of coordination compounds of the transition metals.
9. Define the terms monodentate, bidentate, and polydentate ligands, chelate and chelating agent.
10. Write the form of the formation constant K_f for a reaction in which a hydrated ion reacts with a ligand to form a complex ion.
11. Give the systematic names for complex ions, and write the formulas for complex ions, given their systematic names.
12. Account for the colors and magnetic properties of transition metal compounds.
13. List some important compounds of silver, gold and mercury, and the chief uses of these metals.
14. Give examples of organometallic compounds.
15. Show quantitatively how complex ion formation affects solubility.

PROBLEM SOLVING STRATEGIES

21.3 Coordination Compounds

Formation Constants

We have seen in Chapter 10 that some metal cations from the main group elements can form **complex ions;** for example, aluminum ion in water is $Al(H_2O)_6^{3+}$. In this Chapter, we find that many transition metal cations also form complex ions. The formation of such a complex from the cation and ligands can be described by an equilibrium constant K_f, called the formation constant. For example, $Ag^+(aq) + 2CN^-(aq) \rightleftarrows Ag(CN)_2^-(aq)$

$$K_f = \frac{[Ag(CN)_2^-]}{[Ag^+][CN^-]^2}$$

Problems involving K_f can be solved in the same manner as we saw for solution equilibria in Chapter 14.

Example 21.1

Many transition metal cations form complexes with ammonia. Ag(I) forms $Ag(NH_3)_2^+$, with $K_f = 1.0 \times 10^8\ mol^{-2}\ L^2$. Cu(II) forms $Cu(NH_3)_4^{2+}$, with $K_f = 1.0 \times 10^{12}\ mol^{-4}\ L^4$. Beaker A contains 100 mL of 0.100 M $AgNO_3$, and beaker B contains 100 mL of 0.100 M $Cu(NO_3)_2$. 100 mL of 1.00 M NH_3 solution is added to each. Which beaker has more uncomplexed cation remaining at equilibrium?

Solution 21.1

We will treat the silver complex first.

$$Ag^+(aq) + 2NH_3(aq) \rightleftarrows Ag(NH_3)_2^+(aq)$$

$$K_f = 1.0 \times 10^8 \ mol^{-2} \ L^2 = \frac{[Ag(NH_3)_2{}^+]}{[Ag^+][NH_3]^2}$$

The total volume in each beaker is 200 mL. We first solve for the initial concentrations.

$$[Ag^+] = \frac{\frac{0.100 \ mol}{L} \times 100 \ mL \times \frac{1.0 \ L}{1000 \ mL}}{200 \ mL \times \frac{1.0 \ L}{1000 \ mL}} = 0.050 \ M$$

and similarly $[NH_3] = 0.500$ M.

We will assume that aqueous NH_3 consists **entirely** of dissolved NH_3. This is not exact, since we know there is some small amount of $NH_4{}^+$ and OH^- present. Also we will make another assumption which is at the very heart of the problem. Since we note that K_f is large, we will assume the equilibrium $[Ag(NH_3)_2{}^+]$ equals the **initial** $[Ag^+]$, so as to get $[Ag(NH_3)_2{}^+] = 0.050$ M. This is a good approximation. But realize that $[Ag^+]$ at equilibrium is not **zero,** otherwise there would be no **equilibrium.** In fact $[Ag^+]$ is the **unknown** we will now solve for. It will be the uncomplexed silver cation concentration. Note that $[NH_3] = 0.50 - 2 \times 0.05$, since each Ag^+ reacts with two NH_3's.

$$K_f = 1.0 \times 10^8 = \frac{(0.050)}{[Ag^+] \ (0.500 - 0.100)^2}$$

$$[Ag^+] = \frac{(0.050)}{(0.400)^2(1.0 \times 10^8)} = 3.1 \times 10^{-9} \ M$$

This is the uncomplexed $[Ag^+]$. For $Cu(NH_3)_4{}^{2+}$, the steps are similar.

$$Cu^{2+}(aq) + 4NH_3(aq) \rightleftarrows Cu(NH_3)_4{}^{2+}(aq)$$

At the start, $[Cu^{2+}] = 0.050$ M, $[NH_3] = 0.500$ M, and since K_f is large, at equilibrium

$$[Cu(NH_3)_4{}^{2+}] = 0.05 \ M$$

$$[NH_3] = 0.500 - 4 \times 0.05 = 0.500 - 0.200 = 0.300 \ M$$

Then

$$K_f = 1.0 \times 10^{12} = \frac{(0.050)}{[Cu^{2+}](0.300)^4}$$

$$[Cu^{2+}] = \frac{(0.050)}{(0.300)^4(1.0 \times 10^{12})} = 6.2 \times 10^{-12} \ M$$

This is the uncomplexed $[Cu^{2+}]$. So we see that the beaker originally containing $AgNO_3$, beaker A, has more uncomplexed metal ion left, though the amounts are tiny in both cases.

21.4 Second and Third Series of Transition Metals
Effect of Complex Ion Formation on Solubility

In Chapter 15 we considered the equilibrium between a slightly soluble salt and its ions. We learned how to calculate the solubility of a salt from its K_{sp} value. We also saw that addition of a common ion **decreased** the solubility of the salt. The **formation of a complex ion can also affect solubility**, since one of the ions involved in the K_{sp} equilibrium may also be involved in the formation of a complex. For example, Ag^+ can form a thiosulfate complex, $AgS_2O_3^-$, according to

$$Ag^+(aq) + S_2O_3^{2-}(aq) \rightleftarrows AgS_2O_3^-(aq)$$

$$K_f = \frac{[AgS_2O_3^-]}{[Ag^+][S_2O_3^{2-}]}$$

If the silver ion is in equilibrium with the slightly soluble salt AgBr, for which

$$AgBr(s) \rightleftarrows Ag^+(aq) + Br^-(aq)$$

$$K_{sp} = [Ag^+][Br^-]$$

then $[Ag^+]$ must simultaneously obey two equilibrium expressions. Thus, if $S_2O_3^{2-}$ is added to a saturated solution of AgBr, the $[Ag^+]$ changes. The solubility of the salt **increases.**

In considering these two simultaneous equilibria, we start by rearranging the equations containing K_f and K_{sp} so as to equate the $[Ag^+]$ common to both.

$$[Ag^+] = \frac{K_{sp}}{[Br^-]} = \frac{[AgS_2O_3^-]}{K_f[S_2O_3^{2-}]}$$

By cross-multiplying we get

$$K = K_{sp} \cdot K_f = \frac{[Br^-][AgS_2O_3^-]}{[S_2O_3^{2-}]}$$

We have replaced the **product** of two equilibrium constants by another equilibrium constant K. This K refers to the overall reaction

$$AgBr(s) + S_2O_3^{2-}(aq) \rightleftarrows Br^-(aq) + AgS_2O_3^-(aq)$$

The solid [AgBr] is constant, and does not appear in the K expression, just as it did not appear in K_{sp}. This equilibrium is the starting point for calculations involving the effect of complex ion formation on solubility. The method of attack is just like that for the solubility problems in Chapter 15.

Example 21.2

What is the molar solubility of AgBr in 1.00 M $Na_2S_2O_3$? $K_f = 6.6 \times 10^9$ mol^{-1} L for $AgS_2O_3^-$, and $K_{sp} = 5.0 \times 10^{-13}$ mol^2 L^{-2} for AgBr.

Solution 21.2

Since we have already written the balanced equation involving AgBr and the thiosulfate ion, $S_2O_3^{2-}$, we can start by evaluating $K = K_{sp} \cdot K_f$.

$$K = 5.0 \times 10^{-13} \text{ mol}^2 \text{ L}^{-2} \times 6.6 \times 10^8 \text{ mol}^{-1} \text{ L}$$
$$= 3.3 \times 10^{-4} \text{ mol L}^{-1}$$

Let S = the molar solubility of AgBr in the presence of the complexing agent. From the balanced equation described by K,

$$AgBr(s) + S_2O_3^{2-}(aq) \rightleftarrows Br^-(aq) + AgS_2O_3^-(aq)$$

we see that for each Ag^+ that gets complexed as $AgS_2O_3^-$, there is one Br^- produced and one $S_2O_3^{2-}$ used up.

concentration, mol L^{-1}	$[S_2O_3^{2-}]$	$[Br^-]$	$[AgS_2O_3^-]$
initial	1.00	0	0
change	- S	S	S
equilibrium	1.00 - S	S	S

$$K = 3.3 \times 10^{-4} = \frac{S \times S}{1.00 - S}$$

Let us assume S << 1.00. Then 1.00 - S can be approximated by 1.00, and

$$S^2 = 3.3 \times 10^{-4} \qquad S = 1.82 \times 10^{-2}$$

Thus the solubility of AgBr in 1.00 M $Na_2S_2O_3$ is 1.82×10^{-2} M. For the sake of comparison, let us calculate the solubility of AgBr in **water,** using the method of Chapter 15.

Let T = molar solubility of AgBr in water. Then

$$K_{sp} = 5.0 \times 10^{-13} \text{ mol}^2 \text{ L}^{-2} = T \cdot T$$

$$T = 7.07 \times 10^{-7} \text{ M}$$

So we see that the solubility of AgBr in 1.00 M $Na_2S_2O_3$ is about 10,000 times greater than in water. The addition of the complexing agent has greatly **increased** the solubility.

SELF TEST

1. Which of the following statements about transition metals is(are) true?
   (a) the majority of transition metals form compounds in only one oxidation state, +3.
   (b) transition metals form a large number of complexes.
   (c) all transition metals are solids at 25°C.
   (d) very few transition metal compounds are colored.
   (e) all transition metals contain at least one 3d electron.
2. Choose all the transition metals that are important for their lack of reactivity and resistance to corrosion.
   (a) Ag (b) Au (c) Na (d) Pt (e) Fe
3. Which transition metal has the configuration $1s^22s^22p^63s^23p^64s^23d^{10}$?
   (a) copper (b) lead (c) zinc (d) iron (e) cadmium

4. For the elements with atomic numbers 19 to 30, which has the **smallest** first ionization energy?
(a) potassium (b) zinc (c) scandium (d) vanadium.

5. Which are the three common oxidation states for chromium?
(a) +2 (b) +3 (c) +4 (d) +5 (e) +6

6. When the following redox reaction is balanced, using the smallest possible integers as stoichiometric coefficients, what is the coefficient of Cr_2O_3?

$$Cr_2O_3 + Na_2CO_3 + KNO_3 \rightarrow Na_2CrO_4 + CO_2 + KNO_2$$

(a) 1 (b) 2 (c) 3 (d) (4) (e) greater than 4

7. Titanium dioxide is used as a(n)
(a) reducing agent (b) precipitation agent (c) indicator for acid base titrations
(d) paint pigment (e) major constituent in the alloy Monel

8. A precipitate of zinc hydroxide is
(a) red (b) white (c) blue (d) black (e) green

9. Which of the following are correct for the first transition metal series?
(a) the 3d level is above the 4s level for the neutral atoms
(b) the 3d level is below the 4s level for the neutral atoms
(c) the 3d level is above the 4s level for the metal ions
(d) the 3d level is below the 4s level for the metal ions
(e) the 3d level is at the same energy as the 4s level, both for the neutral atoms and for the metal ions.

10. The complex ion $[Co(NH_3)_4Cl_2]^+$ has a coordination number of
(a) 1 (b) 2 (c) 4 (d) 6 (e) 7

11. In the equilibrium described by $Fe^{3+}(aq) + 6CN^-(aq) \rightleftharpoons Fe(CN)_6{}^{3-}(aq)$, when concentrations are expressed in moles per liter, the units associated with K_f are
(a) mol L^{-6} (b) mol^6 L^{-6} (c) mol^6 L (d) mol^{-6} L^6 (e) none of these

12. Which of the following are correct?
a) The pattern of splitting of 3d levels in a transition metal complex depends on its geometry.
b) Strong field ligands yield high spin complexes whereas weak field ligands yield low spin complexes.
c) Octahedral d^{10} complexes are colored due to transitions between the two types of d orbitals.
d) Paramagnetic compounds are attracted by a magnetic field.

13. The formula for the complex ion hexaaquocobalt(III) is
(a) $[Co_6NH_3]^{3+}$ (b) $[Co_6H_2O]^{3+}$ (c) $[Co(H_2O)_6]^{3+}$ (d) $[Co(NH_3)_6]^{3+}$

14. The copper ammonia complex $[Cu(NH_3)_6]^{2+}$ is
(a) pink (b) green (c) colorless (d) yellow (e) blue

15. The compound $Pb(C_2H_5)_4$ is used in
(a) dentistry (b) gasoline (c) photography
(d) fluorescent lighting (e) coinage

16. What is the molar solubility of silver bromide in 0.1 M NH_3 solution?

$$K_{sp} = 5.0 \times 10^{-13} \text{ mol}^2 \text{ L}^{-2} \text{ for AgBr(s)}$$
$$K_f = 1.0 \times 10^8 \text{ mol}^{-2} \text{ L}^2 \text{ for Ag(NH}_3)_2{}^+\text{(aq)}$$

(a) 7.1×10^{-4} (b) 5.0×10^{-7} (c) 2.2×10^{-3} (d) 5.0×10^{-6}
(e) 7.1×10^{-7}

17. Chromate aqueous solutions, upon addition of acid, produce
(a) CrO_3 (b) $Cr_2O_7{}^{2-}$ (c) O_2 (d) $H_2Cr_2O_4$

18. In acid solution, the manganate ion MnO_4^{2-} disproportionates to
(a) MnO (b) MnO_2 (c) Mn_2O_3 (d) MnO_4^- (e) Mn_2O_7

19. Silver tarnishes in the atmosphere because of the presence of
(a) HCN (b) H_2S (c) H_2O (d) NH_3 (e) O_3

20. The common ions of mercury are
(a) Hg^+ (b) Hg^{2+} (c) Hg^{3+} (d) Hg_2^+ (e) Hg_2^{2+}

<u>Answers to Self Test</u>

1. b,e; 2. a,b,d; 3. c; 4. a; 5. a,b,e; 6. a; 7. d; 8. b; 9. a,d; 10. d; 11. d; 12. a,d; 13. c; 14. e; 15. b; 16. a; 17. b; 18. b,d; 19. b; 20. b,e.

CHAPTER 22

BORON AND SILICON: TWO SEMIMETALS

SUMMARY REVIEW

Silicon, Si, in group 4, and **boron,** B, in group 3 are both semimetals (metalloids). After oxygen, silicon is the most abundant element in the earth's crust; silica, SiO_2, and silicates make up 95% of rock, clay, soil and sand. Boron is rather rare.

Silicon is obtained by reducing SiO_2 at high temperature with coke. Purification is by conversion to $SiCl_4$, which is then reduced with H_2 or Mg. Very high purity Si results from **zone refining.** Silicon is rather unreactive. At high temperature it burns in O_2 to give SiO_2, an acidic oxide which is only slightly soluble in water but which dissolves in bases, such as NaOH and Na_2CO_3, to give silicic acid, $Si(OH)_4$, and its condensed forms. Silicon has the diamond structure; no allotrope analogous to graphite exists. Silica and the silicates all have structures based upon SiO_4 tetrahedra, which can share all four O atoms with adjacent SiO_4 units, as occurs in the various forms of silica (quartz, cristobalite, and tridymite); if less than four O atoms are shared, the structures have 1, 2 or 3 terminal $Si\text{-}O^-$ groups. The structure of cristobalite is that of diamond in which the Si atom is linked to four O atoms, in a 3D infinite covalent network structure. In contrast to the 3D structure of SiO_2, CO_2 is a molecular gas at room temperature containing O=C=O molecules; carbon readily forms multiple bonds while silicon commonly forms only single bonds. Simple silicates, such as **olivines,** have compositions ranging from Mg_2SiO_4 to Fe_2SiO_4 and contain SiO_4^{4-} ions. Condensation of $Si(OH)_4$ molecules gives polysilicic acids, such as $H_6Si_2O_7$, $H_8Si_3O_{10}$, and longer chains or rings. In the cyclic anion $Si_6O_{18}^{12-}$, found in **beryl,** $Be_3Al_2Si_6O_{18}$, six SiO_4 tetrahedra share **two corners** with neighboring SiO_4 units; rings and chains of this kind have the empirical formula $(SiO_3^{2-})_n$. **Pyroxenes,** such as **diopside**, $CaMg(SiO_3)_2$, contain such long chain anions. Linkage of two $(SiO_3^{2-})_n$ chains by sharing O atoms on alternate Si atoms gives an amphibolite with a double silicate chain of empirical formula $Si_4O_{11}^{6-}$, of which the asbestos **tremolite,** $Ca_2Mg_5(Si_4O_{10})_2(OH)_2$, is typical. When SiO_4 tetrahedra share **three corners,** a layered sheet anion of empirical formula $Si_2O_5^{2-}$ results. **Talc,** $Mg_3(Si_2O_5)_2(OH)_2$, and the clay **kaolinite,** $Al_2(Si_2O_5)_2(OH)_2$, are examples of this kind of structure. They consist of $(Si_2O_5^{2-})_n$ anions with sufficient cations and OH^- ions between the layers to make the structure neutral. Such layers readily slide over each other, giving a soapy feel and minerals such as kaolinite can absorb water molecules between the anion layers. SiO_4 tetrahedra share all **four** corners in SiO_2.

In **aluminosilicates,** some of the Si atoms of the SiO_4 tetrahedra are replaced by Al^-. Replacement of one-quarter of the Si atoms of $(SiO_2)_n$ gives the 3D framework anion $(AlSi_3O_8^-)_n$, found in minerals such as **albite,** $NaAlSi_3O_8$. Replacement of one-half of the Si atoms of $(SiO_2)_n$ gives the anion $(AlSiO_4^-)_n$ found in **feldspars.** Replacement of one-quarter of the Si atoms in the amphibole sheet anion $(Si_2O_5^{2-})_n$ gives the anion $(AlSi_3O_{10}^{5-})_n$ found in **micas**, which consist of infinite layer aluminosilicate anions together with Al^{3+}, OH^- and K^+ ions. Micas have negatively-charged sheets held together by cations. The sheets do not slide over each other readily as do those in minerals such as talc; mica cleaves into thin transparent sheets.

Glasses result from cooling molten silica, silicates and aluminosilicates, to form amorphous solids in which the SiO_4 tetrahedra are randomly arranged. Window and bottle glass is a mixture of sodium and calcium silicates (made by heating together Na_2CO_3, CaO or $CaCO_3$, and sand (SiO_2)). Addition of other substances confers special properties or colors: B_2O_3 - a low coefficient of expansion (Pyrex); PbO - optical glass; CoO - blue glass; Cr_2O_3 - green glass; SnO_2 - white opaque glass, and AgCl (or AgBr) - photochromic glass. **Cement** is made by heating powdered limestone, sand and clay, to 1500°C, and adding a little gypsum. Addition of water gives a plastic mass which slowly hardens as interlocking crystals of hydrated aluminosilicates form. **Concrete** is cement mixed with sand, gravel, or crushed rock.

Silicones of general formula $(R_2SiO)_n$, where R is an alkyl or aryl group, are chain polymers formed via the hydroxides $R_2Si(OH)_2$ by reacting organosilicon dichlorides, R_2SiCl_2, with water. Inclusion of R_3SiCl in the reaction mixture provides chain ending groups, while addition of $RSiCl_3$ provides cross-links between silicone chains.

The **silanes**, Si_nH_{2n+2}, are analogs to the alkanes, but form only short chains and are highly reactive, burning spontaneously in air and reacting with OH^- to give H_2. The **halides**, such as $SiCl_4$ and SiF_4 are covalent tetrahedral AX_4 molecules that behave as Lewis acids. With water, $Si(OH)_4$ and HX are formed. With F^- ions, SiF_4 forms trigonal bipyramidal SiF_5^- and octahedral SiF_6^{2-} ions. $(H_3O^+)_2SiF_6^{2-}$ is a strong aqueous acid.

Boron is a black shiny solid with a low electrical conductivity. Boron burns to give weakly acidic B_2O_3. **Boric acid** consists of infinite sheets of $B(OH)_3$ molecules joined by hydrogen bonds. Its acidity is related to its behavior as a Lewis acid rather than as a Bronsted acid. It forms $H_2\overset{+}{O}\text{-}\overset{-}{B}(OH)_3$ which weakly donates a proton to water to give the tetrahedral $B(OH)_4^-$ ion. On heating, boric acid molecules condense to give polymers, such as **metaboric acid**, $(HO\text{-}BO)_3$, a cyclic ring. Complete dehydration gives B_2O_3 with an infinite framework structure based on planar BO_3 groups. A few borates contain the BO_3^{3-} ion; the majority have condensed anions derived from $B(OH)_4^-$ and $B(OH)_3$ and contain both trigonal planar BO_3 groups and tetrahedral BO_4^- groups. **Borax,** $Na_2B_4O_7 \cdot 10H_2O$, contains the $B_4O_5(OH)_4^{2-}$ anion, the conjugate base of a weak acid. The boron halides, BX_3, are covalent molecules and Lewis acids. They react with water to give $B(OH)_3$ and HX. BF_3 forms adducts with Lewis bases such as NH_3, amines, ethers and alcohols. **Boranes** have unexpected formulas, such as B_2H_6, B_4H_{10}, B_5H_9 and $B_{10}H_{14}$. **Diborane,** B_2H_6, has four covalent electron pair B-H bonds and two **three-centered** B-H-B bonds, in which two boron atoms and a hydrogen atom share a single electron pair. Electron deficient B_2H_6 readily adds two hydride ions to give two tetrahedral BH_4^- anions. $LiBH_4$ and $NaBH_4$ are good reducing agents.

Semiconductors such as silicon are much less conducting than metals and their conductivities increase with increasing temperature. One model views them as essentially covalent substances with a few weakly held valence electrons that are free to move through the structure, particularly when the temperature is raised. In a **metal** the valence electrons occupy the **valence band,** a closely spaced set of energy levels. Conduction occurs by promotion of electrons from filled to unfilled energy levels in a **conduction band,** similar in energy to the valence band. In a **nonconductor,** the enregy gap between the valence and conduction bands is too large for conduction to occur. In a **semiconductor** the energy gap between the valence and conduction band allows the few electrons with sufficient thermal energy to cross the gap. However, semiconductor conductivity may be greatly enhanced by introducing small amounts of certain impurities. In the case of silicon, P or As impurity atoms use only four of their five valence electrons to bond to adjacent Si atoms and the extra electrons enters the conduction band. This gives an **n-type** (n = negatively- charged electron) **conductor.** In contrast, impurity atoms such as boron atoms, with only three valence electrons, have insufficient electrons to bond to four adjacent Si atoms, but do so by stealing the necessary electrons from adjacent Si atoms, thereby creating **positive holes** in the structure. Conduction then occurs in these **p-type conductors** (p = positive hole), by electron transfer into the positive holes, thereby transferring the positive hole through the crystal. The combination of n-type and p-type conductors is the basis for modern electronic devices and the silicon chip.

REVIEW QUESTIONS

1. Write the electron configurations of boron and silicon in their ground states. What are their normal valence state electron configurations?
2. Why are boron and silicon clasified as semimetals? Classify each of the elements in group 4 of the periodic table as nonmetal, semimetal or metal.
3. By what reaction is silicon prepared industrially? How is ultra-pure silicon obtained?
4. What is the structure of silicon? Why is there no allotrope of silicon analogous to graphite?
5. What is the basic molecular unit found in the structures of silica and the silicates?
6. How is the structure of the form of silica known as crystobalite related to that of diamond?
7. Discuss why SiO_2 is a hard crystalline solid, while CO_2 is a monomolecular gas.
8. How is the structure of silica glass related to that of quartz?
9. Write equations for the reaction between SiO_2 and NaOH, and SiO_2 and Na_2CO_3.
10. Classify SiO_2 as an acidic, amphoteric or basic oxide.
11. What is the reaction of $SiCl_4$ with water? How does this reaction compare to that of CCl_4 with water?

12. What acid results from the elimination of (a) one water molecule from two molecules of silicic acid, (b) two water molecules from three molecules of silicic acid? What salts would result from complete reaction of each of these acids with NaOH?
13. Given that the mineral beryl contains Al^{3+} ions and silicate ions containing six SiO_4 groups condensed corner to corner to form a ring, what is its empirical formula?
14. An amphibole is formed by joining infinite $(SiO_3{}^{2-})_n$ chains to form a double chain in which O atoms on alternate Si atoms are shared. What is the empirical formula for an amphibole?
15. What feature of the structures of silicates such as talc and kaolinite accounts for empirical formulas such as $Mg_3(Si_2O_5)_2(OH)_2$ and $Al_2Si_2O_5(OH)_4$? Why is talc a soft material with a soapy feel?
16. What accounts for the difference between wet plastic clay and clay that has been fired to form pottery?
17. How are the aluminosilicates related to the silicates?
18. How is silica related to the mineral albite, $NaAlSi_3O_8$?
19. How is common glass made?
20. What oxides are added to common glass to give Pyrex, blue glass, green glass, and opaque white glass?
21. In what way does a glass differ from a true crystalline solid?
22. What are cement and concrete?
23. What silicone polymer results from the reaction of $(CH_3)_2SiCl_2$ with water?
24. How does the addition of small amounts of (a) $(CH_3)_3SiCl$, and (b) CH_3SiCl_3, affect the nature of the polymer formed in the above reaction?
25. How do the structures of ketones differ from those of the corresponding silicon compounds?
26. What is the general formula of a silane, the silicon analogue of an alkane?
27. What reaction occurs when SiH_4 is reacted with a basic aqueous solution?
28. Classify SiH_4, $SiCl_4$ and SiF_4 as ionic or covalent substances.
29. Classify $SiCl_4$ and SiF_4 as Lewis acids or Lewis bases.
30. What are the geometric shapes of SiF_4, $SiF_5{}^-$, $SiF_6{}^{2-}$ and $B(OH)_4{}^-$?
31. What factors account for the great differences between the reactivities of compounds containing C-C and Si-Si bonds?
32. Under what conditions is it possible to make compounds containing Si=Si bonds?
33. What is the molecular structure of boric acid, $(B(OH)_3)_n$?
34. Is boric acid best classified in aqueous solution as a Lewis acid or a Bronsted acid?
35. Is B_2O_3 a basic or an acidic oxide?
36. Draw the structure of the tetraborate ion, $B_4O_5(OH)_4{}^{2-}$.
37. How is boric acid obtained from borax, $Na_2B_4O_7 \cdot 10H_2O$?
38. Write a balanced equation for the complete reaction of BCl_3 with water.
39. Classify BCl_3 and BF_3 as Lewis acids or bases.
40. What is a borane?
41. Draw a diagram showing the positions of all of the atoms and the bonding electron pairs in diborane, B_2H_6. What is a multi-center bond?
42. How could $LiBH_4$ be obtained from diborane?
43. How many electron pairs are available for bonding in the $B_{12}H_{12}{}^{2-}$ anion?

44. What is the valence band and conducting band concept? How does it account for the differences in the conductivities of metals, semiconductors and insulators?
45. Why does addition of small amounts of phosphorus or arsenic to silicon increase its conductivity?
46. What is n-conduction? How do n-conduction and p-conduction come about?

Answers to Selected Review Questions

2. carbon, nonmetal; silicon, semimetal; germanium, semimetal; tin, metal; lead, metal. 5. SiO_4.

```
                              OH
                              |
12. (HO)3Si-O-Si(OH)3; (HO)3Si-O-Si-O-Si(OH)3; Na6Si2O7; Na8Si3O10.
                              |
                              OH
```

12. $(HO)_3Si\text{-}O\text{-}Si(OH)_3$; $(HO)_3Si\text{-}O\text{-}Si(OH)_2\text{-}O\text{-}Si(OH)_3$; $Na_6Si_2O_7$; $Na_8Si_3O_{10}$.

13. $Be_3Al_2(SiO_3)_6$. 14. $Si_4O_{11}^{6-}$.
18. By replacement of 1/4 the Si atoms in silica, $(SiO_2)_n$, by Al^-. 43. 25.

OBJECTIVES

Be able to:

1. Define the term semimetal and indicate where the semimetals are located in the periodic table.
2. State how silicon can be prepared from silica and how ultrapure silicon is obtained.
3. List the physical and chemical properties of silicon.
4. Describe the structures of the crystalline and glassy forms of silicon dioxide.
5. List several reactions by which silicic acid can be prepared, and recognize its polymeric nature.
6. Show how silicates can be constructed from SiO_4 tetrahedra, yielding chains, rings and sheets.
7. Give examples of typical aluminosilicates and list the oxides of the elements that are used in the production of glass.
8. Show how silicones can be prepared, give typical structures and list their properties.
9. List typical reactions of silanes and the silicon halides.
10. Contrast the chemistry of silicon and carbon.
11. Show how boric acid ionizes in water.
12. Give the products of the reaction of boron halides with ammonia, amines, ethers, alcohols and water.
13. Describe the bonding in diborane in terms of three-center bonds.
14. Describe the properties of semiconductors in terms of the band model, and explain how the addition of impurities alters their conductivity.

22.3 Silicic Acid and Silicates

Glass

Ordinary window glass is prepared by cooling a molten complex silicate mixture. The starting materials are SiO_2, Na_2O and CaO, which are all rather inexpensive substances, plus a variety of other oxides, which impart a characteristic property or color to the glass. Although glass is silicate in nature, its composition is often given by quoting the percentages of the oxides used to prepare it. We can convert these compositions to ones based on mass percentage of the elements in the glass. First choose an arbitrary amount of the glass, say 100 grams. Next, calculate how many grams of each element are present in each oxide. This is just a series of mass percentage composition determinations that we saw back in Chapter 2. Third, sum the masses of those elements, like oxygen, that are present in more than one oxide. Finally, **if the calculation is based on a 100 gram sample**, the **total mass in grams of each element will be numerically equal to the percentage of that element in the glass**.

Example 22.1

A batch of optical glass has the following percentage oxide composition: SiO_2, 69.0; CaO, 13.0; K_2O, 12.0; Na_2O, 6.0. Give the percentages of each of the alkali and alkaline earth elements present in the glass.

Solution 22.1

Let's take a 100.0 g sample of the glass, and first consider CaO. By definition,

$$13.0\%\ \text{CaO} = \frac{13.0\ \text{g CaO}}{100.0\ \text{g}} \times 100\%$$

For the 13.0 g CaO, since the molar masses for Ca and O are 40.08 and 16.00 g mol^{-1} respectively, the fraction of CaO that is calcium is

$$\frac{40.08\ \text{g mol}^{-1}\ \text{Ca}}{56.08\ \text{g mol}^{-1}\ \text{CaO}}$$

This **fraction** is the same, whether we consider 1 mole, 1 molecule, or 13.0 g of CaO. The **mass** of Ca, of course, depends on the **mass** of CaO. For the 13.0 g of CaO,

$$\text{mass Ca} = \frac{40.08\ \text{g mol}^{-1}\ \text{Ca}}{56.08\ \text{g mol}^{-1}\ \text{CaO}} \times 13.0\ \text{g CaO} \qquad \text{mass Ca} = 9.29\ \text{g}$$

Similarly, we find: mass K = 9.96 g and mass Na = 4.45 g.

The masses of Si and O present are not needed for the problem. They could have been obtained in the same manner. (Si = 32.25 g, O = 44.05 g).

Since the calcualation is based on 100.0 grams of sample, the mass percentages for Ca, K and Na are readily obtained.

$$\text{percentage Ca} = \frac{9.29 \text{ g Ca}}{100.0 \text{ g}} \times 100\% = 9.29\%$$

$$\text{percentage K} = \frac{9.96 \text{ g K}}{100.0 \text{ g}} \times 100\% = 9.96\%$$

$$\text{percentage Na} = \frac{4.45 \text{ g Na}}{100.0 \text{ g}} \times 100\% = 4.45\%$$

Thus the element from groups 1 and 2 that is present in the greatest abundance is potassium, even though its oxide percentage was less than that for CaO.

SELF TEST

1. Which of the following are semimetals?
   (a) P (b) As (c) Si (d) B (e) Pb
2. Silicon is prepared commercially by heating sand to 3000°C in an electric arc furnace with
   (a) Mg (b) C (c) H_2 (d) NaCl (e) B
3. Silicon reacts with:
   (a) acids at room temperature
   (b) hot concentrated aqueous hydroxide solutions
   (c) oxygen at high temperatures
   (d) sand at room temperature
   (e) molten hydroxide
4. Three different crystalline forms of SiO_2 are found in the minerals
   (a) tridymite (b) olivine (c) diopside (d) cristobalite (e) quartz
5. A concentrated solution of silicic acid can be prepared from water and
   (a) $SiCl_4$ (b) sand (c) quartz (d) Si (e) none of the above
6. SiO_4 tetrahedra can be joined together in many ways, to form silicates. The simplest of these has the empirical formula
   (a) SiO_4^{2-} (b) SiO_4^{4-} (c) $Si_2O_7^{6-}$ (d) $Si_2O_7^{4-}$ (e) $Si_3O_{10}^{8-}$
7. A special type of glass used in surgical implants has the folowing oxide percentage composition: SiO_2, 45; Na_2O, 25; CaO, 24; P_2O_5, 6. Which element is present in the second greatest amount in this glass?
   (a) silicon (b) oxygen (c) sodium (d) calcium (e) phosphorus
8. Which of the following are silicones?
   a) $(HO)_3Si\text{-}O\text{-}Si(OH)_3$ b) $(HO)_2SiCl_2$ c) $[(CH_3)_2SiO]_4$
   d) $SiCl_4$ e) $(CH_3)_3Si\text{-}O\text{-}Si(CH_3)_3$
9. $SiCl_4$ is an example of which type of molecule?
   (a) AX_4 (b) AX_4E (c) AX_4E_2 (d) AX_4E_3 (e) AX_4E_4
10. Which of the following are true?
    (a) the Si-Si bond is stronger than the C-C bond.
    (b) the Si-O bond is stronger than the C-O bond.
    (c) the C=C bond is present in many stable molecules
    (d) the Si=Si bond is present in many stable molecules
    (e) SiF_4 and CF_4 both form complex ions with F^-.

11. $B(OH)_3$ in water is
(a) acidic (b) basic (c) neutral (d) pink (e) polymerized

12. When BF_3 reacts with NH_3
(a) BF_3 is acting as a Lewis base, and NH_3 as a Lewis acid
(b) BF_3 is acting as a Lewis base, and NH_3 as a Bronsted acid
(c) BF_3 is acting as a Lewis acid, and NH_3 as a Lewis base
(d) BF_3 is acting as a Lewis acid, and NH_3 as a Bronsted base
(e) BF_3 is acting as an Arrhenius base, and NH_3 as an Arrhenius acid.

13. Which of the following statements are true?
(a) The conductivity of a semiconductor decreases with decreasing temperature.
(b) In an insulator the valence band is higher in energy than the conduction band.
(c) When silicon has a small amount of boron added, a p-type conductor is created.
(d) In metallic magnesium, the conduction band is widely separated from the valence band.
(e) A p-n junction often serves as a rectifier.

14. The empirical formula of the infinite chain

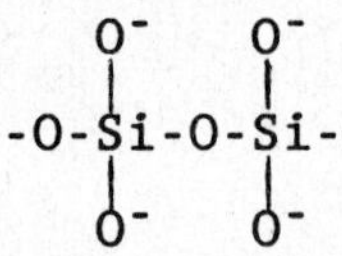

```
      O-    O-
      |     |
  -O-Si-O-Si-O-
      |     |
      O-    O-
```

is

a) SiO_2^{2-} b) SiO_2^{3-} c) SiO_3^{2-} d) SiO_3^{3-} e) SiO_4^{2-}

<u>Answers to Self Test</u>

1. b,c,d; 2. b; 3. b,c,e; 4. a,d,e; 5. a; 6. c; 7. a; 8. c,e; 9. a; 10. b,c; 11. a; 12. c; 13. a,c,e; 14. c.

CHAPTER 23

ORGANIC CHEMISTRY

SUMMARY REVIEW

Organic compounds are hydrocarbons, or can be regarded as derived from them by replacing H atoms with **functional groups.** They occur naturally in large numbers and are essential to life. Many are synthesized on the large scale from simple substances derived from petroleum and natural gas. In the **petrochemical industry,** two basic processes are **thermal cracking** and the manufacture of **synthesis gas.** Thermal cracking takes mixtures of alkanes, heats them at high temperature for short times and breaks them down to shorter-chain alkanes, alkenes and hydrogen. Gasoline and large amounts of ethene and propene are produced in this way. **Synthesis gas** is a mixture of CO and H_2 obtained by heating methane and other hydrocarbons with steam and a nickel catalyst at 700-800°C. It is used as a starting point for many syntheses.

Alkanes have the general formula C_nH_{2n+2} and contain only single bonds. The alkanes with up to four C atoms are gases, those from C_5 to C_{15} are liquids, and the higher alkanes are solids; they are referred to as **saturated** hydrocarbons. **Unsaturated** hydrocarbons contain one or more double and/or triple bonds. The **alkenes** with one C=C double bond have the general formula C_nH_{2n}, and the **alkynes** with one C≡C bond have the general formula C_nH_{2n-2}.

All hydrocarbons burn in air when ignited, to give CO_2 and water. In other respects the alk**anes** are very unreactive; in contrast, the alk**enes** and alk**ynes** readily **add** reagents such as H_2, hydrogen halides, water and halogens to their multiple bonds.

Alkanes contain C-C and C-H bonds and each C atom is at the center of a tetrahedral arrangement of four single bonds. Molecules with the same molecular formula but different structures (different arrangements of the same atoms) are **structural isomers.** For any alkane, the structures of all of its possible isomers can be deduced by starting with the normal (n-) compound, in which all of the C atoms form one continuous chain, and then considering all other possible structures containing shorter continuous chains in which the remaining C atoms form **branched** chains.

The different **conformations** of a molecule are not isomers but are interconvertible by rotating (groups of) atoms around single bonds. For ethane, C_2H_6, two extreme conformations are the **eclipsed** form and the **staggered** conformation; there is also an infinite number of intermediate skew conformations. All the conformations have very similar energies. Free rotation does not occur about C=C or C≡C bonds.

Natural gas is a mixture of methane, ethane, propane and the two butanes. **Petroleum** is a complex mixture of C_5 to C_{40} alkanes, the exact composition of which depends on the source. It is separated into fractions by **fractional distillation.** In the production of **gasoline** (C_6 to C_{10} mixtures), higher alkanes are "cracked" at high temperature.

In **ethene,** $H_2C{=}CH_2$, each C atom is surrounded by four pairs of electrons; two pairs form two CH bonds and the remaining two pairs are shared between the carbon atoms to form the C=C bond. A simple model represents the double bond as two "bent" bonds and describes ethene as having two tetrahedral arrangements of electron pairs sharing an edge. Thus, the two C atoms and all four H atoms lie in the same plane. Otherwise, using VSEPR theory, each carbon atom is described as having AX_3 planar geometry. Reagents such as H_2, hydrogen halides, and halogens readily add to the double bond, one atom to each carbon, to give substituted alkanes.

Because of the restricted rotation about the C=C bond, **butene,** C_4H_{10}, exists as **cis**- and **trans**-positional isomers:

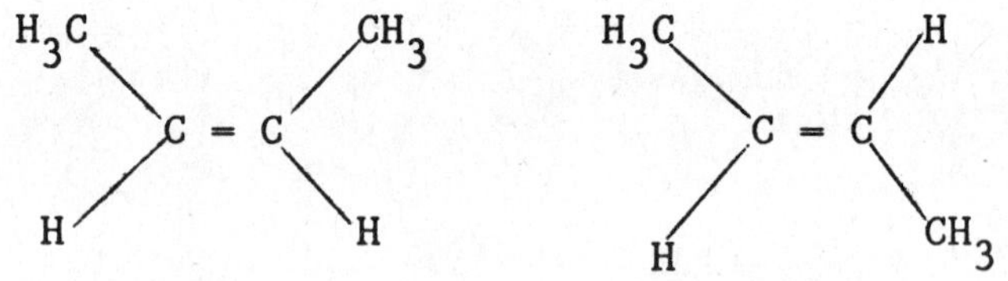

H_3C and CH_3 above, H and H below: $C = C$

cis

H_3C and H above, H and CH_3 below: $C = C$

trans

In **ethyne,** HC≡CH, each C atom is surrounded by a tetrahedral arrangement of four electron pairs; one pair forms the C-H bond and the remaining three pairs are shared with the other C atom to form a **triple** bond; two tetrahedral arrangements of electron pairs share a common face and the H-C-C-H arrangement is thus linear. The triple bond is three "bent" bonds; each C atom has linear AX_2 geometry. Ethyne is manufactured by cracking ethane or from the reaction of water with calcium carbide, $Ca^{2+}C_2^{2-}$. Alkynes readily add H_2, HX or X_2 molecules to give in the first stage substituted alkenes, and in the second substituted alkanes.

Cycloalkanes, $(CH_2)_n$, contain closed rings of carbon atoms. **Cyclopropane** $(CH_2)_3$, and cyclobutane, $(CH_2)_4$, are unusual, in that their CCC bond angles are much smaller than the normal tetrahedral angle of 109.5°. The C-C bonds in these two molecules are "strained" or "bent". Like ethene, $CH_2{=}CH_2$, they react with reagents such as hydrogen or bromine, which open up the ring to give an alkane derivative. Higher cycloalkanes, in which the CCC angles are close to tetrahedral, behave as typical alkanes and are unreactive.

Organic compounds are named using the following IUPAC rules:

1. The longest continuous carbon chain is taken as the parent hydrocarbon.
2. Substituted hydrocarbon groups attached to the parent chain are named by replacing the ending **-ane** by **-yl**; eg., methyl, ethyl, etc.
3. The parent compound is numbered starting from the end of the longest continuous chain, so that the substituent groups are attached at the carbon atoms with the lowest possible numbers.

4. When the molecule contains a double bond, the **-ane** ending is changed to **-ene** and the longest carbon chain numbered so that the lowest possible number is assigned to a carbon atom of the double bond.
5. When the molecule contains a triple bond, the **-ane** ending is changed to **-yne** and the rules for alkenes are followed.
6. **Cycloalkanes** (or alkenes or alkynes) are named by adding the prefix **cyclo-** to the name of the parent hydrocarbon and the C atoms of the ring are numbered to give the lowest possible numbers to the substituted carbon atoms.

Benzene is the simplest **arene** (aromatic hydrocarbon). Industrially it comes from the distillation of coal or the **reformation** of petroleum. The Lewis (Kekulé) structure of benzene, C_6H_6, has a six-membered ring of C atoms with alternate C-C and C=C bonds, which is inconsistent with the observed structure in which all CC bonds are of identical length, intermediate between the lengths of typical single and double CC bonds. The bonds in benzene are described as having a bond order of 1 1/2; this is consistent with treating the two alternative Kekulé structures as resonance structures. Alternatively the ring may be described as having six CC single bonds with an additional three pairs of electrons delocalized around the ring.

Substitution of hydrogen atoms in benzene gives rise to the existence of isomers. The simplest arene containing two fused rings is **naphthalene,** $C_{10}H_8$. A very large number of arenes with more than two fused rings is known. **Graphite** is the final member of the series.

Substituted alkanes are named using the normal alkan- prefix with a suffix that identifies the nature of the substituent (functional group). **Alcohols,** R-OH, employ the suffix -ol, hence: methanol, CH_3OH, etc.

Methanol is synthesized from synthesis gas ($CO + 2H_2 \rightarrow CH_3OH$). A common source of ethanol is the fermentation of sugar, and a general method of synthesis for alcohols is to add water to the double bond of an alkene.

Association of their molecules through hydrogen bonds gives alcohols higher melting points and boiling points than hydrocarbons of similar molar mass. Short chain alcohols are soluble in water in all proportions (miscible), but as the hydrocarbon chain gets longer, solubility decreases; long-chain alcohols are almost insoluble.

Diols and triols have -OH substituents on two or three C atoms; examples include 1,2-ethanediol (ethylene glycol or glycol) and 1,2,3-propanetriol (glycerol or glycerine).

Alcohols behave as weak acids and weak bases. With a very strong base, such as NH_2^- or H^-, the O-H proton is removed to give an **alkoxide** ion, $R\text{-}O^-$. Reactive metals such as sodium also react with alcohols to give **alkoxides** ($2R\text{-}OH + 2Na \rightarrow 2R\text{-}O^-Na^+ + H_2$). Alcohols are weaker bases than water but are acids of approximately the same strength.

Substitution of a ring hydrogen in an aromatic hydrocarbon, such as benzene, by OH gives a **phenol.** Benzenol, C_6H_5OH, is commonly called "phenol" or carbolic acid. Phenols are stronger acids and weaker bases than alcohols; they were among the first successful antiseptics.

Ethers (alkoxy-alkanes), R-O-R′, result from the condensation of two alcohol molecules, using a dehydrating agent, such as H_2SO_4:

$$R\text{-}O\text{-}H + H\text{-}O\text{-}R' \rightarrow R\text{-}O\text{-}R' + H_2O$$

Ethers (R-O-R′), **alcohols** (R-O-H), and **water** (H-O-H), are all bent AX_2E_2 molecules. Unlike alcohols and water, ethers cannot form hydrogen bonds - their boiling points are similar to those of alkanes of similar molar mass and they are immiscible with water but are good solvents for other organic compounds. A general method for preparing ethers is the reaction of an alkali metal alkoxide with an alkyl halide ($R\text{-}O^-Na^+ + R'\text{-}Cl \rightarrow R\text{-}O\text{-}R' + NaCl$). Like water and alcohols, ethers are weak bases.

Dehydrogenation of an alcohol gives an **aldehyde** or **ketone:**

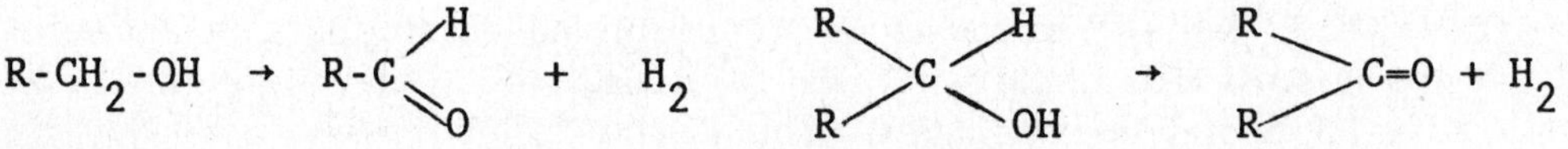

$$R\text{-}CH_2\text{-}OH \rightarrow R\text{-}C(H){=}O + H_2 \qquad R_2C(H)OH \rightarrow R_2C{=}O + H_2$$

Primary alcohol → aldehyde; Secondary alcohol → ketone

They are also prepared by oxidation of alcohols and alkenes:

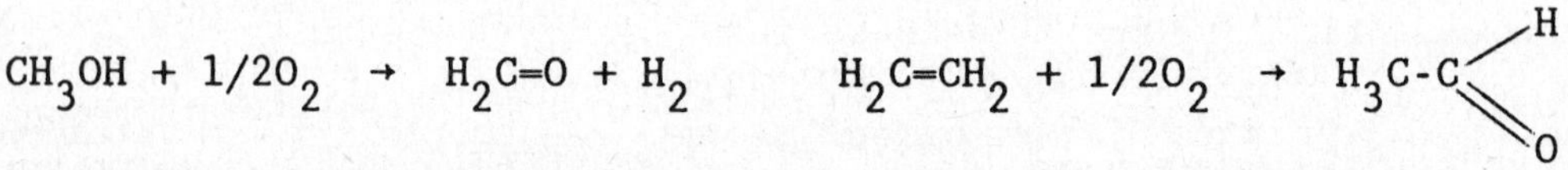

$$CH_3OH + 1/2O_2 \rightarrow H_2C{=}O + H_2 \qquad H_2C{=}CH_2 + 1/2O_2 \rightarrow H_3C\text{-}C(H){=}O$$

Primary alcohols give aldehydes, and secondary alcohols yield ketones. **Aldehydes** are named systematically by replacing the **-e** of the corresponding alkane by **-al,** e.g., $H_2C{=}O$, **methanal** (formaldehyde). **Ketones** are named systematically by replacing the **-e** of the corresponding alkane by **-one,** e.g., $(CH_3)_2C{=}O$, **propanone** (acetone).

Oxidation of an aldehyde gives a **carboxylic acid**, containing the $\text{-}C({=}O)OH$ group: e.g.,

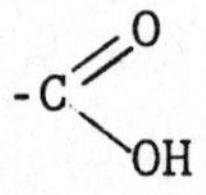

$$2\ H_3C\text{-}C({=}O)H + O_2 \rightarrow 2\ H_3C\text{-}C({=}O)OH$$

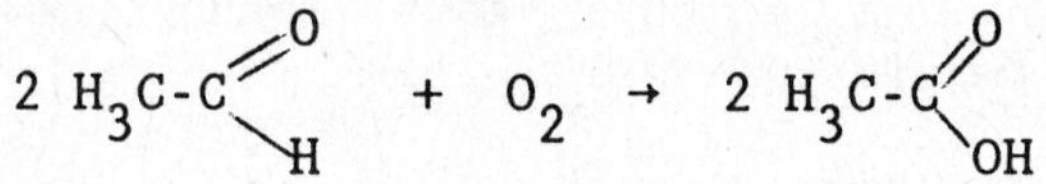

Carboxylic acids are named by replacing the **-e in the name of the parent alkane** by **-oic acid,** e.g., methanoic acid for HCOOH. Carboxylic acids are weak acids, but not as weak as the alcohols. The C=O group is polar, $\rangle C^{\delta+} = O^{\delta-}$.

Formic acid, HCO_2H, results from the reaction of CO with NaOH to give sodium formate, $Na^+HCO_2^-$, followed by acidification and distillation. The simplest aromatic carboxylic acid is **benzoic acid,** $C_6H_5CO_2H$.

Reaction of an alcohol R′OH and a carboxylic acid RCOOH gives an **ester,** RCOOR′, and water. Synthetic esters are important as artificial flavors and perfumes.

Amines are related to ammonia in the same way that alcohols and ethers are related to water. All the amines have the same AX_3E geometry. Amines result from the reaction of an alcohol with ammonia (or another amine) in the presence of Al_2O_3 catalyst.

$$R\text{-}O\text{-}H + NH_3 \rightarrow R\text{-}NH_2 + H_2O$$

$$R\text{-}OH + R'\text{-}NH_2 \rightarrow R\text{-}NH\text{-}R' + H_2O$$

Amines, like ammonia, are weak bases; they react with acids to give salts.

Amides $R\text{-}C(=O)\text{-}NH_2$ result from reaction of a carboxylic acid with ammonia, to give an ammonium salt which decomposes on heating:

$$RCO_2H + NH_3 \rightarrow RCO_2^- NH_4^+ \xrightarrow{\text{heat}} R\text{-}CO\text{-}NH_2 + H_2O$$

The simplest amide is **methanamide,** $H\text{-}CO\text{-}NH_2$ (formamide).

Alkanes are relatively unreactive compounds that react only when C-H bonds are broken; at high temperature, two important reactions are combustion and halogenation. **Chloroalkanes** are formed in chain reactions at high temperature between alkanes and chlorine. Chloroethane, C_2H_5Cl, results from chlorination of ethane or from the addition of HCl to ethene. Addition of chlorine to ethene gives 1,2-dichloroethene, which eliminates HCl on heating to give chloroethene (vinyl chloride), $H_2C{=}CHCl$. Chlorofluoroalkanes are called Freons and have important uses as refrigerator fluids and as aerosol propellants, but are environmental hazards.

The important functional groups and their relationships are shown in the following table:

Functional Group	Name	Functional Group	Name
-H		$-C(=O)H$	aldehyde
-R	alkyl	$-C(=O)R$	ketone
-OH	alcohol	$-C(=O)OH$	carboxylic acid
-OR	ether	$-C(=O)OR$	ester
$-NH_2$	amine	$-C(=O)NH_2$	amide

REVIEW QUESTIONS

1. Write the general formulas for (a) an alkane, (b) an alkene, and (c) an alkyne, containing n carbon atoms.
2. What is the general formula for a cycloalkane with n carbon atoms?
3. What is the difference between a **saturated** and an **unsaturated** hydrocarbon?
4. Name the hydrocarbons with the formulas: CH_4 C_2H_6 C_3H_8 C_4H_{10} C_5H_{12} C_6H_{14}
5. Draw Lewis structures for CH_4, C_2H_6 and C_3H_8.
6. Draw the two structural isomers with the formula C_4H_{10}.
7. Draw the five structural isomers of hexane.
8. What structures are possible for a propyl group, $-C_3H_7$?
9. What four structures are possible for a butyl group, $-C_4H_9$?
10. What are the two possible structures for a hydrocarbon with the formula C_3H_6?
11. Draw diagrams illustrating the relative orientations of the $-CH_3$ groups in the **eclipsed** and **staggered** conformations of ethane.
12. Why are the alkanes regarded as rather unreactive compounds?
13. Write the balanced equation for the combustion of heptane to give (a) carbon monoxide and water, (b) carbon dioxide and water.
14. Draw the Lewis structure for ethene. Why is it a planar molecule?
15. What compound results from the reaction of propene with Br_2?
16. What general term is used to describe reactions in which C=C bonds are changed to C-C bonds?
17. Draw structures for the cis- and trans- isomers of 3-hexene.
18. Draw a diagram illustrating how a triple bond results from the sharing of three common electron pairs between carbon atoms, each of which is at the center of a tetrahedral arrangement of four electron pairs.
19. Write equations illustrating the reaction of HCl with 1-butene and name the products.
20. What products result from the reaction of ethyne with Br_2?
21. What are the products, if any, of treating ethene, cyclopropane and cyclohexane, respectively, with bromine?
22. Draw the structural formula of 2,2-dimethylbutane.
23. What systematic name would be given to a Kekulé structure of benzene?
24. Draw the two Kekulé resonance structures for benzene. What is the carbon-carbon bond order is benzene? Is the bond order consistent with the observed bond length of 140 pm?
25. Draw structures for each of the three isomers of xylene, $C_6H_4(CH_3)_2$, and name each systematically.
26. Draw a Lewis structure for naphthalene.
27. How could the molecular mass of a hydrocarbon that boils below 100°C be determined experimentally?

28 Write general formulas for each of the following substituted alkanes, R-X.
(a) an alcohol (b) an aldehyde (c) an amine (d) a carboxylic acid.

29. Give the common and systematic names of the compounds in question 28 for R = H_3C-, R = H_3C-CH_2-, R = C_6H_5-, and R = $C_6H_5CH_2$-.
30. Write formulas for primary, secondary, and tertiary alcohols with a total of four carbon atoms. Name each of the alcohols.

31. Write the balanced equation for the production of synthesis gas from propane.
32. Starting with synthesis gas, how are (a) methanol, (b) methylamine, obtained?
33. Write an equation for the addition of water to propene and name the possible products.
34. What alcohols of formula C_4H_9OH are possible?
35. Why are the enthalpies of combustion of alcohols less exothermic than those of alkanes with the same number of carbon atoms?
36. Why do alcohols have very much higher melting points and boiling points that the alkanes or ethers of similar molar mass?
37. What are alcohols such as methanol and ethanol miscible with water, while 1-hexanol, for example, has a very low solubility?
38. What compound results from the reaction of ethanol with a very strong base, such as sodium hydride, NaH?
39. How is diethyl ether obtained from ethanol?
40. What is the systematic name for CH_3-O-C_2H_5?
41. What compounds result from the oxidation of 1-propanol and 2-propanol?
42. To what classes of compounds do ethanal and 2-propanone belong?
43. Draw the structure of 2-methyl-3-hexanone.
44. Draw the structure of 3-ethylpentanal.
45. What are the formulas, structures, and systematic names of (a) formaldehyde, (b) acetaldehyde, (c) acetone?
46. Starting with methanol and carbon monoxide, how is acetic acid made commercially?
47. How is ethanoic acid made from ethanol?
48. What are the structures of (a) the carbonyl group, (b) the carboxyl group?
49. What are the bond orders of the CO bonds in the formate ion?
50. What are the formulas, structures and systematic names of (a) formic acid, (b) acetic acid, (c) propionic acid?
51. Write the formulas and structures of (a) benzyl alcohol, (b) benzaldehyde, (c) benzoic acid.
52. Draw a general structure for the ester that results from the reaction of an alcohol R-OH with a carboxylic acid $R'CO_2H$.
53. What is the systematic name and the common name of the ester of formula C_2H_5-O-C-CH_3?

```
C2H5-O-C-CH3
       ||
       O
```

54. Draw the structures for ethylamine, diethylamine, triethylamine, and the tetraethylammonium ion.
55. What product results from reacting aniline with HCl?
56. What are the products of the reaction of methanol and ammonia at 400°C in the presence of Al_2O_3 catalyst?
57. Write the general formula for an amide.
58. What product results from heating ammonium benzoate?
59. What products result from reacting propene with hydrogen bromide?

Answers to Selected Review Questions

2. C_nH_{2n} 8. $H_3C-CH_2-CH_2-$ or $(CH_3)_2CH-$
9. $H_3C-CH_2-CH_2-CH_2-$, $H_3C-CH_2-\overset{|}{C}H-CH_3$, $(CH_3)_3C-$, $(CH_3)_2CH-CH_2-$
10. $CH_3-CH=CH_2$ or $(CH_2)_3$ 15. 1,2-dibromopropane
19. $H_2C=C(H)CH_2CH_3 + HCl \rightarrow H_3CCH(Cl)CH_2CH_3$ (2-chlorobutane) and $ClCH_2CH_2CH_2CH_3$ (1-chlorobutane)

23. 1,3,5-cyclohexatriene   28. (a) R-OH (b) R-C(=O)H (c) $R-NH_2$ (d) R-C(=O)OH

29. (a) methanol (methyl alcohol); ethanol (ethyl alcohol); phenol (carbolic acid); benzyl alcohol.
(b) ethanal (acetaldehyde); propanal (propaldehyde); benzaldehyde; 2-phenylethanal.
(c) methylamine; ethylamine; aminobenzene (aniline); benzylamine.
(d) ethanoic acid (acetic acid); propanoic acid (propionic acid); benzoic acid; benzylic acid.

33. $CH_3-CH=CH_2 + H_2O \rightarrow CH_3-CH(OH)-CH_3$ (2-propanol)
$\rightarrow CH_3-CH_2-CH_2-OH$ (1-propanol)

34. 1-butanol, 2-butanol, 2-methyl-1-propanol, 2-methyl-2-propanol.
40. Methoxyethane.
41. $CH_3CH_2CH_2OH \rightarrow CH_3CH_2CHO$ (propanal) $\rightarrow CH_3CH_2COOH$ (propanoic acid); $CH_3CH(OH)CH_3 \rightarrow (CH_3)_2C=O$ (2-propanone)
49. 1.5.

OBJECTIVES

Be able to:

1. Name and give formulas for the first ten alkanes, and state how boiling points and melting points vary in the series.
2. Draw structural isomers for hydrocarbons.
3. Draw eclipsed and staggered conformations for hydrocarbons.
4. Give names and formulas for cycloalkanes, and use ring strain to account for the reactivity of cycloalkanes compared to alkanes.
5. Write balanced equations for the reactions of alkanes with O_2 and Cl_2, and for typical cracking reactions.
6. List, and give uses of, the various fractions obtained in the distillation of petroleum.
7. Name and give formulas for simple alkenes and alkynes, and give examples of geometric (cis-trans) isomers of alkenes.
8. Classify reactions of hydrocarbons as additions, eliminations, or substitutions.
9. Name hydrocarbons using IUPAC rules.
10. Name simple derivatives of benzene and simple arenes containing fused rings.

11. Describe two modern methods used for identification and analysis of hydrocarbons and their mixtures.
12. Give the general structural formulas for the common functional groups of organic chemistry, and name the classes of compounds formed by joining an alkyl group to each of these functional groups.
13. Give typical examples of products formed from thermal cracking of natural gas or petroleum, and show how alkanes and steam can form synthesis gas.
14. Name, show the preparation, and where appropriate, list the physical properties and typical reactions of alcohols, ethers, aldehydes and ketones, carboxylic acids, esters, amines, amides, and halogen derivatives of alkanes and alkenes.
15. Describe the fermentation process.
16. Show how methane and ethylene can be converted to various classes of organic compounds, and how one class of compounds can be converted to others.

PROBLEM SOLVING STRATEGIES

23.1 Alkanes

Structural Isomers

Alkanes contain carbon-carbon single bonds; they are all of the general formula C_nH_{2n+2}. All the C-C-C, H-C-C and H-C-H bond angles are roughly 109°; however, on paper, the bond angles often are shown as 90° or 180°. For a given value of $n > 3$, there is more than one way to join the carbons; that is, there are **structural isomers,** having different physical and chemical properties. After you have learned a systematic way to name alkanes, you will be able to recognize whether two C_nH_{2n+2} molecules, with the same n, are the same, or are a pair of structural isomers. Even without naming the molecules, you can make such a decision by:

1. counting the longest sequence of carbon atoms;
2. numbering this carbon chain starting at one end;
3. noting the numbers and kinds of substituents, with the substituent numberings as small as possible.

Example 23.1

Are the following two molecules structural isomers?

```
          CH2- CH3
          |
CH3 -     CH          CH3                          CH3        CH3
          |           |                            |          |
          CH - CH2 -  CH - CH3   and   CH3 - CH2-  CH - CH -  C - CH3
          |                                             |     |
CH3 -     CH - CH3                             CH3-    CH2   CH2 - CH3
```

Solution 23.1

Both alkanes are $C_{12}H_{26}$, so they are either identical or are structural isomers. First find the longest continuous chain. By trial and error, a longest chain length of 7 is found for both molecules. Next, number the chains, starting at one end, to keep the substituent numbering as low as possible. We find

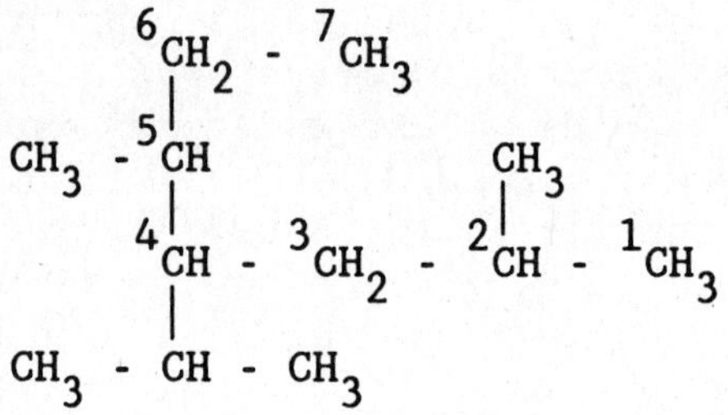

```
             6CH2 - 7CH3
               |
        CH3 - 5CH              CH3
               |                |
              4CH - 3CH2 - 2CH - 1CH3
               |
        CH3 - CH - CH3
```

There is a CH_3 group at carbon 2, and one at carbon 5; a more complex group is at carbon 4. If we had **reversed** the 1 to 7 numbering, the CH_3 substituents would have been at carbons 3 and 6. The second molecule has the following lowest numbering scheme.

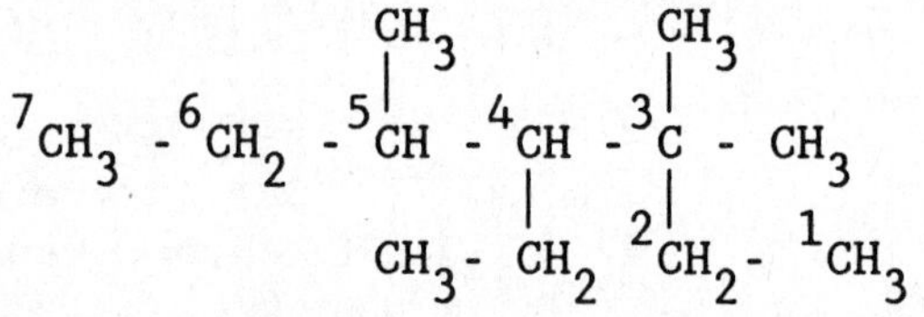

```
                      CH3         CH3
                       |           |
      7CH3 - 6CH2 - 5CH - 4CH - 3C - CH3
                             |     |
                      CH3 - CH2  2CH2 - 1CH3
```

There are two CH_3 groups at carbon 3, 1 at carbon 5, and a CH_2-CH_3 group at carbon 4. So the numbering and the types of substituents are **different.** These 2 molecules are structural isomers.

23.2 Alkenes and Alkynes

Compounds of carbon and hydrogen undergo 3 important types of reactions: **addition, elimination** and **substitution. Addition** reactions involve joining together 2 molecules to form a third molecule. This is the typical reaction of unsaturated molecules like alkenes and alkynes. **Elimination** reactions are just the opposite of addition reactions; a larger molecule is broken into two smaller ones. **Substitution** reactions involve the replacement of one atom or group of atoms by another atom or group of atoms. Alkanes and arenes undergo substitution reactions. If you can systematize organic reactions into these 3 types, it should help you see the similarities between what might seem at first to be very different reactions.

Example 23.2

Give products for the following reactions, if there are any; classify the reactions as addition, elimination or substitution.

```
(a) CH3-CH2-CH -CH2-CH3 + H2 --catalyst-->      (b) CH3CH = CHCH3 + Br2 →
            |
            CH3
```

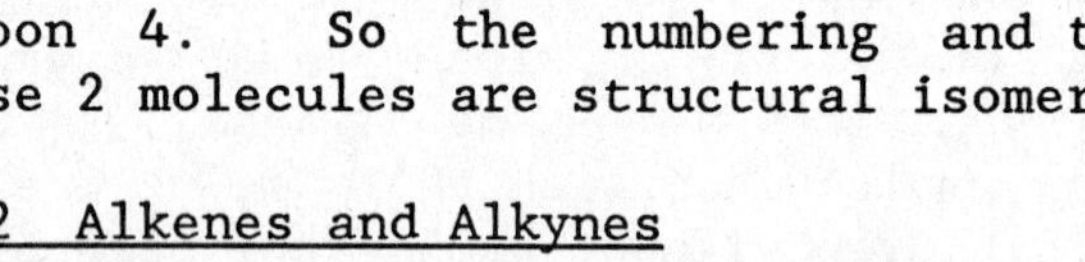

(c) $CH_3 - CH_3 \xrightarrow[\text{heating}]{\text{strong}}$ (d) $CH_3 - CH_3 + Br_2 \xrightarrow[\text{or light}]{\text{heat}}$

Solution 23.2

(a) Alkanes are saturated compounds. They are unreactive towards hydrogen. There is no reaction (under moderate conditions).
(b) Alkenes are unsaturated, and readily undergo addition reactions at the double bond. The product is $CH_3-CHBr-CHBr-CH_3$.
(c) Alkanes break up with strong heating, usually forming alkanes of smaller carbon-content, alkenes, and hydrogen. Ethane forms ethene and hydrogen. This is an elimination reaction. Depending on the temperature chosen, the alkyne $CH\equiv CH$, ethyne, plus another hydrogen molecule could be formed from the ethene. This is also an elimination reaction.
(d) A characteristic reaction of alkanes is substitution. Here a bromine atom replaces a hydrogen atom, and CH_3CH_2Br is formed, along with HBr as the simplest products of this chain reaction.

23.3 Naming of Hydrocarbons

The rules for naming hydrocarbons are presented in the textbook. Although there are common or "trivial" names for some of these, the **systematic** (or **IUPAC**) names are used as the basis for naming all organic compounds. The following examples will test whether you have mastered this topic.

Example 23.3

Name the following hydrocarbons.

(a)
```
                CH3
                |
CH3 - CH2 - CH
                |
        CH3 - CH
                |
                CH2 - CH2 - CH3
```

(b)
```
CH3 - CH2             CH3
          |               |
          CH2 - CH  - CH2
                 |
                 CH3
```

(c)
```
H          CH3
  \        /
   C  =  C
  /        \
CH3        CH3
```

(d)
```
CH3        CH2-CH3
  \        /
   C  =  C
  /        \
H          H
```

Solution 23.3

(a) The longest carbon chain is 7, so this is a heptane. If we start the numbering from the **left**, we get methyl substituents at positions 3 and 4, rather than at positions 4 and 5 if the numbering had started at the right. So the compound is 3,4-dimethylheptane. The numbers are separated by a comma, and a hyphen separates the numbers from the rest of the name. Note that two methyl groups are called dimethyl.

(b) Just because there is a CH_3 group at the right end of the molecule, sticking up, don't consider this to be a substituent. It is a part of the largest carbon chain, which has a length of 6. The compound is 3-methylhexane. If you had incorrectly numbered the molecule starting at the left, you would have named the alkane as 4-methylhexane.

(c) The longest carbon chain **containing the double bond** must be chosen. Thus the compound is a butene, with the double bond located between carbons 2 and 3. The correct name is 2-methyl-2-butene. (Note there is no cis-trans relationship here. The methyl group at carbon 3 is cis to one methyl group at carbon 2, and trans to the other methyl group at carbon 2.)

(d) The longest carbon chain **containing the double bond** is 5, so the molecule is a pentene. The double bond is between carbons 2 and 3, so it is a 2-pentene. The hydrogens are cis to each other. The correct name is cis-2-pentene, not cis-3-pentene.

Example 23.4

Draw a structure for 3,6-dimethyl-1-octene.

Solution 23.4

An octene has a carbon chain length of 8; 1-octene means the double bond is situated between carbons 1 and 2. The methyl groups to be placed at carbons 3 and 6 have no orientational relationship to each other, since there is free rotation about C-C bonds in alkanes. Check that all the hydrogens are present. Remember each carbon must have a total of four bonds. An alkene with 1 double bond has the formula C_nH_{2n}. Thus here we must have $C_{10}H_{20}$.

$$H_2C = CH - \underset{\displaystyle CH_3}{\underset{|}{CH}} - CH_2 - CH_2 - \underset{\displaystyle CH_3}{\underset{|}{CH}} - CH_2 - CH_3$$

Example 23.5

Why do the following names violate IUPAC rules? Give the correct names.

(a) 3-methyl-2-butene (b) 5-methylcyclohexene

Solution 23.5

(a) Let us give a structure corresponding to the proposed name, and then find the error. 2-butene has the structure

$$^1CH_3 - ^2CH = ^3CH - ^4CH_3$$

If we put a methyl group at carbon 3 (replace the H by CH_3) we have the same molecule as we would have by reversing the numbering. That is,

$$CH_3 - CH = \underset{\displaystyle CH_3}{\underset{|}{C}} - CH_3$$

is 2-methyl-2-butene. This is the correct name for the compound, since the methyl group has a lower number than in the alternate name. You have seen this molecule in Example 23.3, part c.

(b) In cyclohexene, the double bond is between carbons 1 and 2. So if we draw 5-methylcyclohexene

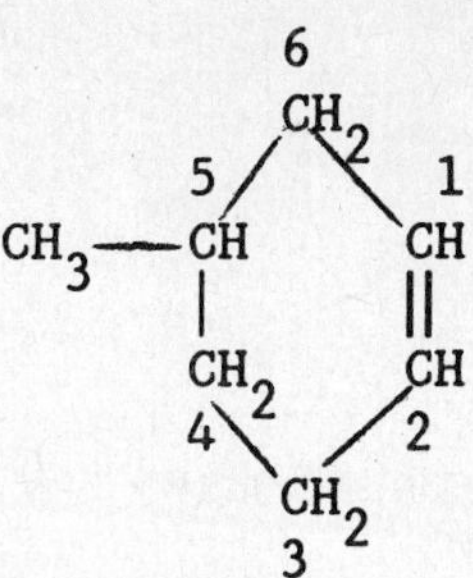

we see that by **renumbering**, starting with carbon 2, and now designating it as carbon 1, and proceeding **counterclockwise**, we get 4-methylcyclohexene.

23.8-23.15 Nomenclature of Organic Compounds

In these sections of the chapter you are introduced to several of the common **functional groups** of organic chemistry, and to some of the typical preparations and reactions of the different classes of compounds containing these functional groups. It is important that you be able to both name an organic compound, given its structural formula, and write its structural formula, given its name. The common functional groups are listed in Table 23.3 of the text. Learn them. You must be able to recognize to what **class of compound** a given organic substance belongs, since the rules of naming organic molecules need that information. The names of several common classes of compounds are also shown in Table 23.3. There can be more than one name for a compound. The **common or trivial name** would have to be memorized for each compound. The **systematic** (IUPAC) **name** on the other hand, requires mastery of only a few rules. For the **systematic** nomenclature, you first must recognize the class to which the compound belongs. Second, name the parent hydrocarbon(s) to which the compound is related. The naming of hydrocarbons has already been covered in Section 23.3. Third, depending on the class of compound, you follow these rules:

Alcohols: replace -e of the alkane by -ol. For example CH_3OH is methanol.

Ethers: replace -yl of the smaller alkyl substituent by -oxy, and join this to the name of the **alkane** corresponding to the other alkyl group. $CH_3CH_2OCH_2CH_2CH_3$ is ethoxypropane.

Aldehyde: replace -e of the alkane by -al. $CH_3CH_2CH_2CHO$ is butanal.

Ketones: replace -e of the alkane by -one. Prefix the name with the lowest possible number to indicate the position of the C=O group. $CH_3CH_2CH_2COCH_3$ is 2-pentanone.

Carboxylic Acids: replace -e of the alkane by -oic acid. CH_3CH_2COOH is propanoic acid.

Esters: use the names of the carboxylic acid and the alcohol that can react to form the ester. Replace -oic acid of the acid with -oate, and prefix with the alkyl

group from the alcohol. $CH_3\underset{\|}{\overset{}{C}}OCH_2CH_2CH_3$ is propylethanoate.
O

Amines: name the alkyl group(s) and add the ending -amine. $CH_3NHCH_2CH_2CH_3$ is methylpropylamine.

Amides: replace -e of the alkane by -amide. $CH_3CH_2CONH_2$ is propanamide.

Haloalkanes and Haloalkenes: already named in Section 23.3. $CH_3CHClCH_3$ is 2-chloropropane.

Example 23.6

Name the following, using the systematic nomenclature:
(a) $CH_3CH_2COCH_2CH_3$ (b) CH_3CONH_2 (c) $CH_3OOCCH_2CH_3$

Solution 23.6

Although an organic compound can contain more than one functional group, these examples contain only one functional group each.

(a) As soon as you notice there is only **one** oxygen atom, and not **two,** in the compound, and no nitrogen, the possible classes of compound are reduced to alcohol, ether, aldehyde or ketone. An alcohol is dismissed, since no -OH group is present. An ether requires just **two** alkyl groups, each singly-bonded to oxygen; here there is an extra carbon, so this choice can be ignored. An aldehyde has a terminal -CHO group, so you are left with a ketone, containing five carbons, with the oxygen doubly-bonded to the carbon at the 3 position. The compound is 3-pentanone.

(b) With nitrogen in the compound, it must be an amine or amide. Since there is not an alkyl group joined to nitrogen, it cannot be an amine. Thus the compound is the 2 carbon amide, ethanamide.

(c) The compound has two oxygens; you are expected to be able to figure out the bonding in the molecule and conclude that the two oxygens are not bonded to each other. (If the two oxygens were bonded to each other, one carbon in the molecule would have a valence of two; carbon has a valence of four.) Thus you should consider either a carboxylic acid or an ester. With no -COOH group, it must be an ester. Remember that you must recognize the carboxylic acid and the alcohol from which the ester is derived. Here the ester is formed from propanoic acid and methanol, and is named methylpropanoate.

Example 23.7

Write structural formulas for each of the following:
(a) 3-bromo-1-propanol (b) 2-methyl-3-phenylpropanal
(c) 2,2-difluoropropanoic acid.

Solution 23.7

(a) This is a substituted 3 carbon alcohol, with the -OH group at the 1 position and the bromine atom at the 3 position. The compound is $HOCH_2CH_2CH_2Br$.

(b) Since the compound is a substituted 3 carbon aldehyde, we know the carbon of -CHO is carbon number 1. As well, there is a methyl group at carbon 2 and a phenyl group at carbon 3. Remember, each carbon must form a total of four bonds. The compound is $C_6H_5CH_2CH(CH_3)CHO$.

(c) This is a substituted 3 carbon carboxylic acid, propanoic acid. The carbon of the carboxylate group is always carbon number 1 in an acid. The compound is CH_3CF_2COOH.

23.16 Review of Functional Groups and Reactions

Organic chemistry, as treated in a general chemistry course, involves a small number of reactions that either insert a functional group into a hydrocarbon, or convert one functional group to another. Once you know the various reactions of a functional group, you can generate **thousands** of **different examples** of those reactions. So mastery of functional group transformations provides you with a wealth of organic reactions. Figures 23.31 and 23.32 in the text show a variety of products that can be obtained from methane, methanol, ethene, and propene. These are **examples** of such functional group transformations. You should appreciate that, for example, alcohols can be obtained from alkenes by addition of H_2O to the double bond. Then the alcohol, if primary, can be oxidized (or dehydrogenated) to an aldehyde or carboxylic acid; if it is a secondary alcohol, it can be oxidized to a ketone. Reduction takes the aldehyde or ketone back to the alcohol. If you are faced with conversion of functional groups which must take several steps, such as the conversion of an alkene to a carboxylic acid, it is often easiest to work backwards from the desired product to the starting material.

Example 23.8

Give the products of the following reactions:

(a) $CH_3CH_2COCH_3 + H_2 \xrightarrow[\text{heat}]{\text{catalyst}}$ (b) $CH_3CH{=}CHCH_3 + H_2O \xrightarrow{\text{catalyst}}$

(c) $NaOOCCOONa + HCl(aq) \longrightarrow$

Solution 23.8

(a) By noting that carbon 2 is joined to only two carbons and thus the oxygen atom must be doubly- bonded to it, it follows that the organic reactant is a **ketone,** 2-butanone. With **reduction** (here by H_2) a ketone is converted to a secondary alcohol; this is just the reverse of the **oxidation** of a secondary alcohol to a ketone. The product is 2-butanol.

$$CH_3CH_2COCH_3 + H_2 \xrightarrow[\text{heat}]{\text{catalyst}} CH_3CH_2CH(OH)CH_3$$

(b) If you remember that the characteristic reaction of alkenes is addition, then it is not difficult to see that 2-butene will form 2-butanol.

$$CH_3CH{=}CHCH_3 + H_2O \xrightarrow{\text{catalyst}} CH_3CH(OH)CH_2CH_3$$

By the way, 2-butanol can be converted back to the alkene by dehydration with concentrated sulfuric acid and heat.

(c) The hardest part in this example is deciding to what class of compound does the organic reactant belong. It is the disodium salt of ethanedioic acid, a weak acid that has 2 replaceable protons. If 1 mole of the acid,

HOOCCOOH, were titrated with 2 moles of NaOH, the salt NaOOCCOONa would be formed. By acidifying the salt, the acid is reformed. So the desired reaction is

$$NaOOCCOONa + 2HCl(aq) \rightarrow HOOCCOOH + 2NaCl(aq)$$

An example in the textbook is very similar:

$HCOONa + HCl(aq) \rightarrow HCOOH + NaCl(aq)$

Example 23.9

Identify the products A, B, C and D in the following sequence.

$$\text{1-propanol} \xrightarrow[K_2Cr_2O_7,\ H_3O^+]{\text{limited amt.}} A \xrightarrow[H_3O^+]{K_2Cr_2O_7} B \xrightarrow[H_3O^+]{CH_3CH_2OH} C$$

$$B \xrightarrow{NH_3,\ \text{heat}} D$$

Solution 23.9

Remember that dichromate ion is a good oxidizing agent. Your should recall that a **primary alcohol** like a 1-propanol can undergo oxidation to an **aldehyde** (with a limited amount of oxidizing agent) or to a **carboxylic acid.** So it is reasonable to expect that A is the aldehyde, propanal, and that B corresponds to the acid CH_3CH_2COOH, propanoic acid. Since the acid B then reacts with alcohol, ethanol, C must be an ester. (The H_3O^+ acts as a catalyst.) Since the alcohol is ethanol, the ester is named ethylpropanoate. Finally the acid B reacts with ammonia to form the **amide,** D. It is named after the parent acid: thus the amide is propanamide.

A: CH_3CH_2CHO, propanal
B: CH_3CH_2COOH, propanoic acid
C: $CH_3CH_2COOCH_2CH_3$, ethylpropanoate
D: $CH_3CH_2CONH_2$, propanamide.

Example 23.10

Compound A, molecular formula C_3H_8O, reacted with Na metal to give $H_2(g)$ and a substance B, C_3H_7ONa. When A was reacted with concentrated sulfuric acid and heat, substance C, C_3H_6, was produced. Oxidation of A with excesss $K_2Cr_2O_7$ and H_3O^+ gave compound D, C_3H_6O. D did not give a positive silver mirror test. Identify the unknown compounds A, B, C and D.

Solution 23.10

Since compound A has one oxygen atom, it could be an alcohol, an ether, an aldehyde, or a ketone. Since A can be **oxidized,** we dismiss an ether and a ketone as choices. Then A must be an alcohol (primary or secondary) or an aldehyde. Since A reacts with Na, giving H_2 gas, A is an **alcohol.** B is the sodium salt of the alcohol, C_3H_7ONa. Since alcohols react with concentrated sulfuric acid and heat to give alkenes, C is propene, $CH_2{=}CHCH_3$. If A is a primary or secondary alcohol, it will yield propene. Compound D is the oxidation product of the alcohol. Since there is only one oxygen present in compound D, that compound **cannot** be a carboxylic acid. Therefore compound A could not have been a primary alcohol. Thus A is $CH_3CH(OH)CH_3$, and D is the ketone CH_3COCH_3, 2-propanone. A ketone does not give a positive silver mirror test.

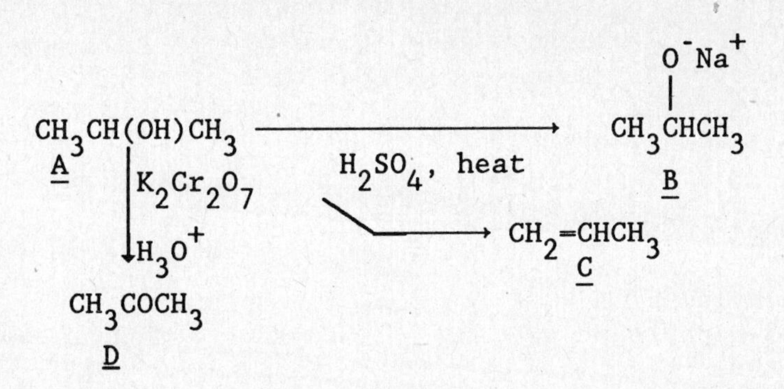

SELF TEST

1. The name of the straight chain hydrocarbon of molecular formula C_9H_{20} is:
(a) cyclonanone (b) nonene (c) nonyl hydride (d) nonane (e) undecane
2. What is the product of the ring-opening addition reaction of cyclopropane with hydrogen?
(a) methylcyclopropane (b) propene (c) 2-butene (d) propane
(e) 2-methylpropane
3. Which of the following has the highest boiling point?
(a) CH_4 (b) C_2H_6 (c) C_3H_8 (d) C_4H_{10} (e) H_2O
4. Which of the following reagents and conditions will yield bromoethane?
(a) bromine and ethane at 25°C
(b) hydrogen bromide and ethane at 25°C
(c) ethane and bromine at high temperatures
(d) ethane and bromine and light at 25°C
(e) ethane and sodium bromide
5. The number of structural isomers that can be written for the molecular formula C_4H_9Cl is
(a) 1 (b) 2 (c) 3 (d) 4 (e) 5
6. Give the correct product(s) for the following reaction. $CH_2{=}CH_2 + Cl_2 \rightarrow$
(a) HCl (b) CH_3CHCl_2 (c) $CH_2Cl\text{-}CH_2Cl$ (d) $CH_2{=}CHCl$ (e) $CH_2 \quad CH_2$ / Cl_2
7. Which of the following could be alkadiene(s)?
(a) C_2H_2 (b) C_2H_4 (c) C_4H_4 (d) C_4H_6 (e) C_6H_6
8. Which of the following is (are) substitution reactions?
(a) $CH_4 + Br_2 \rightarrow CH_3Br + HBr$ (b) $CH_2{=}CH_2 + H_2 \rightarrow CH_3\text{-}CH_3$
(c) $CH_3CH_2OH \xrightarrow{H_2SO_4} CH_2{=}CH_2 + H_2$ (d) diamond → graphite
(e) $SO_3 + H_2O \rightarrow H_2SO_4$
9. A major use for ethyne is
(a) for welding (b) as a lubricating agent (c) for asphalt
(d) as a solvent (e) for nuclear power
10. Give the IUPAC name for

$$\begin{array}{l} CH_2-C-CH_3 \\ \,| \qquad \| \\ CH_2-CH \end{array}$$

(a) 1-methyl-1-cyclobutane (b) 1-methylcyclobutane
(c) cis-1-methylcyclobutene (d) 2-methyl-1-cyclobutene
(e) 1-methylcyclobutene

11. Which of the following names is (are) incorrect?
(a) 4-butene (b) 2-ethylpropane (c) 3-propyl-1-hexene
(d) 2,3-dimethylcyclopropane (d) cyclohexene

12. Give the correct name for

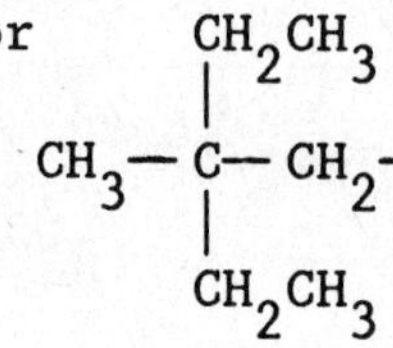

$$\begin{array}{c} CH_2CH_3 \\ | \\ CH_3-C-CH_2-CH_3 \\ | \\ CH_2CH_3 \end{array}$$

(a) 2,2-dimethylbutane (b) 3,3-dimethylbutane
(c) 3-ethyl-3-methylpentane (d) 3,3-dimethyl-3-methylpropane
(e) 2,2-dimethylbutane

13. What is the correct name for CH_3CH_2CHO?
(a) propanol (b) propanoic acid (c) 1-propanone
(d) propanal (e) methylethanoate

14. The common name for $HOOCCH_3$ is
(a) formic acid (b) methylformic acid (c) acetic acid
(d) ethanoic acid (e) none of the above.

15. The major organic product made from synthesis gas is
(a) CH_4 (b) CH_3OH (c) CH_3CH_2OH (d) $CH_2{=}CH_2$ (e) C_6H_6

16. Alcohols can be prepared by which of the following routes?
(a) oxidation of a ketone
(b) reduction of an aldehyde
(c) esterification of a carboxylic acid
(d) hydration of an alkene
(e) burning of an alkane

17. "Alcohol" in beer and wine refers to:
(a) methanol (b) ethanol (c) propanol (d) butanol (e) wood alcohol

18. 2-Butanone may be prepared by the reaction of
(a) $CH_3CH_2CH(OH)CH_3$ + $K_2Cr_2O_7/H_2SO_4$
(b) $CH_3CH_2CH_2CH_3$ + $KMnO_4$
(c) $CH_3CH_2CH_2CH_3$ + $K_2Cr_2O_7/H_2SO_4$
(d) $CH_3CH_2CH_2COOH$ + $K_2Cr_2O_7/H_2SO_4$
(e) $CH_3CH{=}CHCH_3$ + HBr

19. Butanal can be prepared by the reaction of
(a) 1-butene and aqueous H_2SO_4 (b) butane and O_2 at 25°C
(c) $K_2Cr_2O_7/H_2SO_4$ (limited quantities) and 1-butanol
(d) 2-butanol and $K_2Cr_2O_7/H_2SO_4$ (e) $K_2Cr_2O_7/H_2SO_4$ (excess) and 1-butanol

20. Which of the following structural formulas is both a ketone and a tertiary alcohol?

(a) 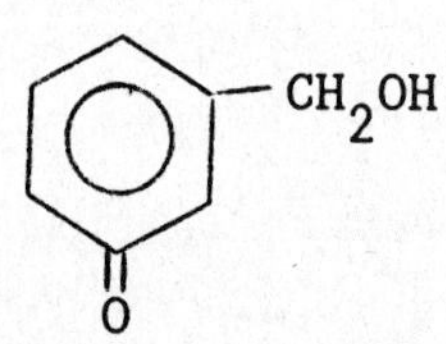 [six-membered ring with CH_2OH substituent and C=O]

(b) 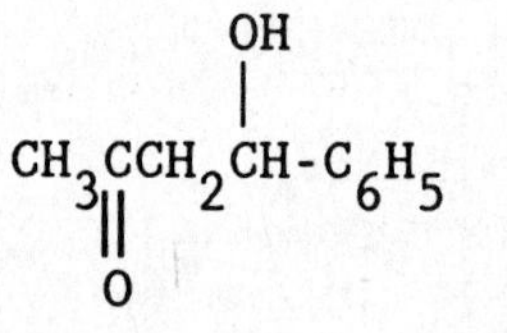 $CH_3C(=O)CH_2CH(OH)-C_6H_5$

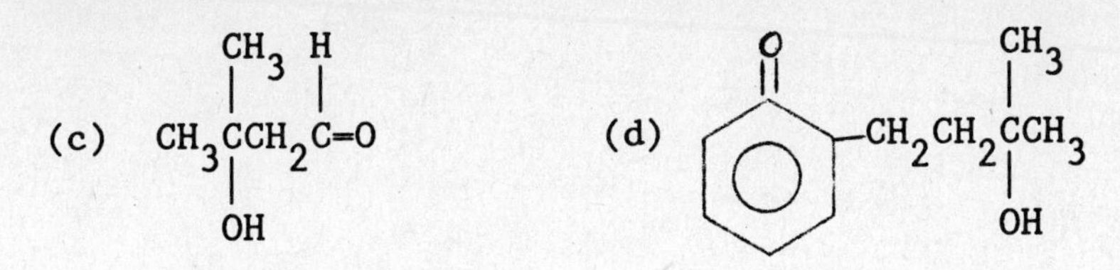

(e) CH_3CH_2COOH

21. Benzoic acid, C_6H_5COOH, will react readily with
    (a) CH_3COCH_3 (b) pentane (c) $K_2Cr_2O_7/H_2SO_4$ at 10°C
    (d) CH_3OH and trace HCl (e) dilute HCl
22. Ethanoic acid (acetic acid) may be prepared by the reaction of
    (a) $Na_2Cr_2O_7$ and CH_3CH_2Cl (b) $KMnO_4$ and $CH_2{=}CH_2$
    (c) CH_3CH_3 and $KMnO_4$ (d) CH_3CHO and H_2 catalyst
    (e) $K_2Cr_2O_7/H_2SO_4$ (excess) and CH_3CH_2OH
23. A reagent that would enable you to carry out the reaction
    $CH_2{=}CH_2 \rightarrow CH_3CH_3$ is
    (a) H_2 and Pt metal (b) Na metal (c) HOOH
    (d) $K_2Cr_2O_7/H_2SO_4$ (e) H_2 and Br_2
24. Propanoic acid, CH_3CH_2COOH, will **not** readily react with which of the following?
    (a) CH_3COOCH (b) $Ba(OH)_2$ (c) CH_3CH_2OH and trace HCl
    (d) aqueous NaCl (e) NH_3
25. The two compounds $CH_3CH{=}CHCH_3$ and $CH_3CH_2CH_2CH_2OH$ can be distinguished by the ready reaction(s) of **one** of them with:
    (a) cyclohexene and light (b) CH_3OCH_3 (c) K metal
    (d) CH_4 (e) $CH_3CH_2COCH_2CH_3$
26. $CH_3CH_2NHCH_3$ is an example of
    (a) a primary amine (b) a primary amide (c) a secondary amine
    (d) a secondary amide (e) a tertiary amine

Answers to Self Test

1. d; 2. d; 3. e; 4. c,d; 5. d; 6. c; 7. d; 8.a; 9. a; 10. e; 11. a,b,d; 12. c; 13. d; 14. c; 15. b; 16. b,d; 17. b; 18. a; 19. c; 20. d; 21. d; 22. e; 23. a; 24. a,d; 25. c; 26. c.

CHAPTER 24

POLYMERS: SYNTHETIC AND NATURAL

SUMMARY REVIEW

Polymers are molecules, usually of large molar mass, in which there are repeating groups of atoms. Very large molecules are also called **macromolecules.** Many occur naturally in living systems and are referred to as **biopolymers;** others are made synthetically by combining large numbers of small molecules **(monomers).** Polymers may have structures that extend in one, two or three dimensions.

Addition polymers are formed by the successive **addition** of monomer units, with no loss of atoms, normally through a free radical mechanism. Important classes of addition polymers are based on **ethene** and the substituted ethenes, and on **butadiene** and substituted butadienes (the rubbers). In vulcanization of rubber, polymer chains are linked by chains of sulfur atoms.

Condensation polymers differ from addition polymers in that they are formed from the **condensation** of monomers and do not contain all of the atoms of the original monomers. **Polyamides,** from condensation of a dicarboxylic acid and diamine, and **polyesters,** both formed from condensation of a dicarboxylic acid and a diol, with elimination of water, are two important examples.

Among important condensation **biopolymers** are the **carbohydrates, proteins** and **nucleic acids.** Green plants synthesize carbohydrates photochemically from CO_2 and water and have the empirical formula $C_x(H_2O)_y$; they are classified as **monosaccharides, disaccharides** and **polysaccharides.** Some monosaccharides occur naturally as sugars; glucose, $C_6H_{12}O_6$, occurs in two forms, both with six-membered rings, while **fructose** with the same formula has a five-membered ring. Sucrose, $C_{12}H_{22}O_{11}$, is a common disaccharide formed by condensing together glucose and fructose. Polymerization of α-glucose gives **starch** which is a mixture of polysaccharides in which **amylose** (a unbranched condensation polymer) predominates. **Cellulose** is a straight chain polysaccharide obtained from β-glucose, with an average of 3000 glucose units. It has a very large number of -OH groups that can be reacted with carboxylic acids to give **polyesters.** Replacement of all three -OH groups of each glucose unit by acetate groups gives an acetate polymer (e.g., Arnel). Replacement by nitrate groups gives nitrocellulose (guncotton and celluloid). (See text for diagrams.)

Proteins of many different kinds are found in living cells; they are **polyamides** formed from α-amino acids (with the $-NH_2$ group on the same C atom as the $-CO_2H$ group). Condensation of α-amino acids gives polymers containing the **peptide** link $-\underset{\underset{O}{\|}}{C}-NH$. Polyamides are also called **polypeptides.** There are twenty α-amino acids commonly found in proteins. The **primary structure** is the order in which the amino acids occur in the structure. The **secondary structure** is due to hydrogen bonding between N-H and O=C groups. When these are between separate polymer chains, numerous parallel chains can be bonded to form a sheet and the sheets can be stacked to form a three-dimensional structure, as is found, for example, in silk. In other proteins, such as those found in wool or hair, hydrogen bonding between C=O and N-H groups in the same chain forms a spiral known as an **α-helix.** α-Helices can then be twisted together into a **protofibril.** These in turn can be packed in parallel bundles to form a fiber. Alternatively, α-helices can be folded to give a compact structure called the **tertiary structure** and held together by disulfide bridges and by interactions between polar or ionic side-groups. Such structures are found in the **globular proteins,** such as myoglobin and hemoglobin, and the many **enzymes** that catalyze biochemical reactions. A simple description of the mechanism of enzyme behavior is the **lock-and-key** theory which supposes that reactant molecules (the **substrate)** fit into a cavity in the enzyme and are held in such an orientation that reaction can occur rapidly.

Organisms synthesize characteristic proteins using information that is stored in the **DNA** (deoxyribonucleic acid) found in the nuclei of all cells. **Nucleic acids** are polymers formed from **nucleotides,** which result from condensation of a molecule of **phosphoric acid,** a molecule of the sugar **deoxyribose** and a molecule of a **nitrogen base** (adenine, guanine, cytosine or thymine). The four different nucleotides condense together to form the DNA, which has a sugar-phosphate backbone with one of the nitrogen bases attached to each sugar, leading to an enormous number of possible different nitrogen base sequences. The unique feature of its structure is that it consists of **two** polynucleotide chains held together by hydrogen bonds in the form of a **double helix.** Hydrogen bonds are only possible between adenine and thymine, or between guanine and cytosine, so that the sequence of bases on one chain completely determines that on the other. If the hydrogen bonds are broken, each strand of the DNA can act as the template for the formation of the other.

REVIEW QUESTIONS

1. Name five examples of common substances that are made of synthetic polymers and five that are made of natural polymers.
2. How is the reaction between ethene molecules to give polyethylene initiated? Why is polyethylene referred to as an addition polymer?
3. What monomers are the starting materials for the preparation of the following addition polymers? Teflon, Lucite, Styrofoam, Orlon, Acrilan, PVC.
4. How many monomer units are there in a single molecule of polystyrene, if its molar mass is 2.08×10^5 g mol^{-1}?
5. Write a general formula for polybutadiene polymer (synthetic rubber).
6. What happens to rubber during the process known as vulcanization?
7. Give an example of a condensation reaction. Give two examples of condensation polymers.
8. What monomers are required to form a polyamide?
9. What are the specific monomers used in preparing Nylon-66?

10. What monomers are required to form a polyester?
11. What are the specific monomers used in preparing Dacron?
12. Write an equation indicating the formation of sucrose by photosynthesis.
13. Both α-glucose and β-glucose have the molecular formula $C_6H_{12}O_6$. In what way are the structures different?
14. Draw the structure for fructose.
15. By what linkage are the two monosaccharide units in sucrose joined? Draw the structure of sucrose. Is sucrose an addition or a condensation polymer?
16. What are the monomers from which starch and cellulose are formed? To what general class of condensation polymers do starch and cellulose belong?
17. How is nitrocellulose prepared?
18. To what general class of condensation polymers do proteins belong?
19. Draw a general structure for an α-amino acid.
20. What is the peptide link?
21. What are (a) the primary, (b) the secondary, (c) the tertiary, structure of a protein?
22. In what structural way is silk different from wool and hair?
23. How many α-amino acid molecules are there in one turn of an α-helix?
24. What is a protofibril?
25. How does a globular protein differ in structure from a protofibril?
26. What is an enzyme? According to the "lock and key" theory, how does an enzyme work?
27. What is a nucleotide? What four nitrogen bases are found in nucleotides?
28. What is the name and the structure of the sugar found in the nucleotides?
29. What is the full chemical name of DNA? What kind of polymer is DNA?
30. What was the revolutionary proposal made by Watson and Crick concerning the structure of DNA? Why in DNA is adenine always found in association with thymine and guanine always found in association with cytosine?
31. How is it supposed that DNA can replicate itself?
32. What is a gene?

Answers to Selected Review Questions

4. 2000.

OBJECTIVES

Be able to:

1. Define the terms monomer, polymer and macromolecule.
2. Give examples of common addition polymers and their monomer units.
3. Show the structure of polyethylene and the involvement of a free radical in the polymerization of ethene.
4. Give the monomer units for natural and synthetic rubber, and write the general formulas for the polymers.
5. Show how condensation polymers, such as polyamides and polyesters, are formed.
6. State typical examples and functions of the four classes of lipids.

7. State the empirical formula for a carbohydrate, and give examples of monosaccharides, disaccharides and polysaccharides.
8. Give the functional groups present in amino acids and show how proteins can be formed from amino acids.
9. Differentiate between the terms primary, secondary, and tertiary structure of a protein.
10. Define the terms globular protein, enzyme and substrate, and state the lock-and-key theory of enzyme action.
11. Give the structural formulas and names of the four nitrogen bases found in DNA.
12. List the three types of molecules needed to form a nucleotide.
13. Explain how DNA can replicate itself.

PROBLEM SOLVING STRATEGIES

24.1 Synthetic Polymers

Condensation Polymers

Although **addition** polymers are prepared by successive **addition** of monomer units, which normally contain C=C bonds, **condensation** polymers are formed by **eliminating** a small molecule like H_2O, HCl or CH_3OH when two monomers combine. Each of the monomer units contains **two** functional groups. For example, one end of a dicarboxylic acid $HO\overset{O}{\overset{\|}{C}}{\sim\sim\sim}\overset{O}{\overset{\|}{C}}OH$ can react with one end of a diol $HOCH_2{\sim\sim\sim}CH_2OH$ to form water and an ester $HO\overset{O}{\overset{\|}{C}}{\sim\sim\sim}\overset{O}{\overset{\|}{C}}\text{-}OCH_2{\sim\sim\sim}CH_2OH$. (The symbol ~~~ indicates parts of the molecule that are not important for the discussion here.) Note that the ester formed still has a carboxylic acid group -COOH and an alcohol group -OH for further reaction. With continued ester formation, accompanied by elimination of water, a polyester is formed. In order to understand the chemistry involved in polymerization, you need to have mastered the functional group chemistry in Chapter 23. Once this has been accomplished, you should be able to recognize the monomer units used to form a condensation polymer.

Example 24.1

Name and draw the structures of the monomers used to prepare the following condensation polymer.

$$\text{-[-O-}CH_2\text{-}CH_2\text{-O-}\overset{O}{\overset{\|}{C}}\text{-}\overset{O}{\overset{\|}{C}}\text{-O-}CH_2\text{-}CH_2\text{-O-}\overset{O}{\overset{\|}{C}}\text{-}\overset{O}{\overset{\|}{C}}\text{]-}_n$$

Solution 24.1

As noted in the text, most common condensation polymers are either polyesters or polyamides. Since the polymer in this example does not contain nitrogen, it cannot be a polyamide. So let's try to identify a dicarboxylic acid and a diol that could serve as starting materials. By the way, when a carboxylic acid and an alcohol react to give an ester and water, the **oxygen** atom in the water molecule comes from the **acid.** So when a dicarboxylic acid reacts with a diol, we look for an acid fragment that is missing two -OH groups, leaving $-\overset{O}{\overset{\|}{C}}\sim\sim\sim\overset{O}{\overset{\|}{C}}-$, and an alcohol fragment that has lost two hydrogen atoms (one from each OH group), leaving $-OCH_2\sim\sim\sim CH_2O-$. Thus we can pick out $-OCH_2CH_2O-$, which is that part of 1,2-ethanediol, $HOCH_2CH_2OH$, left after the two terminal hydrogens have been removed. For the carboxylic acid, we recognize $-\overset{O}{\overset{\|}{C}}-\overset{O}{\overset{\|}{C}}-$, which comes from $HO\overset{O}{\overset{\|}{C}}-\overset{O}{\overset{\|}{C}}OH$, ethanedioic acid. The alcohol fragment is underlined with a single line, and the acid fragment with a double line in the diagram below.

$$-[\underline{-O-CH_2-CH_2-O-}\underline{\underline{\overset{O}{\overset{\|}{C}}-\overset{O}{\overset{\|}{C}}}}\underline{-O-CH_2-CH_2-O-}\underline{\underline{\overset{O}{\overset{\|}{C}}-\overset{O}{\overset{\|}{C}}}}-]_n$$

SELF TEST

1. For the sequence ACCGTCT, what is the complimentary sequence of nitrogen bases that would be found along the second strand of a DNA molecule?
   (a) CCGTCTA (b) CAATGAG (c) GTTACTC (d) TGGCAGA (e) CAGCCTA
2. Which of the following contains at least one nitrogen atom?
   (a) chloroprene (b) glycine (c) DNA (d) nylon (e) starch
3. Which of the following is(are) examples of addition polymers?
   (a) Teflon (b) polyacrylonitrile (c) a polyamide (d) histidine (e) DNA
4. What is the structure of the polymer formed from 2-butene?
   (a) $-[-CH_2-CH_2-]_n$ (b) $-[-\underset{\displaystyle CH_3}{\underset{|}{CH_2}}-CH_2-]_n$
   (c) $-[-\underset{\displaystyle H_3C}{\underset{|}{CH}}=\underset{\displaystyle CH_3}{\underset{|}{CH}}-]_n$ (d) $-[-CH-\overset{\displaystyle CH}{\overset{|}{\underset{\displaystyle CH_3}{\underset{|}{CH}}}}-]_n$
   (e) none of these.
5. Which of the following are monosaccharides?
   (a) sucrose (b) glucose (c) fructose (d) amylose (e) cellulose
6. When starch is hydrolyzed, the product(s) is(are)
   (a) amylose (b) sucrose (c) fructose (d) glucose (e) DNA
7. A fat is:
   (a) a nucleotide (b) a phospholipid (c) a form of rubber
   (d) an ester of glycol (e) a steroid
8. Select the monomer units needed to form the following condensation polymer

$$-O-CH_2-CH_2-O-\overset{O}{\overset{\|}{C}}-C_6H_4-\overset{O}{\overset{\|}{C}}-O-CH_2-CH_2-O-\overset{O}{\overset{\|}{C}}-C_6H_4-\overset{O}{\overset{\|}{C}}-$$

   (a) $CH_3-O-\overset{O}{\overset{\|}{C}}OH$ (b) $HO-(CH_2)_2-OH$

(c) C_6H_5COOH (d) $C_6H_4(COOH)_2$ (e)

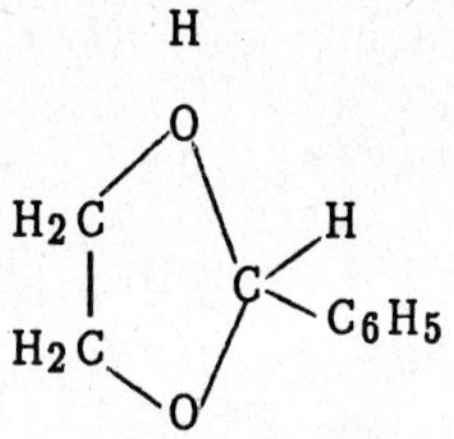

9. All amino acids contain
   (a) DNA (b) the -COOH functional group
   (c) the $-\overset{O}{\overset{\|}{C}}-CH_3$ functional group (d) the $-NH_2$ functional group
   (e) the -NH- functional group

10. The structural formula for adenine is

   (a)

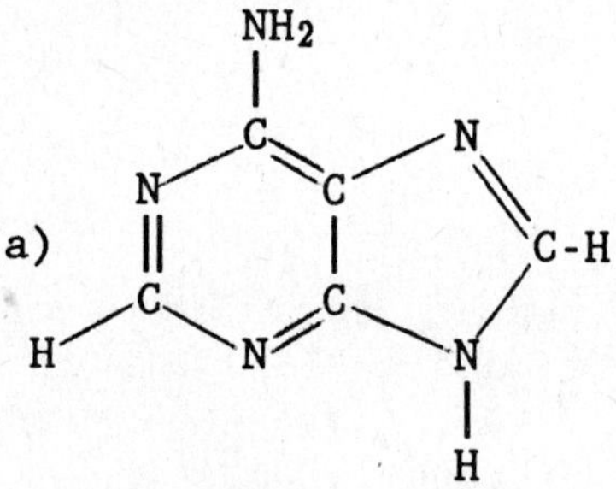

   (b)

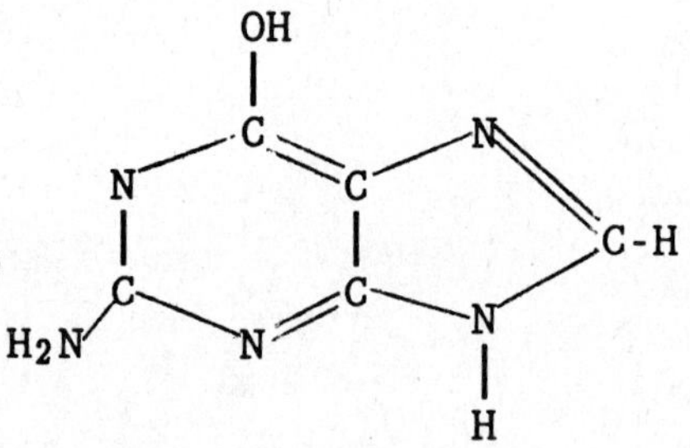

   (c)

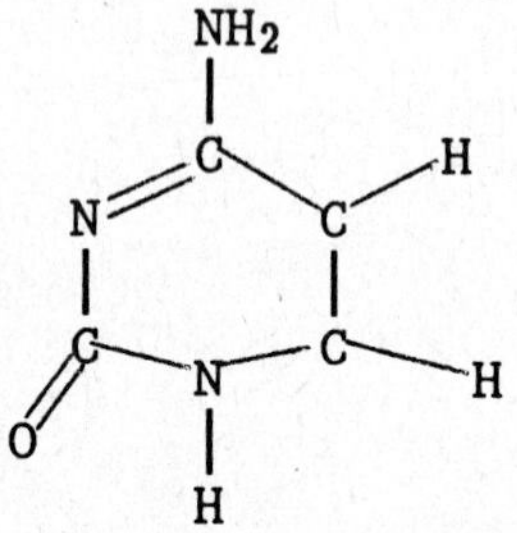

   (d)

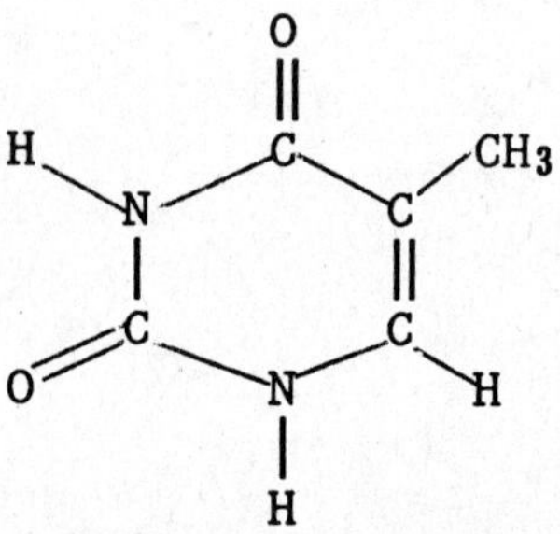

   (e) none of these

11. Which of the following statements is (are) true?
   (a) the α-glucose ring contains six carbon atoms
   (b) cellulose is a disaccharide
   (c) in DNA the two polynucleotides form a double helix joined by carbon-carbon double bonds.
   (d) polyethylene polymers are generally amorphous, but they may be at least partially crystalline.
   (e) a protein adopts a very definite conformation called the secondary structure.

12. A synthetic polymer of glucose is found to have a molar mass of about 1.0 x 10^5 g mol^{-1}. How many glucose units does one molecule contain?
   (a) 550 (b) 620 (c) 5600 (d) 10^5

<u>Answers to Self Test</u>

1. d; 2. b,c,d; 3. a,b; 4. d; 5. b,c; 6. d; 7. d; 8. b,d; 9. b,d; 10. a; 11. d,e; 12. b.

CHAPTER 25

NUCLEAR AND RADIOCHEMISTRY

SUMMARY REVIEW

Nuclear chemistry deals with **nuclear** reactions: the spontaneous radioactive decay of unstable nuclei, or those brought about by bombarding nuclei with high energy subatomic particles. In **radioactive nuclear decay,** various particles can be emitted, but there is always an overall conservation of mass number and atomic number:

particle emitted	nuclear reaction
α-particle (He nucleus)	${}^{A}_{Z}X \rightarrow {}^{A-4}_{Z-2}X + {}^{4}_{2}He$
β-particle (electron)	${}^{A}_{Z}X \rightarrow {}^{A}_{Z+1}X + {}^{0}_{-1}e$
position	${}^{A}_{Z}X \rightarrow {}^{A}_{Z-1}X + {}^{0}_{1}e$
electron capture	${}^{A}_{Z}X + {}^{0}_{-1}e \rightarrow {}^{A}_{Z-1}X$
γ-ray photon	${}^{A}_{Z}X^{*} \rightarrow {}^{A}_{Z}X + \gamma$

Nuclear stability is associated with attractive nuclear forces between protons and neutrons, which have to outweigh the repulsive forces between the protons for a nucleus to be stable. Stable nuclei fall in a narrow band of neutron to proton ratios. After Z = 83, no increase in the number of neutrons is sufficient to keep a nucleus stable. For the lighter elements, the most stable isotopes have neutron/proton ratios close to 1.0. Thus, ${}^{14}_{6}C$ lies above **the band of stability** and decays by electron emission to ${}^{14}_{7}N$, while ${}^{11}_{6}C$ lies below the band of stability and decays by positron emission to ${}^{11}_{5}B$. The series of radioactive disintegrations that continues until a stable isotope is formed is called a **radioactive-decay series.**

The rate of decay of a radioactive substance is first order; the number of nuclear disintegrations in a given time is proportional to the number of atoms present. For N_0 nuclei at time zero, and N nuclei at time t

$$\ln N_0/N = kt$$

where k is the first order rate constant. The time for half the nuclei to decay, $t_{1/2}$, the half-life, is given by $kt_{1/2} = \ln 2 = 0.693$.

In **radiochemical dating,** it is assumed that a radioactive isotope present at the time of formation of a rock, fossil or archeological object, has decayed with first-order kinetics. From its initial amount, its current amount, and the known half-life of the isotope, the age of the object may be determined; alternatively, the current rate of disintegration, R, may be compared with the initial rate, R_0.

$$t = \frac{t_{1/2}}{0.693} \ln \frac{N_0}{N} = \frac{t_{1/2}}{0.693} \ln \frac{R_0}{R}$$

Stable and unstable (radioactive) isotopes can be made by bombarding other isotopes with highly energetic particles, such as α-particles or neutrons. The transuranium elements have been made in this way. The biological hazards of radiation are mainly due to the formation of free radicals.

The **binding energy** of a nucleus is the amount of energy required to decompose it into its constituent neutrons and protons. Binding energies of different nuclei are compared on the basis of binding energy per nucleon. The most stable nucleus is ^{56}Fe. Energy is released when heavy nuclei split into lighter nuclei. The **fission** of ^{235}U is initiated by neutrons and an average of 2.4 neutrons are produced together with lighter nuclei, so that for a **critical mass** of ^{235}U the nuclear reaction is sustained. For a greater mass than the critical mass, an explosive **branching chain reaction** occurs, which is the basis of the atomic bomb. In **nuclear reactors** fission is carried out in a controlled manner and the heat produced is transferred to a liquid and used to produce steam to drive turbines. The radioactive fission products are hazardous and many problems are associated with their disposal. Nuclear **fusion** is the fusion of very light nuclei into heavier nuclei with the release of very large amounts of energy. Because of the very high temperatures involved, controlled nuclear fusion has not yet been achieved industrially.

REVIEW QUESTIONS

1. What is a nuclear reaction?
2. What is radioactivity? Why are uranium-238 nuclei described as radioactive?
3. If an $^{238}_{92}U$ nucleus emits an α-particle, what is the atomic number and the mass number of the resulting nucleus?
4. What is a β-particle?
5. What is a positron? What happens when a positron is emitted from a nucleus?
6. What is a nuclear capture process?
7. To what group in the periodic table would the product nucleus belong if a group 4 nucleus emitted (a) an α-particle, (b) a β-particle, (c) a positron?
8. What is a Geiger counter?

9. What balance of nuclear forces is necessary in order for a nucleus to be stable? What is meant by the nuclear band of stability for nuclei?
10. What is meant by a radioactive decay series?
11. What is radiochemical dating?
12. What isotopes are suitable for dating (a) the ages of rocks, (b) wooden objects from Roman times?
13. On what assumption does the dating of objects containing carbon rest?
14. What isotope must be bombarded with α-particles to obtain $^{17}_{8}O$ and $^{1}_{1}H$ as the products?
15. What is a synthetic isotope?
16. Give an example of the use of radioactive isotopes in studying the mechanisms of reactions.
17. What is the binding energy of a nucleus? How can binding energy be calculated?
18. Which is the most stable nucleus?
19. What is a fission reaction?
20. What is a branching chain reaction?
21. Why is a critical mass necessary to sustain fission?
22. What is a fusion reaction? What problems are associated with the practial use of fusion reactions as a source of useful energy?

Answers to Selected Review Questions

3. 90, 234 7. (a) group 2, (b) group 5, (c) group 3

OBJECTIVES

Be able to:

1. Define the terms radioactivity, radioactive, and positron.
2. Write nuclear equations involving emission of a helium nucleus (α-particle), an electron (β-particle), a positron and capture of an electron.
3. Calculate the energy of radiation accompanying a radioactive decay process.
4. Explain how a Geiger counter works.
5. Sketch the relationship (the band of stability) between the number of neutrons and the number of protons for the stable, naturally-occurring nuclei.
6. Give examples how nuclei lying outside the band of stability decompose.
7. Use the half-life expression for radioactive decay to calculate disintegration rates, half-lives, and the mass of a radioactive sample remaining or used up after a given time.
8. Perform calculations involving radiochemical dating.
9. Give examples illustrating how artificial radioisotopes can be synthesized.
10. List several uses of radioisotopes.
11. Calculate the binding energy for a nucleus and the binding energy per nucleon.

12. Define the terms fission, branching chain reaction, critical mass, and fusion.
13. Describe the operation of a nuclear reactor and the problems associated with radioactive waste.
14. Give examples of fusion reactions.

PROBLEM SOLVING STRATEGIES

25.1 Radioactivity

Some **nuclei** decompose into smaller **nuclei spontaneously;** the process can be accompanied by emission of a helium nucleus (an α-particle), an electron (a β-particle), a positron, or high energy radiation. This phenomenon is called **radioactivity.** In addition to these **spontaneous** nuclear reactions, other nuclear reactions involve one nucleus **capturing** another particle, such as an electron, a positron, a neutron, an α-particle, or other particle. All these nuclear reactions can be written in equation form. Several rules must be followed.

1. The sum of mass numbers for reactants must equal the sum of mass numbers for products. **Mass numbers appear as left superscripts** to atomic or particle symbols.
2. Atomic numbers usually change in nuclear reactions. However, the sum of atomic numbers for reactants must equal the sum of atomic numbers for products. **Atomic numbers appear as left subscripts.**
3. A neutron is symbolized by ${}^{0}_{1}n$; an electron is ${}^{0}_{-1}e$; a positron ${}^{0}_{1}e$.
4. In nuclear reactions, **nuclei** undergo changes. The symbols normally used for atoms and molecules in **chemical** reactions in this book are also used for **nuclear** reactions. For example, in a **nuclear** reaction, we use ${}^{1}_{1}H$ for a hydrogen **nucleus** ${}^{1}_{1}H^{+}$; we use ${}^{4}_{2}He$ to represent ${}^{4}_{2}He^{2+}$, a helium nucleus (an alpha particle), i.e., we ignore net charges on atoms due to missing orbital electrons.
5. The energy changes accompanying nuclear reactions, which are much larger than for **chemical** reactions, can be exothermic or endothermic. If we are only interested in the reactants and products, the energy change can be omitted.

Example 25.1

Give the symbols for the missing product or reactant in the following.

(a) ${}^{14}_{7}N + {}^{1}_{1}H \rightarrow {}^{4}_{2}He +$ (b) ${}^{10}_{5}B + {}^{4}_{2}He \rightarrow$ (c) ${}^{106}_{47}Ag \rightarrow {}^{106}_{48}Cd +$

(d) ${}^{27}_{14}Si \rightarrow {}^{27}_{13}Al +$ (e) $3{}^{1}_{1}H + 4{}^{1}_{0}n \rightarrow$ (f) $\rightarrow {}^{13}_{6}C + {}^{0}_{1}e$

Solution 25.1

(a) Since the mass numbers sum to 15 on the left side of the equation, the mass number of the missing product must be 11. As well, the atomic numbers on the right must sum to 8. Thus the product nucleus has atomic number 6, which is carbon. The product is $^{11}_{6}C$.

$$^{14}_{7}N + ^{1}_{1}H \rightarrow ^{4}_{2}He + ^{11}_{6}C$$

(b) This is an example of an α-particle, $^{4}_{2}He$, combining with a boron nucleus to produce a heavier nucleus. This product nucleus must have a mass number of 14 and an atomic number 7. Since we know atomic number 7 refers to nitrogen, we write the product as $^{14}_{7}N$.

$$^{10}_{5}B + ^{4}_{2}He \rightarrow ^{14}_{7}N$$

(c) In this example the mass number doesn't change, but the atomic number increases. This is consistent with an electron (a β-particle) being emitted; the missing product is $^{0}_{-1}e$.

$$^{106}_{47}Ag \rightarrow ^{106}_{48}Cd + ^{0}_{-1}e$$

(d) Just as in example (c), the mass number doesn't change. Here, the atomic number of the listed product is one less than that for the reactant nucleus. Thus a positron, $^{0}_{1}e$, is emitted.

$$^{27}_{14}Si \rightarrow ^{27}_{13}Al + ^{0}_{1}e$$

(e) In this example, 3 protons and 4 neutrons combine to form a nucleus that has a mass number of 3 x 1 + 4 x 1 = 7 and an atomic number of 3 x 1 + 4 x 0 = 3. The product is $^{7}_{3}Li$, since lithium is the element with atomic number 3.

$$3\,^{1}_{1}H + 4\,^{1}_{0}n \rightarrow ^{7}_{3}Li$$

(f) For this example we can obtain the mass number of the reactant by adding the mass numbers of the products, and get the atomic number of the reactant by adding the atomic number of the products. We find that the reactant nucleus has a mass number of 13 and atomic number equal to 7. Thus the reactant is $^{13}_{7}N$.

$$^{13}_{7}N \rightarrow ^{13}_{6}C + ^{0}_{1}e$$

25.3 Radioactive-Decay Rates

Radioactive decay follows first order kinetics; thus the half-life, $t_{1/2}$, for a given radioactive isotope is a **constant.** From the rate law for a first order process, we have

$$-\ln\frac{N}{N_0} = kt \qquad \text{or} \qquad \ln\frac{N_0}{N} = kt.$$

To get the $t_{1/2}$ expression, just let $t = t_{1/2}$ and $N = N_0/2$. Then

$$\ln\frac{N_0}{N_0/2} = kt_{1/2} \qquad \text{or} \qquad t_{1/2} = \frac{0.693}{k}$$

Also remember that the amount of a radioactive sample **used up** after a given time equals the initial amount minus the amount remaining.

Example 25.2

$^{147}_{61}Pm$ is a β emitter with a half-life of 2.65 years. If a battery for a wrist watch uses this decay as the source of electrons, how many years would it take for the rate of electron production to fall to 5% of its initial value? What product nucleus is produced by this decay?

Solution 25.2

The plan of attack is as follows. Use $t_{1/2}$ to solve for the rate constant k. Then use k and $N = 0.05\ N_0$ to solve for t in the expression

$$\ln \frac{N_0}{N} = kt$$

Since we know $t_{1/2} = 2.65 \text{ years} = \frac{0.693}{k}$, $k = \frac{0.693}{2.65}$ years

and

$$\ln \frac{N_0}{0.05\ N_0} = \frac{0.693}{2.65} \text{ years}^{-1} \times t$$

Note that since the left side is dimensionless, the units of t are years. Rearranging, we have

$$t = \ln \left(\frac{1}{0.05}\right) \times \frac{2.65}{0.693} \text{ years} \qquad t = 11.5 \text{ years}$$

The product has nuclear charge = 61 + 1 = 62, and this is Sm; its mass number is the same as that for Pm, 147. Thus the nuclear reaction is $^{147}_{61}Pm \rightarrow {}^{0}_{-1}e + {}^{147}_{62}Sm$.

The half-life expression is also useful for problems involving radiochemical dating. No new equations are involved. However, you may need to know the nuclear reaction. Also it may be necessary to know the mass relationship between the product nucleus formed and the reactant nucleus, as in Example 25.4 of the text.

Example 25.3

A uranium-containing rock has been undergoing radioactive decay since its formation by a complex set of steps that can be summarized by the net equation

$$^{238}_{92}U \rightarrow {}^{206}_{82}Pb + 8\,{}^{4}_{2}He + 6\,{}^{0}_{-1}e$$

When the rock was crushed and analyzed, a 1.00 g sample yielded 3.50×10^{-7} g of uranium-238 and 4.0×10^{-5} mL of helium gas at STP (the $^{4}_{2}He$ nuclei have captured electrons to form He gas). What was the age of the rock? The half-life of $^{238}_{92}U$ is 4.5×10^{9} years.

Solution 25.3

We begin by calculating the mass of uranium that produced the helium gas. Using the ideal gas equation,

$$n = \frac{PV}{RT} = \frac{1.00 \text{ atm} \times 4.0 \times 10^{-5} \text{ mL} \times \dfrac{1.0 \text{ L}}{10^{3} \text{ mL}}}{0.0821 \text{ L atm mol}^{-1} \text{ K} \times 273 \text{ K}} = 1.78 \times 10^{-9} \text{ mol He}$$

From the balanced nuclear equation

$$\text{mol U} = \text{mol He} \times \frac{1 \text{ mol U}}{8 \text{ mol He}} = 2.22 \times 10^{-10} \text{ mol U}$$

Since the molar mass of ${}^{238}_{92}U$ is, to 3 significant figures, 238 g mol^{-1}, the mass of ${}^{238}_{98}U$ that produced the helium gas is

$$\text{mass } {}^{238}_{92}U = 2.22 \times 10^{-10} \text{ mol} \times 238 \text{ g mol}^{-1} = 5.28 \times 10^{-8} \text{ g}$$

The total amount of ${}^{238}_{92}U$ present per gram of rock, at its formation, assuming no products have escaped, is

$$3.50 \times 10^{-7} \text{g} + 0.53 \times 10^{-7} \text{ g} = 4.03 \times 10^{-7} \text{ g}$$

We use

$$N_0 = 4.03 \times 10^{-7} \text{ g} \quad \text{and} \quad N = 3.50 \times 10^{-7} \text{ g}$$

$$t_{1/2} = 4.51 \times 10^{9} \text{ years} = \frac{0.693}{k} \qquad k = \frac{0.693}{4.51 \times 10^{9}} \text{ years}^{-1}$$

$$\ln \frac{N_0}{N} = kt \quad t = \frac{\ln \dfrac{N_0}{N}}{k} = \frac{\ln \left(\dfrac{4.03 \times 10^{-7}}{3.50 \times 10^{-7}}\right)}{\dfrac{0.693}{4.51 \times 10^{9}}} \text{ years} \quad t = 9.18 \times 10^{8} \text{ years.}$$

25.5 Nuclear Energy

Mass and energy are related by the Einstein equation $E = mc^2$. **Mass is not constant for nuclear reactions.** For a nuclear reaction $\Delta E = c^2 \Delta m$. This means that if the sum of the masses of the products is greater than that of the reactants, both Δm and ΔE are positive. In other words, for Δm and ΔE positive, energy must be supplied to the reaction. If the reaction involves a nucleus being dissociated to neutrons and protons, this energy is called the **binding energy.** If the total number of nucleons (neutrons and protons) is n, the binding energy per nucleon is $\Delta E/n$.

Example 25.4

Calculate the binding energy and the binding energy per nucleon for ${}^{12}_{6}C$.

Solution 25.4

The reaction we are interested in is the **nuclear** reaction

$${}^{12}_{6}C \rightarrow 6\,{}^{1}_{1}H + 6\,{}^{1}_{0}n$$

In Table 2.1 we find the mass of a ${}^{12}_{6}C$ **atom** to be 12.00000 u; as well, the proton mass is 1.00728 u, the neutron mass is 1.00866 u, and the electron mass is 0.00055 u. Thus

$$\Delta m = 6(1.00728) + 6(1.00866) + 6(0.00055) - 12.00000 = 0.09894 \text{ u}$$

We must convert this mass to kg and use the velocity of light, c, in m s^{-1} to obtain an answer in Joules.

$$\Delta E = (2.998 \times 10^8 \text{ m s}^{-1})^2 \times 0.09894 \text{ u} \times \frac{1.000 \text{ g}}{6.022 \times 10^{23} \text{ u}} \times \frac{1 \text{ kg}}{10^3 \text{ g}}$$

$$\Delta E = 1.477 \times 10^{-11} \text{ J}$$

This is the binding energy. Since 6 protons and 6 neutrons are produced, the binding energy per nucleon is

$$\frac{\Delta E}{12 \text{ nucleons}} = 1.231 \times 10^{-12} \text{ J nucleon}^{-1}$$

SELF TEST

1. In the following nuclear reaction, supply the missing product.

$$^{238}_{92}U \rightarrow \alpha\text{-particle} +$$

(a) $^{234}_{91}Pa$ (b) $^{234}_{90}Th$ (c) $^{230}_{90}Th$ (d) $^{208}_{82}Pb$ (e) $^{236}_{88}Ra$

2. Choose all the statements that are correct. In radioactive decay there are examples where
(a) the mass number of the nucleus does not change.
(b) the atomic number of the nucleus does not change.
(c) the difference between the mass number of the radioactive nucleus and the product nucleus is four.
(d) the atomic number of the product nucleus is greater than the atomic number of the radioactive nucleus.
(e) no particle is emitted.

3. A sample of matter consisting of a new radioactive nucleus is monitored over time. At the end of 3.2 years of study, it is found that three-quarters of the original mass of the isotope has disintegrated. What is the half-life of the nucleus?
a) 1.6 y b) 2.1 y c) 2.4 y d) 6.4 y e) 9.6 y

4. The nuclear reaction

$$^{14}_{6}C \rightarrow {}^{14}_{7}N + {}^{0}_{-1}e$$

is classified as:
a) fission b) fusion c) radioactive decay d) electron capture
e) β particle decay

5. ^{232}Th has a half-life of 1.4×10^{10} years. If the age of the earth is 4.5×10^9 years, what % of the ^{232}Th originally present has now decayed?
(a) 10.8 (b) 20.0 (c) 2.23 (d) 97.77 (e) 0.108

6. Experimentally one can measure the [^{14}C] in a sample from the rate of disintegration of ^{14}C nuclei. A piece of wood from an ancient coffin has a disintegration rate of 7.50 disintegrations per min per g of carbon-14. How many years old is the wood? The half-life of $^{14}_{6}C$ is 5730 years. In living organisms there are 15.3 disintegrations per min per g of $^{14}_{6}C$.
(a) 6000 (b) 5900 (c) 5700 (d) 5500 (e) 5000

7. Give the missing product:

$$^{6}_{3}Li + ^{1}_{0}n \rightarrow \alpha\text{-particle} +$$

(a) $^{5}_{1}H$ (b) $^{2}_{1}H$ (c) $^{4}_{2}He$ (d) $^{3}_{1}H$ (e) $^{7}_{3}Li$

8. What is the binding energy, in Joules nucleon^{-1}, for $^{27}_{13}Al$? Use Table 2.1 in the text.

(a) 3.6×10^{-11} (b) 1.2×10^{-19} (c) 1.3×10^{-12} (d) 4.5×10^{-21}

(e) -3.6×10^{-11}

9. Which of the following are examples of fusion reactions?

(a) $^{6}_{3}Li + ^{1}_{0}n \rightarrow ^{4}_{2}He + ^{3}_{1}H$ (b) $^{1}_{1}H + ^{2}_{1}H \rightarrow ^{3}_{2}He$ (c) $^{1}_{1}H + ^{1}_{1}H \rightarrow ^{2}_{1}H + ^{0}_{1}e$

(d) $^{3}_{1}H + ^{2}_{1}H \rightarrow ^{4}_{2}He + ^{1}_{0}n$ (e) $^{1}_{0}n + ^{0}_{1}e \rightarrow ^{1}_{1}H$

Answers to Self Test

1. b; 2. a,b,c,d,e; 3. a; 4. e; 5. b; 6. b; 7. d; 8. c; 9. b,c,d.